Universitext

For other books in this series:
http://www.springer.com/series/223

Ravi P. Agarwal
Donal O'Regan

Ordinary and Partial Differential Equations

With Special Functions, Fourier Series, and Boundary Value Problems

Ravi P. Agarwal
Florida Institute of Technology
Department of Mathematical Sciences
150 West University Blvd.
Melbourne, FL 32901
agarwal@fit.edu

Donal O'Regan
National University of Ireland, Galway
Mathematics Department
University Road
Galway, Ireland
donal.oregan@nuigalway.ie

ISBN: 978-0-387-79145-6   e-ISBN: 978-0-387-79146-3
DOI 10.1007/978-0-387-79146-3

Library of Congress Control Number: 2008938952

Mathematics Subject Classification (2000): 00-01, 34-XX, 35-XX

Printed on acid-free paper

springer.com

Dedicated to our Sons

Hans Agarwal   and   Daniel Joseph O'Regan

Preface

This book comprises 50 class-tested lectures which both the authors have given to engineering and mathematics major students under the titles *Boundary Value Problems* and *Methods of Mathematical Physics* at various institutions all over the globe over a period of almost 35 years. The main topics covered in these lectures are power series solutions, special functions, boundary value problems for ordinary differential equations, Sturm–Liouville problems, regular and singular perturbation techniques, Fourier series expansion, partial differential equations, Fourier series solutions to initial-boundary value problems, and Fourier and Laplace transform techniques. The prerequisite for this book is calculus, so it can be used for a senior undergraduate course. It should also be suitable for a beginning graduate course because, in undergraduate courses, students do not have any exposure to various intricate concepts, perhaps due to an inadequate level of mathematical sophistication. The content in a particular lecture, together with the problems therein, provides fairly adequate coverage of the topic under study. These lectures have been delivered in one year courses and provide flexibility in the choice of material for a particular one-semester course. Throughout this book, the mathematical concepts have been explained very carefully in the simplest possible terms, and illustrated by a number of complete workout examples. Like any other mathematical book, it does contain some theorems and their proofs.

A detailed description of the topics covered in this book is as follows: In Lecture 1 we find explicit solutions of the first-order linear differential equations with variable coefficients, second-order homogeneous differential equations with constant coefficients, and second-order Cauchy–Euler differential equations. In Lecture 2 we show that if one solution of the homogeneous second-order differential equation with variable coefficients is known, then its second solution can be obtained rather easily. Here we also demonstrate the method of variation of parameters to construct the solutions of nonhomogeneous second-order differential equations.

In Lecture 3 we provide some basic concepts which are required to construct power series solutions to differential equations with variable coefficients. Here through various examples we also explain ordinary, regular singular, and irregular singular points of a given differential equation. In Lecture 4 first we prove a theorem which provides sufficient conditions so that the solutions of second-order linear differential equations can be expressed as power series at an ordinary point, and then construct power series solutions of Airy, Hermite, and Chebyshev differential equations. These equations occupy a central position in mathematical physics, engineering, and approximation theory. In Lectures 5 and 6 we demonstrate the method of Frobenius to construct the power series solutions of second-order linear differential equations at a regular singular point. Here we prove a general result which provides three possible different forms of the power series solution. We illustrate this result through several examples, including Laguerre's equation, which arises in quantum mechanics. In Lecture 7 we study Legendre's differential equation, which arises in problems such as the flow of an ideal fluid past a sphere, the determination of the electric field due to a charged sphere, and the determination of the temperature distribution in a sphere given its surface temperature. Here we also develop the polynomial solution of the Legendre differential equation. In Lecture 8 we study polynomial solutions of the Chebyshev, Hermite, and Laguerre differential equations. In Lecture 9 we construct series solutions of Bessel's differential equation, which first appeared in the works of Euler and Bernoulli. Since many problems of mathematical physics reduce to the Bessel equation, we investigate it in somewhat more detail. In Lecture 10 we develop series solutions of the hypergeometric differential equation, which finds applications in several problems of mathematical physics, quantum mechanics, and fluid dynamics.

Mathematical problems describing real world situations often have solutions which are not even continuous. Thus, to analyze such problems we need to work in a set which is bigger than the set of continuous functions. In Lecture 11 we introduce the sets of piecewise continuous and piecewise smooth functions, which are quite adequate to deal with a wide variety of applied problems. Here we also define periodic functions, and introduce even and odd extensions. In Lectures 12 and 13 we introduce orthogonality of functions and show that the Legendre, Chebyshev, Hermite, and Laguerre polynomials and Bessel functions are orthogonal. Here we also prove some fundamental properties about the zeros of orthogonal polynomials.

In Lecture 14 we introduce boundary value problems for second-order ordinary differential equations and provide a necessary and sufficient condition for the existence and uniqueness of their solutions. In Lecture 15 we formulate some boundary value problems with engineering applications, and show that often solutions of these problems can be written in terms of Bessel functions. In Lecture 16 we introduce Green's functions of homogeneous boundary value problems and show that the solution of a given nonhomogeneous boundary value problem can be explicitly expressed in terms of Green's function of the corresponding homogeneous equation.

In Lecture 17 we discuss the regular perturbation technique which relates the unknown solution of a given initial value problem to the known solutions of the infinite initial value problems. In many practical problems one often meets cases where the methods of regular perturbations cannot be applied. In the literature such problems are known as singular perturbation problems. In Lecture 18 we explain the methodology of singular perturbation technique with the help of some examples.

If the coefficients of the homogeneous differential equation and/or of the boundary conditions depend on a parameter, then one of the pioneer problems of mathematical physics is to determine the values of the parameter (eigenvalues) for which nontrivial solutions (eigenfunctions) exist. In Lecture 19 we explain some of the essential ideas involved in this vast field, which is continuously growing.

In Lectures 20 and 21 we show that the sets of orthogonal polynomials and functions we have provided in earlier lectures can be used effectively as the basis in the expansions of general functions. This in particular leads to Fourier's cosine, sine, trigonometric, Legendre, Chebyshev, Hermite and Bessel series. In Lectures 22 and 23 we examine pointwise convergence, uniform convergence, and the convergence in the mean of the Fourier series of a given function. Here the importance of Bessel's inequality and Parseval's equality are also discussed. In Lecture 24 we use Fourier series expansions to find periodic particular solutions of nonhomogeneous differential equations, and solutions of nonhomogeneous self-adjoint differential equations satisfying homogeneous boundary conditions, which leads to the well-known Fredholm's alternative.

In Lecture 25 we introduce partial differential equations and explain several concepts through elementary examples. Here we also provide the most fundamental classification of second-order linear equations in two independent variables. In Lecture 26 we study simultaneous differential equations, which play an important role in the theory of partial differential equations. Then we consider quasilinear partial differential equations of the Lagrange type and show that such equations can be solved rather easily, provided we can find solutions of related simultaneous differential equations. Finally, we explain a general method to find solutions of nonlinear first-order partial differential equations which is due to Charpit. In Lecture 27 we show that like ordinary differential equations, partial differential equations with constant coefficients can be solved explicitly. We begin with homogeneous second-order differential equations involving only second-order terms, and then show how the operator method can be used to solve some particular nonhomogeneous differential equations. Then, we extend the method to general second and higher order partial differential equations. In Lecture 28 we show that coordinate transformations can be employed successfully to reduce second-order linear partial differential equations to some standard forms, which are known as canonical forms. These transformed equations sometimes can be solved rather easily. Here the concept of characteristic of second-order partial differential equations plays an important role.

The method of separation of variables involves a solution which breaks up into a product of functions each of which contains only one of the variables. This widely used method for finding solutions of linear homogeneous partial differential equations we explain through several simple examples in Lecture 29. In Lecture 30 we derive the one-dimensional heat equation and formulate initial-boundary value problems, which involve the heat equation, the initial condition, and homogeneous and nonhomogeneous boundary conditions. Then we use the method of separation of variables to find the Fourier series solutions to these problems. In Lecture 31 we construct the Fourier series solution of the heat equation with Robin's boundary conditions. In Lecture 32 we provide two different derivations of the one-dimensional wave equation, formulate an initial-boundary value problem, and find its Fourier series solution. In Lecture 33 we continue using the method of separation of variables to find Fourier series solutions to some other initial-boundary value problems related to one-dimensional wave equation. In Lecture 34 we give a derivation of the two-dimensional Laplace equation, formulate the Dirichlet problem on a rectangle, and find its Fourier series solution. In Lecture 35 we discuss the steady-state heat flow problem in a disk. For this, we consider the Laplace equation in polar coordinates and find its Fourier series solution. In Lecture 36 we use the method of separation of variables to find the temperature distribution of rectangular and circular plates in the transient state. Again using the method of separation of variables, in Lecture 37 we find vertical displacements of thin membranes occupying rectangular and circular regions. The three-dimensional Laplace equation occurs in problems such as gravitation, steady-state temperature, electrostatic potential, magnetostatics, fluid flow, and so on. In Lecture 38 we find the Fourier series solution of the Laplace equation in a three-dimensional box and in a circular cylinder. In Lecture 39 we use the method of separation of variables to find the Fourier series solutions of the Laplace equation in and outside a given sphere. Here, we also discuss briefly Poisson's integral formulas. In Lecture 40 we demonstrate how the method of separation of variables can be employed to solve nonhomogeneous problems.

The Fourier integral is a natural extension of Fourier trigonometric series in the sense that it represents a piecewise smooth function whose domain is semi-infinite or infinite. In Lecture 41 we develop the Fourier integral with an intuitive approach and then discuss Fourier cosine and sine integrals which are extensions of Fourier cosine and sine series, respectively. This leads to Fourier cosine and sine transform pairs. In Lecture 42 we introduce the complex Fourier integral and the Fourier transform pair and find the Fourier transform of the derivative of a function. Then, we state and prove the Fourier convolution theorem, which is an important result. In Lectures 43 and 44 we consider problems in infinite domains which can be effectively solved by finding the Fourier transform, or the Fourier sine or cosine transform of the unknown function. For such problems usually the method of separation of variables does not work because the Fourier series are not adequate to yield complete solutions. We illustrate the method by considering several examples, and obtain the famous Gauss–Weierstrass, d'Alembert's, and Poisson's integral formulas.

In Lecture 45 we introduce some basic concepts of Laplace transform theory, whereas in Lecture 46 we prove several theorems which facilitate the computation of Laplace transforms. The method of Laplace transforms has the advantage of directly giving the solutions of differential equations with given initial and boundary conditions without the necessity of first finding the general solution and then evaluating from it the arbitrary constants. Moreover, the ready table of Laplace transforms reduces the problem of solving differential equations to mere algebraic manipulations. In Lectures 47 and 48 we employ the Laplace transform technique to find solutions of ordinary and partial differential equations, respectively. Here we also develop the famous Duhamel's formula.

A given problem consisting of a partial differential equation in a domain with a set of initial and/or boundary conditions is said to be well-posed if it has a unique solution which is stable. In Lecture 49 we demonstrate that problems considered in earlier lectures are well-posed. Finally, in Lecture 50 we prove a few theorems which verify that the series or integral form of the solutions we have obtained in earlier lectures are actually the solutions of the problems considered.

Two types of exercises are included in the book, those which illustrate the general theory, and others designed to fill out text material. These exercises form an integral part of the book, and every reader is urged to attempt most, if not all of them. For the convenience of the reader we have provided answers or hints to almost all the exercises.

In writing a book of this nature no originality can be claimed, only a humble attempt has been made to present the subject as simply, clearly, and accurately as possible. It is earnestly hoped that *Ordinary and Partial Differential Equations* will serve an inquisitive reader as a starting point in this rich, vast, and ever-expanding field of knowledge.

We would like to express our appreciation to Professors M. Bohner, S.K. Sen, and P.J.Y. Wong for their suggestions and criticisms. We also want to thank Ms. Vaishali Damle at Springer New York for her support and cooperation.

Ravi P. Agarwal
Donal O'Regan

Contents

Lecture 1
Solvable Differential Equations

In this lecture we shall show that first-order linear differential equations with variable coefficients, second-order homogeneous differential equations with constant coefficients, and second-order Cauchy–Euler differential equations can be solved in terms of the known quantities.

First-order equations. Consider the differential equation (DE)

$$y' + p(x)y = q(x), \quad ' = \frac{d}{dx} \tag{1.1}$$

where the functions $p(x)$ and $q(x)$ are continuous in some interval J. The corresponding homogeneous equation

$$y' + p(x)y = 0 \tag{1.2}$$

obtained by taking $q(x) \equiv 0$ in (1.1) can be solved by separating the variables, i.e.,

$$\frac{1}{y}y' + p(x) = 0$$

and now integrating it, to obtain

$$\ln y(x) + \int^x p(t)dt = \ln c,$$

or

$$y(x) = c \exp\left(-\int^x p(t)dt\right). \tag{1.3}$$

In dividing (1.2) by y we have lost the solution $y(x) \equiv 0$, which is called the *trivial solution* (for a linear homogeneous DE $y(x) \equiv 0$ is always a solution). However, it is included in (1.3) with $c = 0$.

If $x_0 \in J$, then the function

$$y(x) = y_0 \exp\left(-\int_{x_0}^x p(t)dt\right) \tag{1.4}$$

clearly satisfies the DE (1.2) and passes through the point (x_0, y_0). Thus, this is the solution of the *initial value problem*: DE (1.2) together with the *initial condition*

$$y(x_0) = y_0. \tag{1.5}$$

R.P. Agarwal, D. O'Regan, *Ordinary and Partial Differential Equations*,
Universitext, DOI 10.1007/978-0-387-79146-3_1,
© Springer Science+Business Media, LLC 2009

To find the solution of the DE (1.1) we shall use the *method of variation of parameters* due to Lagrange. In (1.3) we assume that c is a function of x, i.e.,

$$y(x) = c(x) \exp\left(-\int^x p(t)dt\right) \tag{1.6}$$

and search for $c(x)$ so that (1.6) becomes a solution of the DE (1.1). For this, setting (1.6) into (1.1), we find

$$c'(x) \exp\left(-\int^x p(t)dt\right) - c(x)p(x) \exp\left(-\int^x p(t)dt\right)$$

$$+ c(x)p(x) \exp\left(-\int^x p(t)dt\right) = q(x),$$

which is the same as

$$c'(x) = q(x) \exp\left(\int^x p(t)dt\right). \tag{1.7}$$

Integrating (1.7), we obtain the required function

$$c(x) = c_1 + \int^x q(t) \exp\left(\int^t p(s)ds\right) dt.$$

Now, substituting this $c(x)$ in (1.6), we find the solution of (1.1) as

$$y(x) = c_1 \exp\left(-\int^x p(t)dt\right) + \int^x q(t) \exp\left(-\int^x_t p(s)ds\right) dt. \tag{1.8}$$

This solution $y(x)$ is of the form $c_1 u(x) + v(x)$. It is to be noted that $c_1 u(x)$ is the general solution of (1.2). Hence, the general solution of (1.1) is obtained by adding any particular solution of (1.1) to the general solution of (1.2).

From (1.8) the solution of the initial value problem (1.1), (1.5), where $x_0 \in J$, is easily obtained as

$$y(x) = y_0 \exp\left(-\int^x_{x_0} p(t)dt\right) + \int^x_{x_0} q(t) \exp\left(-\int^x_t p(s)ds\right) dt. \tag{1.9}$$

Example 1.1. Consider the initial value problem

$$xy' - 4y + 2x^2 + 4 = 0, \quad x \neq 0, \quad y(1) = 1. \tag{1.10}$$

Since $x_0 = 1$, $y_0 = 1$, $p(x) = -4/x$ and $q(x) = -2x - 4/x$ from (1.9) the solution of (1.10) can be written as

$$\begin{aligned}
y(x) &= \exp\left(\int^x_1 \frac{4}{t}dt\right) + \int^x_1 \left(-2t - \frac{4}{t}\right) \exp\left(\int^x_t \frac{4}{s}ds\right) dt \\
&= x^4 + \int^x_1 \left(-2t - \frac{4}{t}\right) \frac{x^4}{t^4}dt \\
&= x^4 + x^4\left(\frac{1}{x^2} + \frac{1}{x^4} - 2\right) = -x^4 + x^2 + 1.
\end{aligned}$$

Alternatively, instead of using (1.9) we can find the solution of (1.10) as follows: For the corresponding homogeneous DE $y' - (4/x)y = 0$ the general solution is cx^4, and a particular solution of the DE (1.10) is

$$\int^x \left(-2t - \frac{4}{t}\right) \exp\left(\int_t^x \frac{4}{s}ds\right) dt = x^2 + 1$$

and hence the general solution of the DE (1.10) is $y(x) = cx^4 + x^2 + 1$. Now, in order to satisfy the initial condition $y(1) = 1$, it is necessary that $1 = c + 1 + 1$, or $c = -1$. The solution of (1.10) is, therefore, $y(x) = -x^4 + x^2 + 1$.

Second-order equations with constant coefficients. We shall find solutions of the second-order DE

$$y'' + ay' + by = 0, \tag{1.11}$$

where a and b are constants.

As a first step toward finding a solution to this DE we look back at the equation $y' + ay = 0$ (a is a constant) for which all solutions are constant multiples of e^{-ax}. Thus, for (1.11) also some form of exponential function would be a reasonable choice and would utilize the property that the differentiation of an exponential function e^{rx} always yields a constant multiplied by e^{rx}.

Thus, we try $y = e^{rx}$ and find the value(s) of r. We have

$$r^2 e^{rx} + are^{rx} + be^{rx} = 0,$$

or

$$(r^2 + ar + b)e^{rx} = 0,$$

or

$$r^2 + ar + b = 0. \tag{1.12}$$

Hence, e^{rx} is a solution of (1.11) if r is a solution of (1.12). Equation (1.12) is called the *characteristic polynomial* of (1.11). For the roots of (1.12) we have the following three cases:

1. Distinct real roots. If r_1 and r_2 are real and distinct roots of (1.12), then $e^{r_1 x}$ and $e^{r_2 x}$ are two solutions of (1.11), and its general solution can be written as

$$y(x) = c_1 e^{r_1 x} + c_2 e^{r_2 x}.$$

In the particular case when $r_1 = r$, $r_2 = -r$ (then the DE (1.11) is $y'' -$

$r^2 y = 0$), we have

$$y(x) = c_1 e^{rx} + c_2 e^{-rx} = \left(\frac{A+B}{2}\right) e^{rx} + \left(\frac{A-B}{2}\right) e^{-rx}$$

$$= A\left(\frac{e^{rx} + e^{-rx}}{2}\right) + B\left(\frac{e^{rx} - e^{-rx}}{2}\right)$$

$$= A \cosh rx + B \sinh rx.$$

2. Repeated real roots. If $r_1 = r_2 = r$ is a repeated root of (1.12), then e^{rx} is a solution. To find the second solution, we let $y(x) = u(x)e^{rx}$ and substitute it in (1.11), to get

$$e^{rx}(u'' + 2ru' + r^2 u) + ae^{rx}(u' + ru) + bue^{rx} = 0,$$

or

$$u'' + (2r + a)u' + (r^2 + ar + b)u = u'' + (2r + a)u' = 0.$$

Now since r is a repeated root of (1.12), it follows that $2r + a = 0$ and hence $u'' = 0$, i.e., $u(x) = c_1 + c_2 x$. Thus,

$$y(x) = (c_1 + c_2 x)e^{rx} = c_1 e^{rx} + c_2 x e^{rx}.$$

Hence, the second solution of (1.11) is $x e^{rx}$.

3. Complex conjugate roots. Let $r_1 = \mu + i\nu$ and $r_2 = \mu - i\nu$, where $i = \sqrt{-1}$, so that

$$e^{(\mu \pm i\nu)x} = e^{\mu x}(\cos \nu x \pm i \sin \nu x).$$

Since for the DE (1.11) the real part (i.e., $e^{\mu x} \cos \nu x$) and the imaginary part (i.e., $e^{\mu x} \sin \nu x$) both are solutions, the general solution of (1.11) can be written as

$$y(x) = c_1 e^{\mu x} \cos \nu x + c_2 e^{\mu x} \sin \nu x.$$

In the particular case when $r_1 = i\nu$ and $r_2 = -i\nu$ (then the DE (1.11) is $y'' + \nu^2 y = 0$) we have $y(x) = c_1 \cos \nu x + c_2 \sin \nu x$.

Cauchy–Euler equations. For the *Cauchy–Euler equation*

$$t^2 y'' + at y' + by = 0, \quad t > 0 \ (t \text{ is the independent variable}), \quad ' = \frac{d}{dt} \quad (1.13)$$

which occurs in studying the temperature distribution generated by a heat source such as the sun or a nuclear reactor, we assume $y(t) = t^m$ to obtain

$$t^2 m(m-1)t^{m-2} + atmt^{m-1} + bt^m = 0,$$

or

$$m(m-1) + am + b = 0. \tag{1.14}$$

This is the characteristic equation for (1.13), and as earlier for (1.12) the nature of its roots determines the general solution:

Real, distinct roots $m_1 \neq m_2$: $y(t) = c_1 t^{m_1} + c_2 t^{m_2}$,

Real, repeated roots $m = m_1 = m_2$: $y(t) = c_1 t^m + c_2 (\ln t) t^m$,

Complex conjugate roots $m_1 = \mu + i\nu$, $m_2 = \mu - i\nu$: $y(t) = c_1 t^\mu \cos(\nu \ln t) + c_2 t^\mu \sin(\nu \ln t)$.

In the particular case

$$t^2 y'' + ty' - \lambda^2 y = 0, \quad t > 0, \quad \lambda > 0 \tag{1.15}$$

the characteristic equation is $m(m-1) + m - \lambda^2 = 0$, or $m^2 - \lambda^2 = 0$. Thus, the roots are $m = \pm\lambda$, and hence the general solution of (1.15) appears as

$$y(t) = c_1 t^\lambda + c_2 t^{-\lambda}. \tag{1.16}$$

Problems

1.1. (Principle of Superposition). If $y_1(x)$ and $y_2(x)$ are solutions of $y' + p(x)y = q_i(x)$, $i = 1, 2$ respectively, then show that $c_1 y_1(x) + c_2 y_2(x)$ is a solution of the DE $y' + p(x)y = c_1 q_1(x) + c_2 q_2(x)$, where c_1 and c_2 are constants.

1.2. Find general solutions of the following DEs:

(i) $y' - (\cot x)y = 2x \sin x$

(ii) $y' + y + x + x^2 + x^3 = 0$

(iii) $(y^2 - 1) + 2(x - y(1 + y)^2)y' = 0$

(iv) $(1 + y^2) = (\tan^{-1} y - x)y'$.

1.3. Solve the following initial value problems:

(i) $y' + 2y = \begin{cases} 1, & 0 \leq x \leq 1 \\ 0, & x > 1 \end{cases}$, $y(0) = 0$

(ii) $y' + p(x)y = 0$, $y(0) = 1$ where $p(x) = \begin{cases} 2, & 0 \leq x \leq 1 \\ 1, & x > 1. \end{cases}$

1.4. Let $q(x)$ be continuous in $[0, \infty)$ and $\lim_{x \to \infty} q(x) = L$. For the DE $y' + ay = q(x)$ show that

(i) if $a > 0$, every solution approaches L/a as $x \to \infty$

(ii) if $a < 0$, there is one and only one solution which approaches L/a as $x \to \infty$.

1.5. Let $y(x)$ be the solution of the initial value problem (1.1), (1.5) in $[x_0, \infty)$, and let $z(x)$ be a continuously differentiable function in $[x_0, \infty)$ such that $z' + p(x)z \le q(x)$, $z(x_0) \le y_0$. Show that $z(x) \le y(x)$ for all x in $[x_0, \infty)$. In particular, for the problem $y' + y = \cos x$, $y(0) = 1$ verify that $2e^{-x} - 1 \le y(x) \le 1$, $x \in [0, \infty)$.

1.6. Certain nonlinear first-order DEs can be reduced to linear equations by an appropriate change of variables. For example, this is always possible for the *Bernoulli equation:*

$$y' + p(x)y = q(x)y^n, \quad n \ne 0, 1.$$

Indeed this equation is equivalent to the DE

$$y^{-n}y' + p(x)y^{1-n} = q(x)$$

and now the substitution $v = y^{1-n}$ (used by Leibniz in 1696) leads to the first-order linear DE

$$\frac{1}{1-n}v' + p(x)v = q(x).$$

In particular, show that the general solution of the DE $xy' + y = x^2y^2$, $x \ne 0$ is $y(x) = (cx - x^2)^{-1}$, $x \ne 0, c$.

1.7. Find general solutions of the following homogeneous DEs:

(i) $y'' + 7y' + 10y = 0$

(ii) $y'' - 8y' + 16y = 0$

(iii) $y'' + 2y' + 3y = 0$.

1.8. Show that if the real parts of the roots of (1.12) are negative, then $\lim_{x \to \infty} y(x) = 0$ for every solution $y(x)$ of (1.11).

1.9. Show that the solution of the initial value problem

$$y'' - 2(r + \beta)y' + r^2y = 0, \quad y(0) = 0, \quad y'(0) = 1$$

can be written as

$$y_\beta(x) = \frac{1}{2\sqrt{\beta(2r + \beta)}} \left[e^{[r+\beta+\sqrt{\beta(2r+\beta)}]x} - e^{[r+\beta-\sqrt{\beta(2r+\beta)}]x} \right].$$

Further, show that $\lim_{\beta \to 0} y_\beta(x) = xe^{rx}$.

1.10. The following fourth order DEs occur in applications as indicated:

(i) $y'''' - k^4y = 0$ (vibration of a beam)

(ii) $y'''' + 4k^4y = 0$ (beam on an elastic foundation)

(iii) $y'''' - 2k^2 y'' + k^4 y = 0$ (bending of an elastic plate),

where $k \neq 0$ is a constant. Find their general solutions.

Answers or Hints

1.1. Use the definition of a solution.

1.2. (i) $c \sin x + x^2 \sin x$ (ii) $ce^{-x} - x^3 + 2x^2 - 5x + 5$ (iii) $x(y-1)/(y+1) = y^2 + c$ (iv) $x = \tan^{-1} y - 1 + ce^{-\tan^{-1} y}$.

1.3. (i) $y(x) = \begin{cases} \frac{1}{2}(1 - e^{-2x}), & 0 \leq x \leq 1 \\ \frac{1}{2}(e^2 - 1)e^{-2x}, & x > 1 \end{cases}$ (ii) $y(x) = \begin{cases} e^{-2x}, & 0 \leq x \leq 1 \\ e^{-(x+1)}, & x > 1. \end{cases}$

1.4. (i) In $y(x) = y(x_0)e^{-a(x-x_0)} + [\int_{x_0}^{x} e^{at} q(t)dt]/e^{ax}$ take the limit $x \rightarrow \infty$ (ii) In $y(x) = e^{-ax} \left[y(x_0)e^{ax_0} + \int_{x_0}^{\infty} e^{at} q(t)dt - \int_{x}^{\infty} e^{at} q(t)dt \right]$ choose $y(x_0)$ so that $y(x_0)e^{ax_0} + \int_{x_0}^{\infty} e^{at} q(t)dt = 0$ $(\lim_{x \rightarrow \infty} q(x) = L)$. Now in $y(x) = -[\int_{x}^{\infty} e^{at} q(t)dt]/e^{ax}$ take the limit $x \rightarrow \infty$.

1.5. There exists a continuous function $r(x) \geq 0$ such that $z' + p(x)z = q(x) - r(x)$, $z(x_0) \leq y_0$. Thus, for the function $\phi(x) = y(x) - z(x)$, $\phi' + p(x)\phi = r(x) \geq 0$, $\phi(x_0) = y_0 - z(x_0) \geq 0$.

1.6. Using the substitution $v = y^{-1}$ the given equation reduces to $-xv' + v = x^2$.

1.7. (i) $c_1 e^{-2x} + c_2 e^{-5x}$ (ii) $(c_1 + c_2 x)e^{4x}$ (iii) $c_1 e^{-x} \cos \sqrt{2}x + c_2 e^{-x} \times \sin \sqrt{2}x$.

1.8. Use explicit forms of the solution.

1.9. Note that $\sqrt{\beta(\beta + 2r)} \rightarrow 0$ as $\beta \rightarrow 0$.

1.10. (i) $c_1 e^{kx} + c_2 e^{-kx} + c_3 \cos kx + c_4 \sin kx$ (ii) $e^{kx}(c_1 \cos kx + c_2 \sin kx) + e^{-kx}(c_3 \cos kx + c_4 \sin kx)$ (iii) $e^{kx}(c_1 + c_2 x) + e^{-kx}(c_3 + c_4 x)$.

Lecture 2
Second-Order Differential Equations

Generally, second-order differential equations with variable coefficients cannot be solved in terms of the known functions. In this lecture we shall show that if one solution of the homogeneous equation is known, then its second solution can be obtained rather easily. Further, by employing the method of variation of parameters, the general solution of the nonhomogeneous equation can be constructed provided two solutions of the corresponding homogeneous equation are known.

Homogeneous equations. For the homogeneous linear DE of second-order with variable coefficients

$$y'' + p_1(x)y' + p_2(x)y = 0, \tag{2.1}$$

where $p_1(x)$ and $p_2(x)$ are continuous in J, there does not exist any method to solve it. However, the following results are well-known.

Theorem 2.1. There exist exactly two solutions $y_1(x)$ and $y_2(x)$ of (2.1) which are linearly independent (essentially different) in J, i.e., there does not exist a constant c such that $y_1(x) = cy_2(x)$ for all $x \in J$.

Theorem 2.2. Two solutions $y_1(x)$ and $y_2(x)$ of (2.1) are linearly independent in J if and only if their *Wronskian* defined by

$$W(x) = W(y_1, y_2)(x) = \begin{vmatrix} y_1(x) & y_2(x) \\ y_1'(x) & y_2'(x) \end{vmatrix} \tag{2.2}$$

is different from zero for some $x = x_0$ in J.

Theorem 2.3. For the Wronskian defined in (2.2) the following Abel's identity holds:

$$W(x) = W(x_0) \exp\left(- \int_{x_0}^{x} p_1(t)dt \right), \quad x_0 \in J. \tag{2.3}$$

Thus, if Wronskian is zero at some $x_0 \in J$, then it is zero for all $x \in J$.

Theorem 2.4. If $y_1(x)$ and $y_2(x)$ are solutions of (2.1) and c_1 and c_2 are arbitrary constants, then $c_1 y_1(x) + c_2 y_2(x)$ is also a solution of (2.1).

R.P. Agarwal, D. O'Regan, *Ordinary and Partial Differential Equations*,
Universitext, DOI 10.1007/978-0-387-79146-3_2,
© Springer Science+Business Media, LLC 2009

Further, if $y_1(x)$ and $y_2(x)$ are linearly independent, then any solution $y(x)$ of (2.1) can be written as $y(x) = \bar{c}_1 y_1(x) + \bar{c}_2 y_2(x)$, where $\bar{c}_1$ and $\bar{c}_2$ are suitable constants.

Now we shall show that, if one solution $y_1(x)$ of (2.1) is known (by some clever method) then we can employ *variation of parameters* to find the second solution of (2.1). For this, we let $y(x) = u(x)y_1(x)$ and substitute this in (2.1), to get

$$(uy_1)'' + p_1(uy_1)' + p_2(uy_1) = 0,$$

or

$$u''y_1 + 2u'y_1' + uy_1'' + p_1 u'y_1 + p_1 uy_1' + p_2 uy_1 = 0,$$

or

$$u''y_1 + (2y_1' + p_1 y_1)u' + (y_1'' + p_1 y_1' + p_2 y_1)u = 0.$$

However, since y_1 is a solution of (2.1), the above equation with $v = u'$ is the same as

$$y_1 v' + (2y_1' + p_1 y_1)v = 0, \tag{2.4}$$

which is a first-order equation, and it can be solved easily provided $y_1 \neq 0$ in J. Indeed, multiplying (2.4) by y_1, we find

$$(y_1^2 v' + 2y_1' y_1 v) + p_1 y_1^2 v = 0,$$

which is the same as

$$(y_1^2 v)' + p_1(y_1^2 v) = 0;$$

and hence

$$y_1^2 v = c \exp\left(-\int^x p_1(t)dt\right),$$

or, on taking $c = 1$,

$$v(x) = \frac{1}{y_1^2(x)} \exp\left(-\int^x p_1(t)dt\right).$$

Hence, the second solution of (2.1) is

$$y_2(x) = y_1(x)\int^x \frac{1}{y_1^2(t)} \exp\left(-\int^t p_1(s)ds\right) dt. \tag{2.5}$$

Example 2.1. It is easy to verify that $y_1(x) = x^2$ is a solution of the DE

$$x^2 y'' - 2xy' + 2y = 0, \quad x \neq 0.$$

For the second solution we use (2.5), to obtain

$$y_2(x) = x^2 \int^x \frac{1}{t^4} \exp\left(-\int^t \left(-\frac{2s}{s^2}\right) ds\right) dt = x^2 \int^x \frac{1}{t^4} t^2 dt = -x.$$

We note that the substitution $w = y'/y$ converts (2.1) into a first-order nonlinear DE

$$w' + p_1(x)w + p_2(x) + w^2 = 0. \tag{2.6}$$

This DE is called *Riccati's equation.* In general it is not integrable, but if a particular solution, say, $w_1(x)$ is known, then by the substitution $z = w - w_1(x)$ it can be reduced to Bernoulli's equation (see Problem 1.6). In fact, we have

$$z' + w_1'(x) + p_1(x)(z + w_1(x)) + p_2(x) + (z + w_1(x))^2 = 0,$$

which is the same as

$$z' + (p_1(x) + 2w_1(x))z + z^2 = 0. \tag{2.7}$$

Since this equation can be solved easily to obtain $z(x)$, the solution of (2.6) takes the form $w(x) = w_1(x) + z(x)$.

Example 2.2. It is easy to verify that $w_1(x) = x$ is a particular solution of the Riccati equation

$$w' = 1 + x^2 - 2xw + w^2.$$

The substitution $z = w - x$ in this equation gives the Bernoulli equation

$$z' = z^2,$$

whose general solution is $z(x) = 1/(c-x)$, $x \neq c$. Thus, the general solution of the given Riccati's equation is $w(x) = x + 1/(c-x)$, $x \neq c$.

Nonhomogeneous equations. Now we shall find a particular solution of the nonhomogeneous equation

$$y'' + p_1(x)y' + p_2(x)y = r(x). \tag{2.8}$$

For this also we shall apply the method of *variation of parameters.* Let $y_1(x)$ and $y_2(x)$ be two solutions of (2.1). We assume $y(x) = c_1(x)y_1(x) + c_2(x)y_2(x)$ is a solution of (2.8). Note that $c_1(x)$ and $c_2(x)$ are two unknown functions, so we can have two sets of conditions which determine $c_1(x)$ and $c_2(x)$. Since

$$y' = c_1 y_1' + c_2 y_2' + c_1' y_1 + c_2' y_2$$

as a first condition we assume that

$$c_1' y_1 + c_2' y_2 = 0. \tag{2.9}$$

Thus, we have

$$y' = c_1 y_1' + c_2 y_2'$$

and on differentiation

$$y'' = c_1 y_1'' + c_2 y_2'' + c_1' y_1' + c_2' y_2'.$$

Substituting these in (2.8), we get

$$c_1(y_1'' + p_1 y_1' + p_2 y_1) + c_2(y_2'' + p_1 y_2' + p_2 y_2) + (c_1' y_1' + c_2' y_2') = r(x).$$

Clearly, this equation, in view of $y_1(x)$ and $y_2(x)$ being solutions of (2.1), is the same as

$$c_1' y_1' + c_2' y_2' = r(x). \tag{2.10}$$

Solving (2.9), (2.10), we find

$$c_1' = -\frac{r(x)y_2(x)}{\begin{vmatrix} y_1(x) & y_2(x) \\ y_1'(x) & y_2'(x) \end{vmatrix}}, \quad c_2' = \frac{r(x)y_1(x)}{\begin{vmatrix} y_1(x) & y_2(x) \\ y_1'(x) & y_2'(x) \end{vmatrix}};$$

and hence a particular solution of (2.8) is

$$
\begin{aligned}
y_p(x) &= c_1(x)y_1(x) + c_2(x)y_2(x) \\
&= -y_1(x)\int^x \frac{r(t)y_2(t)}{\begin{vmatrix} y_1(t) & y_2(t) \\ y_1'(t) & y_2'(t) \end{vmatrix}}\,dt + y_2(x)\int^x \frac{r(t)y_1(t)}{\begin{vmatrix} y_1(t) & y_2(t) \\ y_1'(t) & y_2'(t) \end{vmatrix}}\,dt \\
&= \int^x H(x,t)r(t)\,dt,
\end{aligned}
$$

$$\tag{2.11}$$

where

$$H(x,t) = \begin{vmatrix} y_1(t) & y_2(t) \\ y_1(x) & y_2(x) \end{vmatrix} \Bigg/ \begin{vmatrix} y_1(t) & y_2(t) \\ y_1'(t) & y_2'(t) \end{vmatrix}. \tag{2.12}$$

Thus, the general solution of (2.8) is

$$y(x) = c_1 y_1(x) + c_2 y_2(x) + y_p(x). \tag{2.13}$$

The following properties of the function $H(x,t)$ are immediate:

(i). $H(x,t)$ is defined for all $(x,t) \in J \times J$;

(ii). $\partial^j H(x,t)/\partial x^j$, $j = 0,1,2$ are continuous for all $(x,t) \in J \times J$;

(iii). for each fixed $t \in J$ the function $z(x) = H(x,t)$ is a solution of the homogeneous DE (2.1) satisfying $z(t) = 0$, $z'(t) = 1$; and

(iv). the function

$$v(x) = \int_{x_0}^x H(x,t)r(t)\,dt$$

is a particular solution of the nonhomogeneous DE (2.8) satisfying $y(x_0) = y'(x_0) = 0$.

Example 2.3.   Consider the DE

$$y'' + y = \cot x.$$

For the corresponding homogeneous DE $y'' + y = 0$, $\sin x$ and $\cos x$ are solutions. Thus, its general solution can be written as

$$
y(x) = c_1 \sin x + c_2 \cos x + \int^x \frac{\begin{vmatrix} \sin x & \cos x \\ \sin t & \cos t \end{vmatrix}}{\begin{vmatrix} \sin t & \cos t \\ \cos t & -\sin t \end{vmatrix}} \frac{\cos t}{\sin t} dt
$$

$$
= c_1 \sin x + c_2 \cos x - \int^x (\sin t \cos x - \sin x \cos t)\frac{\cos t}{\sin t} dt
$$

$$
= c_1 \sin x + c_2 \cos x - \cos x \sin x + \sin x \int^x \frac{1 - \sin^2 t}{\sin t} dt
$$

$$
= c_1 \sin x + c_2 \cos x - \cos x \sin x - \sin x \int^x \sin t\, dt + \sin x \int^x \frac{1}{\sin t} dt
$$

$$
= c_1 \sin x + c_2 \cos x + \sin x \int^x \frac{\operatorname{cosec} t(\operatorname{cosec} t - \cot t)}{(\operatorname{cosec} t - \cot t)} dt
$$

$$
= c_1 \sin x + c_2 \cos x + \sin x \ \ln[\operatorname{cosec} x - \cot x].
$$

Finally, we remark that if the functions $p_1(x)$, $p_2(x)$ and $r(x)$ are continuous on J and $x_0 \in J$, then the DE (2.8) together with the *initial conditions*

$$y(x_0) = y_0, \quad y'(x_0) = y_1 \tag{2.14}$$

has a unique solution. The problem (2.8), (2.14) is called an *initial value problem*. Note that in (2.14) conditions are prescribed at the same point, namely, x_0.

Problems

2.1.   Given the solution $y_1(x)$, find the second solution of the following DEs:

(i) $(x^2 - x)y'' + (3x - 1)y' + y = 0$ $(x \neq 0, 1)$, $y_1(x) = (x - 1)^{-1}$
(ii) $x(x - 2)y'' + 2(x - 1)y' - 2y = 0$ $(x \neq 0, 2)$, $y_1(x) = (1 - x)$
(iii) $xy'' - y' - 4x^3 y = 0$ $(x \neq 0)$, $y_1(x) = \exp(x^2)$
(iv) $(1 - x^2)y'' - 2xy' + 2y = 0$ $(|x| < 1)$, $y_1(x) = x$.

2.2.   The differential equation

$$xy'' - (x+n)y' + ny = 0$$

is interesting because it has an exponential solution and a polynomial solution.

(i)   Verify that one solution is $y_1(x) = e^x$.

(ii)   Show that the second solution has the form $y_2(x) = ce^x \int^x t^n e^{-t} dt$. Further, show that with $c = -1/n!$,

$$y_2(x) = 1 + \frac{x}{1!} + \frac{x^2}{2!} + \cdots + \frac{x^n}{n!}.$$

Note that $y_2(x)$ is the first $n+1$ terms of the Taylor series about $x = 0$ for e^x, that is, for $y_1(x)$.

2.3.   The differential equation

$$y'' + \delta(xy' + y) = 0$$

occurs in the study of the turbulent flow of a uniform stream past a circular cylinder. Verify that $y_1(x) = \exp(-\delta x^2/2)$ is one solution. Find its second solution.

2.4.   Let $y_1(x) \neq 0$ and $y_2(x)$ be two linearly independent solutions of the DE (2.1). Show that $y(x) = y_2(x)/y_1(x)$ is a nonconstant solution of the DE

$$y_1(x)y'' + (2y_1'(x) + p_1(x)y_1(x))y' = 0.$$

2.5.   Let the function $p_1(x)$ be differentiable in J. Show that the substitution $y(x) = z(x)\exp\left(-\frac{1}{2}\int^x p_1(t)dt\right)$ transforms (2.1) to the differential equation

$$z'' + \left(p_2(x) - \frac{1}{2}p_1'(x) - \frac{1}{4}p_1^2(x)\right)z = 0.$$

In particular show that the substitution $y(x) = z(x)/\sqrt{x}$ transforms *Bessel's DE*

$$x^2 y'' + xy' + (x^2 - a^2)y = 0, \tag{2.15}$$

where a is a constant (parameter), into a simple DE

$$z'' + \left(1 + \frac{1 - 4a^2}{4x^2}\right)z = 0. \tag{2.16}$$

2.6.   Let $v(x)$ be the solution of the initial value problem

$$y'' + p_1 y' + p_2 y = 0, \quad y(0) = 0, \quad y'(0) = 1$$

where p_1 and p_2 are constants. Show that the function

$$y(x) = \int_{x_0}^{x} v(x-t)r(t)dt$$

is the solution of the nonhomogeneous DE

$$y'' + p_1 y' + p_2 y = r(x)$$

satisfying $y(x_0) = y'(x_0) = 0$.

2.7. Find general solutions of the following nonhomogeneous DEs:

(i) $y'' + 4y = \sin 2x$

(ii) $y'' + 4y' + 3y = e^{-3x}$

(iii) $y'' + 5y' + 4y = e^{-4x}$.

2.8. Verify that $y_1(x) = x$ and $y_2(x) = 1/x$ are solutions of

$$x^3 y'' + x^2 y' - xy = 0.$$

Use this information and the variation of parameters method to find the general solution of

$$x^3 y'' + x^2 y' - xy = x/(1+x).$$

Answers or Hints

2.1. (i) $\ln x/(x-1)$ (ii) $(1/2)(1-x)\ln[(x-2)/x]-1$ (iii) e^{-x^2} (iv) $(x/2)\times \ln[(1+x)/(1-x)] - 1$.

2.2. (i) Verify directly (ii) Use (2.5).

2.3. $e^{-\delta x^2/2} \int^x e^{\delta t^2/2} dt$.

2.4. Use $y_2(x) = y_1(x)y(x)$ and the fact that $y_1(x)$ and $y_2(x)$ are solutions.

2.5. Verify directly.

2.6. Use Leibniz's formula:

$\frac{d}{dx}\int_{\alpha(x)}^{\beta(x)} f(x,t)dt = f(x,\beta(x))\frac{d\beta}{dx} - f(x,\alpha(x))\frac{d\alpha}{dx} + \int_{\alpha(x)}^{\beta(x)} \frac{\partial f}{\partial x}(x,t)dt.$

2.7. (i) $c_1\cos 2x + c_2\sin 2x - \frac{1}{4}x\cos 2x$ (ii) $c_1 e^{-x} + c_2 e^{-3x} - \frac{1}{2}xe^{-3x}$ (iii) $c_1 e^{-x} + c_2 e^{-4x} - \frac{1}{3}xe^{-4x}$.

2.8. $c_1 x + (c_2/x) + (1/2)[(x-(1/x))\ln(1+x) - x\ln x - 1]$.

Lecture 3
Preliminaries to Series Solutions

In our previous lecture we have remarked that second-order differential equations with variable coefficients cannot be solved in terms of the known functions. In fact, the simple DE $y'' + xy = 0$ defies all our efforts. However, there is a fairly large class of DEs whose solutions can be expressed either in terms of power series, or as simple combination of power series and elementary functions. It is this class of DEs that we shall study in the next several lectures. Here we introduce some basic concepts which will be needed in our later discussion.

Power series. A *power series* is a series of functions of the form

$$\sum_{m=0}^{\infty} c_m(x - x_0)^m = c_0 + c_1(x - x_0) + c_2(x - x_0)^2 + \cdots + c_m(x - x_0)^m + \cdots$$

in which the coefficients c_m, $m = 0, 1, \cdots$ and the point x_0 are independent of x. The point x_0 is called the *point of expansion* of the series.

A function $f(x)$ is said to be *analytic* at $x = x_0$ if it can be expanded in a power series in powers of $(x - x_0)$ in some interval of the form $|x - x_0| < \mu$, where $\mu > 0$. If $f(x)$ is analytic at $x = x_0$, then

$$f(x) = \sum_{m=0}^{\infty} c_m(x - x_0)^m, \quad |x - x_0| < \mu,$$

where $c_m = f^{(m)}(x_0)/m!$, $m = 0, 1, \cdots$ which is the same as Taylor's expansion of $f(x)$ at $x = x_0$.

The following properties of power series will be needed later:

1. A power series $\sum_{m=0}^{\infty} c_m(x - x_0)^m$ is said to *converge* at a point x if $\lim_{n\to\infty} \sum_{m=0}^{n} c_m(x - x_0)^m$ exists. It is clear that the series converges at $x = x_0$; it may converge for all x, or it may converge for some values of x and not for others.

2. A power series $\sum_{m=0}^{\infty} c_m(x - x_0)^m$ is said to converge *absolutely* at a point x if the series $\sum_{m=0}^{\infty} |c_m(x - x_0)^m|$ converges. If the series converges absolutely, then the series also converges; however, the converse is not necessarily true.

R.P. Agarwal, D. O'Regan, *Ordinary and Partial Differential Equations*,
Universitext, DOI 10.1007/978-0-387-79146-3_3,
© Springer Science+Business Media, LLC 2009

3. If the series $\sum_{m=0}^{\infty} c_m(x-x_0)^m$ converges absolutely for $|x-x_0| < \mu$ and diverges for $|x-x_0| > \mu$, then μ is called the *radius of convergence*. For a series that converges nowhere except at x_0, we define μ to be zero; for a series that converges for all x, we say μ is infinite.

4. Ratio Test. If for a fixed value of x,

$$\lim_{m \to \infty} \left| \frac{c_{m+1}(x-x_0)^{m+1}}{c_m(x-x_0)^m} \right| = L,$$

then the power series $\sum_{m=0}^{\infty} c_m(x-x_0)^m$ converges absolutely at values of x for which $L < 1$, and diverges where $L > 1$. If $L = 1$, the test is inconclusive.

5. Comparison Test. If we have two power series $\sum_{m=0}^{\infty} c_m(x-x_0)^m$ and $\sum_{m=0}^{\infty} C_m(x-x_0)^m$ where $|c_m| \leq C_m$, $m = 0, 1, \cdots$, and if the series $\sum_{m=0}^{\infty} C_m(x-x_0)^m$ converges for $|x-x_0| < \mu$, then the series $\sum_{m=0}^{\infty} c_m(x-x_0)^m$ also converges for $|x-x_0| < \mu$.

6. If a series $\sum_{m=0}^{\infty} c_m(x-x_0)^m$ is convergent for $|x-x_0| < \mu$, then for any x, $|x-x_0| = \mu_0 < \mu$ there exists a constant M such that $|c_m|\mu_0^m \leq M$, $m = 0, 1, \cdots$.

7. The derivative of a power series is obtained by term by term differentiation; i.e., if $f(x) = \sum_{m=0}^{\infty} c_m(x-x_0)^m$, then

$$
\begin{aligned}
f'(x) &= c_1 + 2c_2(x-x_0) + 3c_3(x-x_0)^2 + \cdots \\
&= \sum_{m=1}^{\infty} mc_m(x-x_0)^{m-1} = \sum_{m=0}^{\infty} (m+1)c_{m+1}(x-x_0)^m.
\end{aligned}
$$

Further, the radii of convergence of these two series are the same. Similarly, the second derivative of $f(x)$ can be written as

$$f''(x) = 2c_2 + 3.2c_3(x-x_0) + \cdots = \sum_{m=0}^{\infty} (m+1)(m+2)c_{m+2}(x-x_0)^m.$$

8. Consider two power series $f(x) = \sum_{m=0}^{\infty} c_m(x-x_0)^m$ and $g(x) = \sum_{m=0}^{\infty} d_m(x-x_0)^m$, which converge for $|x-x_0| < \mu_1$ and $|x-x_0| < \mu_2$ respectively. If $\mu = \min\{\mu_1, \mu_2\}$, then

$$f(x) \pm g(x) = \sum_{m=0}^{\infty} (c_m \pm d_m)(x-x_0)^m,$$

and

$$f(x)g(x) = \sum_{m=0}^{\infty} a_m(x-x_0)^m,$$

where

$$a_m = \sum_{k=0}^{m} c_k d_{m-k} = \sum_{k=0}^{m} c_{m-k} d_k$$

converge for $|x - x_0| < \mu$.

9. Gauss Test. If at the end points of the interval $|x - x_0| < \mu$, the successive terms of the series $\sum_{m=0}^{\infty} c_m (x - x_0)^m$ are of fixed sign, and if the ratio of the $(m+1)$th term to the mth term can be written in the form $1 - (c/m) + O(1/m^2)$, where c is independent of m, then the series converges if $c > 1$ and diverges if $c \leq 1$.

Gamma and Beta functions. It is possible to write long expressions in very compact form using Gamma and Beta functions which we shall define now. The *Gamma function*, denoted by $\Gamma(x)$, is defined by

$$\Gamma(x) = \int_0^{\infty} t^{x-1} e^{-t} dt, \quad x > 0. \tag{3.1}$$

This improper integral can be shown to converge only for $x > 0$; thus the Gamma function is defined by this formula only for the positive values of its arguments. However, later we shall define it for the negative values of its arguments as well.

From the definition (3.1), we find

$$\Gamma(1) = \int_0^{\infty} e^{-t} dt = 1. \tag{3.2}$$

Also, we have

$$\Gamma(x+1) = \int_0^{\infty} t^x e^{-t} dt = \left[-t^x e^{-t} \right]\Big|_0^{\infty} + x \int_0^{\infty} t^{x-1} e^{-t} dt = x\Gamma(x),$$

which is the *recurrence formula*

$$\Gamma(x+1) = x\Gamma(x). \tag{3.3}$$

From (3.3) and (3.2) it is immediate that for any nonnegative integer n the function $\Gamma(n+1) = n!$, and hence the Gamma function, can be considered as a generalization of the factorial function.

Now we rewrite (3.3) in the form

$$\Gamma(x) = \frac{\Gamma(x+1)}{x}, \tag{3.4}$$

which holds only for $x > 0$. However, we can use (3.4) to define $\Gamma(x)$ in the range $-1 < x < 0$ since the right-hand side of (3.4) is well defined for x in this range. Also since

$$\Gamma(x+1) = \frac{\Gamma(x+2)}{(x+1)}$$

when $x > -1$, we may write

$$\Gamma(x) = \frac{\Gamma(x+2)}{x(x+1)} \tag{3.5}$$

for $x > 0$. But since $\Gamma(x+2)$ is defined for $x > -2$, we can use (3.5) to define $\Gamma(x)$ for $-2 < x < 0$, $x \neq -1$. Continuing this process, we have

$$\Gamma(x) = \frac{\Gamma(x+k)}{x(x+1)\cdots(x+k-1)}$$

for any positive integer k and for $x > 0$. By this formula the function $\Gamma(x)$ is defined for $-k < x < 0$, $x \neq -1, -2, \cdots, -k+1$. Hence, $\Gamma(x)$ is defined for all values of x other than $0, -1, -2, \cdots$, and at these points it becomes infinite.

The *Beta function* $B(x,y)$ is defined as

$$B(x,y) = \int_0^1 t^{x-1}(1-t)^{y-1}dt, \tag{3.6}$$

which converges for $x > 0$, $y > 0$.

Gamma and Beta functions are related as follows:

$$B(x,y) = \frac{\Gamma(x)\,\Gamma(y)}{\Gamma(x+y)}. \tag{3.7}$$

Oscillatory equations.   A nontrivial solution of the DE

$$y'' + q(x)y = 0 \tag{3.8}$$

is said to be *oscillatory* if it has no last zero, i.e., if $y(x_1) = 0$, then there exists a $x_2 > x_1$ such that $y(x_2) = 0$. Equation (3.8) itself is said to be oscillatory if every solution of (3.8) is oscillatory. A solution which is not oscillatory is called *nonoscillatory*. For example, the DE $y'' + y = 0$ is oscillatory, whereas $y'' - y = 0$ is nonoscillatory in $J = [0, \infty)$. The following easily verifiable oscillation criterion for the equation (3.8) is well known.

Theorem 3.1.   If the function $q(x)$ is continuous in $J = (0, \infty)$, and

$$\int^{\infty} q(x)dx = \infty, \tag{3.9}$$

then the DE (3.8) is oscillatory in J.

This result can be used easily to show that solutions of Bessel's DE (2.15) for all a, are oscillatory. For this, in Problem 2.5 we have noted that the substitution $y(x) = z(x)/\sqrt{x}$ transforms this equation into a simple DE (2.16). Clearly, this transformation does not alter the oscillatory behavior of two equations; moreover, for all a, there exists a sufficiently large x_0 such that for all $x \geq x_0$,

$$1 + \frac{1 - 4a^2}{4x^2} > \frac{1}{2}$$

and hence

$$\int^{\infty} \left(1 + \frac{1 - 4a^2}{4x^2}\right) dx = \infty.$$

Thus, Theorem 3.1 implies that the equation (2.16) is oscillatory.

Ordinary and singular points. If at a point $x = x_0$ the functions $p_1(x)$ and $p_2(x)$ are analytic, then the point x_0 is said to be an *ordinary point* of the DE (2.1). Further, if at $x = x_0$ the functions $p_1(x)$ and/or $p_2(x)$ are not analytic, then x_0 is said to be a *singular point* of (2.1).

Example 3.1. If in the DE (2.1), $p_1(x)$ and $p_2(x)$ are constants, then every point is an ordinary point.

Example 3.2. Since the function $p_2(x) = x$ is analytic at every point, for the DE $y'' + xy = 0$ every point is an ordinary point.

Example 3.3. In Euler's equation

$$x^2 y'' + a_1 xy' + a_2 y = 0$$

$x = 0$ is a singular point, but every other point is an ordinary point.

A singular point x_0 at which the functions $p(x) = (x - x_0)p_1(x)$ and $q(x) = (x - x_0)^2 p_2(x)$ are analytic is called a *regular singular point* of the DE (2.1). Thus, a second-order DE with a regular singular point x_0 has the form

$$y'' + \frac{p(x)}{(x - x_0)} y' + \frac{q(x)}{(x - x_0)^2} y = 0, \tag{3.10}$$

where the functions $p(x)$ and $q(x)$ are analytic at $x = x_0$. Hence, in Example 3.3 the point $x_0 = 0$ is a regular singular point.

If a singular point x_0 is not a regular singular point, then it is called an *irregular singular point*.

Example 3.4. Consider the DE

$$y'' + \frac{1}{(x - 1)^2} y' + \frac{8}{x(x - 1)} y = 0. \tag{3.11}$$

For the equation (3.11) the singular points are 0 and 1. At the point 0, we have

$$xp_1(x) = \frac{x}{(x-1)^2} = x(1-x)^{-2}$$

and

$$x^2 p_2(x) = \frac{8x^2}{x(x-1)} = -8x(1-x)^{-1},$$

which are analytic at $x = 0$, and hence the point 0 is a regular singular point. At the point 1, we have

$$(x-1)p_1(x) = \frac{(x-1)}{(x-1)^2} = \frac{1}{(x-1)},$$

which is not analytic at $x = 1$, and hence the point 1 is an irregular singular point.

Problems

3.1. Show that

(i) $\Gamma\left(\dfrac{1}{2}\right) = \sqrt{\pi}$

(ii) for all $p > -1$ and $q > -1$ the following holds:

$$\int_0^{\pi/2} \sin^p x \cos^q x\, dx = \frac{1}{2} B\left(\frac{p+1}{2}, \frac{q+1}{2}\right) = \frac{\Gamma\left(\frac{p+1}{2}\right)\Gamma\left(\frac{q+1}{2}\right)}{2\,\Gamma\left(\frac{p+q+2}{2}\right)}.$$

3.2. Locate and classify the singular points of the following DEs:

(i) $x^2(x+2)y'' + xy' - (2x-1)y = 0$

(ii) $(x-1)^2(x+3)y'' + (2x+1)y' - y = 0$

(iii) $(1-x^2)^2 y'' + x(1-x)y' + (1+x)y = 0$

(iv) $(x^2 - x - 2)y'' + (x-2)y' + xy = 0.$

3.3. Show that $x_0 = 0$ is a regular singular point of the *Riccati–Bessel equation*

$$x^2 y'' - (x^2 - k)y = 0, \quad -\infty < k < \infty.$$

3.4. Show that $x_0 = 0$ is a regular singular point of the *Coulomb wave equation*

$$x^2 y'' + [x^2 - 2\ell x - k]y = 0, \quad \ell \text{ fixed}, \quad -\infty < k < \infty.$$

3.5. Let the point $x = x_0$, where $x_0 \neq 0$ be an ordinary point of the DE (2.1). Show that the change of the independent variable $t = x - x_0$ leads to the DE

$$\frac{d^2 y}{dt^2} + \bar{p}_1(t)\frac{dy}{dt} + \bar{p}_2(t)y = 0 \tag{3.12}$$

for which the point $t = 0$ is an ordinary point. Further, show that the function $y(t) = \sum_{m=0}^{\infty} c_m t^m$, $|t| < \mu$ is a solution of the DE (3.12) if and only if the corresponding function $y(x) = \sum_{m=0}^{\infty} c_m (x - x_0)^m$, $|x - x_0| < \mu$ is a solution of the DE (2.1).

3.6. Let the DE (2.1) have a regular singular point at $x = x_0$, where $x_0 \neq 0$. Verify that the change of the independent variable $t = x - x_0$ leads to the DE (3.12) which has a regular singular point at $t = 0$.

3.7. Show that the substitution $x = 1/t$ transforms the DE (2.1) into the form

$$\frac{d^2 y}{dt^2} + \left(\frac{2}{t} - \frac{1}{t^2}p_1\left(\frac{1}{t}\right)\right)\frac{dy}{dt} + \frac{1}{t^4}p_2\left(\frac{1}{t}\right)y = 0. \tag{3.13}$$

Thus, the nature of the point $x = \infty$ of (2.1) is the same as the nature of the point $t = 0$ of (3.13). Use this substitution to show that for the DE.

$$y'' + \frac{1}{2}\left(\frac{1}{x^2} + \frac{1}{x}\right)y' + \frac{1}{2x^3}y = 0$$

the point $x = \infty$ is a regular singular point.

3.8. Show that for Bessel's DE (2.15) the point $x = \infty$ is an irregular singular point.

3.9. Examine the nature of the point at infinity for the following DEs:

Airy's DE: $y'' - xy = 0$ $\hfill (3.14)$

Chebyshev's DE: $(1 - x^2)y'' - xy' + a^2 y = 0$ $\hfill (3.15)$

Hermite's DE: $y'' - 2xy' + 2ay = 0$ $\hfill (3.16)$

Hypergeometric DE: $x(1 - x)y'' + [c - (a + b + 1)x]y' - aby = 0$ $\hfill (3.17)$

Laguerre's DE: $xy'' + (a + 1 - x)y' + by = 0$ $\hfill (3.18)$

Legendre's DE: $(1 - x^2)y'' - 2xy' + a(a + 1)y = 0.$ $\hfill (3.19)$

3.10. The *Schrödinger wave equation* for a simple harmonic oscillator is

$$-\frac{h^2}{8\pi^2 m}\frac{d^2\psi}{dz^2} + \frac{K}{2}z^2\psi = E\psi, \tag{3.20}$$

where h is Planck's constant; E, K, and m are positive real numbers, and $\psi(x)$ is the Schrödinger wave function. Show that the change to dimensionless coordinate $x = \alpha z$ reduces (3.20) to

$$\frac{d^2\psi}{dx^2} + (2a + 1 - x^2)\psi = 0, \tag{3.21}$$

where $\alpha^4 = 4\pi^2 mK/h^2$ and $2a + 1 = (4\pi E/h)\sqrt{m/K}$. Further, show that the second change of variables $\psi = ye^{-x^2/2}$ reduces (3.21) to the Hermite equation (3.16).

Answers or Hints

3.1. (i) $\Gamma\left(\frac{1}{2}\right) = \int_0^\infty t^{-1/2}e^{-t}dt = 2\int_0^\infty e^{-u^2}du$, $t = u^2$ (ii) Use the substitution $t = \sin^2 x$.

3.2. (i) 0, -2 regular singular points (ii) 1 irregular singular point, -3 regular singular point (iii) 1 regular singular point, -1 irregular singular point (iv) $2, -1$ regular singular points.

3.3. Use definition.

3.4. Use definition.

3.5. The change of the independent variable $x = t + x_0$ gives $\frac{dy}{dx} = \frac{dy}{dt}$, $\frac{d^2y}{dx^2} = \frac{d^2y}{dt^2}$, $p_1(x) = p_1(t + x_0) = \bar{p}_1(t)$, and $p_2(x) = p_2(t + x_0) = \bar{p}_2(t)$, thus, it reduces (2.1) to (3.12). Further, since this transformation shifts every point by $-x_0$, if x_0 is an ordinary point of (2.1), then $t = 0$ is an ordinary point of (3.12).

3.6. The proof is similar to that of Problem 3.5.

3.7. The transformed equation is $\frac{d^2y}{dt^2} + \left(\frac{3-t}{2t}\right)\frac{dy}{dt} + \frac{1}{2t}y = 0$.

3.8. Since $\frac{1}{t}p\left(\frac{1}{t}\right) = 1$, and $\frac{1}{t^2}q\left(\frac{1}{t}\right) = \frac{1-t^2a^2}{t^2}$ is not analytic at $t = 0$, the point $x = \infty$ is an irregular singular point.

3.9. Irregular singular, regular singular, irregular singular, regular singular, irregular singular, regular singular.

3.10. Verify directly.

Lecture 4
Solution at an Ordinary Point

In this lecture we shall construct power series solutions of Airy, Hermite and Chebyshev DEs. These equations occupy a central position in mathematical physics, engineering, and approximation theory.

We begin by proving the following theorem, which provides sufficient conditions so that the solutions of (2.1) can be expressed as power series at an ordinary point.

Theorem 4.1. Let the functions $p_1(x)$ and $p_2(x)$ be analytic at $x = x_0$; hence these can be expressed as power series in $(x - x_0)$ in some interval $|x - x_0| < \mu$. Then, the DE (2.1) together with the initial conditions

$$y(x_0) = c_0, \quad y'(x_0) = c_1 \tag{4.1}$$

possesses a unique solution $y(x)$ that is analytic at x_0, and hence can be expressed as

$$y(x) = \sum_{m=0}^{\infty} c_m(x - x_0)^m \tag{4.2}$$

in some interval $|x - x_0| < \mu$. The coefficients c_m, $m \geq 2$ in (4.2) can be obtained by substituting it in the DE (2.1) directly.

Proof. In view of Problem 3.5 we can assume that $x_0 = 0$. Let

$$p_1(x) = \sum_{m=0}^{\infty} \bar{p}_m x^m, \quad p_2(x) = \sum_{m=0}^{\infty} \tilde{p}_m x^m, \quad |x| < \mu \tag{4.3}$$

and

$$y(x) = \sum_{m=0}^{\infty} c_m x^m, \tag{4.4}$$

where c_0 and c_1 are the same constants as in (4.1). Then,

$$y'(x) = \sum_{m=0}^{\infty} (m+1)c_{m+1}x^m, \quad y''(x) = \sum_{m=0}^{\infty} (m+1)(m+2)c_{m+2}x^m$$

and

$$p_1(x)y'(x) = \sum_{m=0}^{\infty} \left(\sum_{k=0}^{m} (k+1)c_{k+1}\bar{p}_{m-k} \right) x^m,$$

R.P. Agarwal, D. O'Regan, *Ordinary and Partial Differential Equations*,
Universitext, DOI 10.1007/978-0-387-79146-3_4,
© Springer Science+Business Media, LLC 2009

$$p_2(x)y(x) = \sum_{m=0}^{\infty} \left(\sum_{k=0}^{m} c_k \tilde{p}_{m-k} \right) x^m.$$

Substituting these expressions in the DE (2.1), we obtain

$$\sum_{m=0}^{\infty} \left[(m+1)(m+2)c_{m+2} + \sum_{k=0}^{m} (k+1)c_{k+1}\bar{p}_{m-k} + \sum_{k=0}^{m} c_k \tilde{p}_{m-k} \right] x^m = 0.$$

Hence, $y(x)$ is a solution of the DE (2.1) if and only if the constants c_m satisfy the recurrence relation

$$c_{m+2} = -\frac{1}{(m+1)(m+2)} \left[\sum_{k=0}^{m} \left\{ (k+1)c_{k+1}\bar{p}_{m-k} + c_k \tilde{p}_{m-k} \right\} \right], \quad m \geq 0$$

which is the same as

$$c_m = -\frac{1}{m(m-1)} \left[\sum_{k=0}^{m-2} \left\{ (k+1)c_{k+1}\bar{p}_{m-k-2} + c_k \tilde{p}_{m-k-2} \right\} \right], \quad m \geq 2.$$
(4.5)

By this relation $c_2, c_3, \cdots$ can be determined successively as linear combinations of c_0 and c_1.

Now we shall show that the series with these coefficients converges for $|x| < \mu$. Since the series for $p_1(x)$ and $p_2(x)$ converge for $|x| < \mu$, for any $|x| = \mu_0 < \mu$ there exists a constant $M > 0$ such that

$$|\bar{p}_j|\mu_0^j \leq M \quad \text{and} \quad |\tilde{p}_j|\mu_0^j \leq M, \quad j = 0, 1, \cdots.$$
(4.6)

Using (4.6) in (4.5), we find

$$|c_m| \leq \frac{M}{m(m-1)} \left[\sum_{k=0}^{m-2} \left\{ \frac{(k+1)|c_{k+1}|}{\mu_0^{m-k-2}} + \frac{|c_k|}{\mu_0^{m-k-2}} \right\} \right] + \frac{M|c_{m-1}|\mu_0}{m(m-1)}, \quad m \geq 2$$
(4.7)

where the term $M|c_{m-1}|\mu_0/m(m-1)$ has been included, the purpose of which will be clear later.

Now we define positive constants C_m by the equations $C_0 = |c_0|$, $C_1 = |c_1|$,

$$C_m = \frac{M}{m(m-1)} \left[\sum_{k=0}^{m-2} \left\{ \frac{(k+1)C_{k+1}}{\mu_0^{m-k-2}} + \frac{C_k}{\mu_0^{m-k-2}} \right\} \right] + \frac{MC_{m-1}\mu_0}{m(m-1)}, \quad m \geq 2.$$
(4.8)

From (4.7) and (4.8) it is clear that $|c_m| \leq C_m$, $m = 0, 1, \cdots$.

Next we replace m by $m + 1$ in (4.8), to obtain

$$C_{m+1} = \frac{M}{m(m+1)} \left[\sum_{k=0}^{m-1} \left\{ \frac{(k+1)C_{k+1}}{\mu_0^{m-k-1}} + \frac{C_k}{\mu_0^{m-k-1}} \right\} \right] + \frac{MC_m\mu_0}{m(m+1)}$$

and hence

$$\mu_0 C_{m+1} = \frac{M\mu_0}{m(m+1)} \left[\sum_{k=0}^{m-2} \left\{ \frac{(k+1)C_{k+1}}{\mu_0^{m-k-1}} + \frac{C_k}{\mu_0^{m-k-1}} \right\} \right]$$
$$+ \frac{M\mu_0}{m(m+1)} [mC_m + C_{m-1}] + \frac{MC_m\mu_0^2}{m(m+1)}. \tag{4.9}$$

Combining (4.8) and (4.9), we get

$$\mu_0 C_{m+1} = \frac{M}{m(m+1)} \left[\frac{m(m-1)}{M} C_m - \mu_0 C_{m-1} \right]$$
$$+ \frac{M\mu_0}{m(m+1)} [mC_m + C_{m-1}] + \frac{MC_m\mu_0^2}{m(m+1)},$$

which is the same as

$$\mu_0 C_{m+1} = \frac{(m-1)}{(m+1)} C_m + \frac{mM\mu_0 C_m}{m(m+1)} + \frac{MC_m\mu_0^2}{m(m+1)}. \tag{4.10}$$

Thus, the addition of $M|c_{m-1}|\mu_0/m(m-1)$ in (4.7) has led to a two-term recurrence relation (4.10) from which we have

$$\left| \frac{C_{m+1}x^{m+1}}{C_m x^m} \right| = \frac{m(m-1) + mM\mu_0 + M\mu_0^2}{\mu_0 m(m+1)} |x|$$

and hence

$$\lim_{m \to \infty} \left| \frac{C_{m+1}x^{m+1}}{C_m x^m} \right| = \frac{|x|}{\mu_0}.$$

Thus, the ratio test establishes that the series $\sum_{m=0}^{\infty} C_m x^m$ converges for $|x| < \mu_0$, and by the comparison test it follows that the series $\sum_{m=0}^{\infty} c_m x^m$ converges absolutely in $|x| < \mu_0$. Since $\mu_0 \in (0, \mu)$ is arbitrary, the series converges absolutely in the interval $|x| < \mu$.

Hence, we have shown that a function which is analytic at $x = x_0$ is a solution of the initial value problem (2.1), (4.1) if and only if the coefficients in its power series expansion satisfy the relation (4.5). Also, from the uniqueness of the solutions of (2.1), (4.1) it follows that this will be the only solution. ∎

Airy's equation.   Solutions of Airy's DE (3.14) are called *Airy functions*, which have applications in the theory of diffraction. Clearly, for (3.14) hypotheses of Theorem 4.1 are satisfied for all x, and hence its solutions have power series expansion about any point $x = x_0$. In the case $x_0 = 0$, we assume that $y(x) = \sum_{m=0}^{\infty} c_m x^m$ is a solution of (3.14). A direct substitution of this in (3.14) gives

$$\sum_{m=0}^{\infty} (m+1)(m+2)c_{m+2}x^m - x \sum_{m=0}^{\infty} c_m x^m = 0,$$

which is the same as

$$2c_2 + \sum_{m=1}^{\infty} [(m+1)(m+2)c_{m+2} - c_{m-1}]x^m = 0.$$

Hence, it follows that

$$c_2 = 0, \quad c_m = \frac{1}{m(m-1)}c_{m-3}, \quad m \geq 3. \tag{4.11}$$

If $m = 3k + 2$, then (4.11) becomes

$$c_{3k+2} = \frac{1}{(3k+2)(3k+1)}c_{3k-1} = \frac{1.2.3.6.9\cdots(3k)}{(3k+2)!}c_2 = 0, \quad k = 1, 2, \cdots.$$

If $m = 3k + 1$, then (4.11) is the same as

$$c_{3k+1} = \frac{1}{(3k+1)(3k)}c_{3k-2} = \frac{2.5\cdots(3k-1)}{(3k+1)!}c_1, \quad k = 1, 2, \cdots.$$

If $m = 3k$, then (4.11) reduces to

$$c_{3k} = \frac{1}{(3k)(3k-1)}c_{3k-3} = \frac{1.4.7\cdots(3k-2)}{(3k)!}c_0, \quad k = 1, 2, \cdots.$$

Since

$$y(x) = c_0 + c_1 x + \sum_{k=1}^{\infty} c_{3k}x^{3k} + \sum_{k=1}^{\infty} c_{3k+1}x^{3k+1} + \sum_{k=1}^{\infty} c_{3k+2}x^{3k+2},$$

Airy functions are given by

$$y(x) = c_0 \left[1 + \sum_{k=1}^{\infty} \frac{1.4\cdots(3k-2)}{(3k)!}x^{3k} \right] + c_1 \left[x + \sum_{k=1}^{\infty} \frac{2.5\cdots(3k-1)}{(3k+1)!}x^{3k+1} \right]$$
$$= c_0 y_1(x) + c_1 y_2(x). \tag{4.12}$$

Finally, since $y_1(0) = 1$, $y_1'(0) = 0$ and $y_2(0) = 0$, $y_2'(0) = 1$ functions $y_1(x)$ and $y_2(x)$ are linearly independent solutions of Airy's equation (cf. Theorem 2.2).

Hermite's equation. Solutions of Hermite's DE (3.16) are called *Hermite functions*. This equation is used in quantum mechanics to study the spatial position of a moving particle that undergoes simple harmonic motion in time. In quantum mechanics the exact position of a particle at a given time cannot be predicted, as in classical mechanics. It is possible to determine only the probability of the particle's being at a given location at a given time. The unknown function $y(x)$ in (3.16) is then related to the probability of finding the particle at the position x. The constant a is related to the energy of the particle. Clearly, for (3.16) also hypotheses of Theorem 4.1 are satisfied for all x, and hence its solutions have power series expansion about any point $x = x_0$. In the case $x_0 = 0$, we again assume that $y(x) = \sum_{m=0}^{\infty} c_m x^m$ is a solution of (3.16), and obtain the recurrence relation

$$c_m = \frac{2(m-2-a)}{m(m-1)} c_{m-2}, \quad m = 2, 3, \cdots. \qquad (4.13)$$

From (4.13) it is easy to find

$$c_{2m} = \frac{(-1)^m 2^{2m} \Gamma\left(\frac{1}{2}a + 1\right)}{(2m)!\, \Gamma\left(\frac{1}{2}a - m + 1\right)} c_0, \quad m = 0, 1, \cdots$$

and

$$c_{2m+1} = \frac{(-1)^m 2^{2m+1} \Gamma\left(\frac{1}{2}a + \frac{1}{2}\right)}{2\,(2m+1)!\, \Gamma\left(\frac{1}{2}a - m + \frac{1}{2}\right)} c_1, \quad m = 0, 1, \cdots.$$

Hence, Hermite functions can be written as

$$\begin{aligned}
y(x) &= c_0\, \Gamma\left(\frac{1}{2}a + 1\right) \sum_{m=0}^{\infty} \frac{(-1)^m (2x)^{2m}}{(2m)!\, \Gamma\left(\frac{1}{2}a - m + 1\right)} \\
&\quad + c_1 \frac{1}{2}\, \Gamma\left(\frac{1}{2}a + \frac{1}{2}\right) \sum_{m=0}^{\infty} \frac{(-1)^m (2x)^{2m+1}}{(2m+1)!\, \Gamma\left(\frac{1}{2}a - m + \frac{1}{2}\right)} \qquad (4.14) \\
&= c_0 y_1(x) + c_1 y_2(x).
\end{aligned}$$

Obviously, $y_1(x)$ and $y_2(x)$ are linearly independent solutions of Hermite's equation.

Chebyshev's equation. The Chebyshev DE (3.15), where a is a real constant (parameter), arises in approximation theory. Since the functions

$$p_1(x) = -\frac{x}{1-x^2} \quad \text{and} \quad p_2(x) = \frac{a^2}{1-x^2}$$

are analytic for $|x| < 1$, $x = x_0 = 0$ is an ordinary point. Thus, Theorem 4.1 ensures that its series solution $y(x) = \sum_{m=0}^{\infty} c_m x^m$ converges for $|x| < 1$. To find this solution, we substitute it directly in (3.15), to find the recurrence relation

$$c_{m+2} = \frac{(m^2 - a^2)}{(m+2)(m+1)} c_m, \quad m \geq 0 \tag{4.15}$$

which can be solved to obtain

$$c_{2m} = \frac{(-a^2)(2^2 - a^2) \cdots ((2m-2)^2 - a^2)}{(2m)!} c_0, \quad m \geq 1$$

$$c_{2m+1} = \frac{(1^2 - a^2)(3^2 - a^2) \cdots ((2m-1)^2 - a^2)}{(2m+1)!} c_1, \quad m \geq 1.$$

Hence, the solution of (3.15) can be written as

$$\begin{aligned}
y(x) &= c_0 \left[1 + \sum_{m=1}^{\infty} \frac{(-a^2)(2^2 - a^2) \cdots ((2m-2)^2 - a^2)}{(2m)!} x^{2m} \right] \\
&\quad + c_1 \left[x + \sum_{m=1}^{\infty} \frac{(1^2 - a^2)(3^2 - a^2) \cdots ((2m-1)^2 - a^2)}{(2m+1)!} x^{2m+1} \right] \\
&= c_0 y_1(x) + c_1 y_2(x).
\end{aligned}$$

(4.16)

It is easy to verify that $y_1(x)$ and $y_2(x)$ are linearly independent solutions of Chebyshev's equation.

Problems

4.1. Verify that for each of the following DEs the given point is an ordinary point and express the general solution of each equation in terms of power series about this point:

(i) $y'' + xy' + y = 0$, $x = 0$

(ii) $y'' + x^2 y' + xy = 0$, $x = 0$

(iii) $y'' + x^2 y = 0$, $x = 0$

(iv) $(x^2 - 1)y'' - 6xy' + 12y = 0$, $x = 0$

(v) $(x^2 - 1)y'' + 8xy' + 12y = 0$, $x = 0$

(vi) $y'' - 2(x + 3)y' - 3y = 0$, $x = -3$

(vii) $y'' + (x - 2)^2 y' - 7(x - 2)y = 0$, $x = 2$

(viii) $(x^2 - 2x)y'' + 5(x - 1)y' + 3y = 0$, $x = 1$.

4.2. For each of the power series obtained in Problem 4.1 find the radius of convergence and the *interval of convergence* (the interval centered at x_0 in which the power series converges).

4.3.   Find series solutions of the following initial value problems:

(i) $y'' + xy' - 2y = 0$, $y(0) = 1$, $y'(0) = 0$

(ii) $x(2 - x)y'' - 6(x - 1)y' - 4y = 0$, $y(1) = 1$, $y'(1) = 0$

(iii) $y'' + e^x y' + (1 + x^2)y = 0$, $y(0) = 1$, $y'(0) = 0$

(iv) $y'' - (\sin x)y = 0$, $y(\pi) = 1$, $y'(\pi) = 0$.

4.4.   If the hypotheses of Theorem 4.1 are satisfied, then the solution $y(x)$ of (2.1), (4.1) possesses a unique Taylor's series expansion at x_0, i.e.,

$$y(x) = \sum_{m=0}^{\infty} \frac{y^{(m)}(x_0)}{m!}(x - x_0)^m. \tag{4.17}$$

For many problems it is easy to find $y^{(m)}(x_0)$ for all m, and hence we can start directly with (4.17). We call this procedure *Taylor's series method*. Use this method to solve the following problems:

(i) $y'' + y = 2x - 1$, $y(1) = 1$, $y'(1) = 3$

(ii) $y'' + 4y' + 3y = 0$, $y(0) = 1$, $y'(0) = -1$.

4.5.   *Van der Pol's equation,*

$$y'' + \mu(y^2 - 1)y' + y = 0, \tag{4.18}$$

finds applications in physics and electrical engineering. It first arose as an idealized description of a spontaneously oscillating circuit. Find first three nonzero terms of the power series solution about $x = 0$ of (4.18) with $\mu = 1$ subject to the conditions $y(0) = 0$, $y'(0) = 1$.

4.6.   *Rayleigh's equation,*

$$my'' + ky = ay' - b(y')^3, \tag{4.19}$$

models the oscillation of a clarinet reed. Find first three nonzero terms of the power series solution about $x = 0$ of (4.19) with $m = k = a = 1$, $b = 1/3$ subject to the conditions $y(0) = 1$, $y'(0) = 0$.

Answers or Hints

4.1.   (i) $c_0 \sum_{m=0}^{\infty} \frac{(-1)^m}{2^m \, m!} x^{2m} + c_1 \sum_{m=0}^{\infty} \frac{(-1)^m \, 2^m \, m!}{(2m+1)!} x^{2m+1}$

(ii) $c_0 \left[1 + \sum_{m=1}^{\infty}(-1)^m \frac{1^2 \cdot 4^2 \cdot 7^2 \cdots (3m-2)^2}{(3m)!} x^{3m} \right]$

$\qquad + c_1 \left[x + \sum_{m=1}^{\infty}(-1)^m \frac{2^2 \cdot 5^2 \cdot 8^2 \cdots (3m-1)^2}{(3m+1)!} x^{3m+1} \right]$

(iii) $c_0 \left[1 + \sum_{m=1}^{\infty} \frac{(-1)^m x^{4m}}{3 \cdot 7 \cdots (4m-1) \, 4^m \, m!} \right] + c_1 \sum_{m=0}^{\infty} \frac{(-1)^m x^{4m+1}}{1 \cdot 5 \cdot 9 \cdots (4m+1) \, 4^m \, m!}$

(iv) $c_0(1 + 6x^2 + x^4) + c_1(x + x^3)$

(v) $c_0 \sum_{m=0}^{\infty}(m+1)(2m+1)x^{2m} + c_1 \sum_{m=0}^{\infty} \frac{(m+1)(2m+3)}{3}x^{2m+1}$

(vi) $c_0 \left[1 + \sum_{m=1}^{\infty} \frac{3\cdot7\cdot11\cdots(4m-1)}{(2m)!}(x+3)^{2m}\right]$

$\quad + c_1 \sum_{m=0}^{\infty} \frac{1\cdot5\cdot9\cdots(4m+1)}{(2m+1)!}(x+3)^{2m+1}$

(vii) $c_0 \left[1 + \sum_{m=1}^{\infty} \frac{(-1)^{m+1}28(x-2)^{3m}}{3^m\, m!(3m-1)(3m-4)(3m-7)}\right]$

$\quad + c_1 \left[(x-2) + \frac{1}{2}(x-2)^4 + \frac{1}{28}(x-2)^7\right]$

(viii) $c_0 \sum_{m=0}^{\infty} \frac{1\cdot3\cdot5\cdots(2m+1)}{2^m\, m!}(x-1)^{2m} + c_1 \sum_{m=0}^{\infty} \frac{2^m\,(m+1)!}{1\cdot3\cdot5\cdots(2m+1)}(x-1)^{2m+1}.$

4.2. (i) For $\sum_{m=0}^{\infty} \frac{(-1)^m}{2^m\, m!}x^{2m}$, $\left|\frac{C_{m+1}}{C_m}\right| = \frac{2^m\, m!}{2^{m+1}\,(m+1)!}|x|^2 = \frac{|x|^2}{2(m+1)} \to$ 0, and hence the interval of convergence is the whole real line $\mathbb{R}$. For $\sum_{m=0}^{\infty} \frac{(-1)^m\, 2^m\, m!}{(2m+1)!}x^{2m+1}$, $\left|\frac{C_{m+1}}{C_m}\right| = \frac{|x|^2}{2m+3} \to$ 0, and hence again the interval of convergence is $\mathbb{R}$ (ii) $\mathbb{R}$ (iii) $\mathbb{R}$ (iv) $\mathbb{R}$ (v) 1, $(-1,1)$ (vi) $\mathbb{R}$ (vii) $\mathbb{R}$ (viii) 1, $(0,2)$.

4.3. (i) $1 + x^2$ (ii) $\sum_{m=0}^{\infty}(m+1)(x-1)^{2m}$ (iii) $1 - \frac{1}{2}x^2 + \frac{1}{6}x^3 - \frac{1}{120}x^5 + \frac{11}{720}x^6 \cdots$ (iv) $1 - \frac{1}{6}(x-\pi)^3 + \frac{1}{120}(x-\pi)^5 + \frac{1}{180}(x-\pi)^6 + \cdots$.

4.4. (i) $y^{(2m)}(1) = 0$, $m = 1, 2, \cdots$; $y^{(2m+1)}(1) = 1$, $m = 2, 4, \cdots$; $y^{(2m+1)}(1) = -1$, $m = 1, 3, \cdots$; $(2x-1) + \sin(x-1)$ (ii) e^{-x}.

4.5. $x + \frac{1}{2}x^2 - \frac{1}{8}x^4 - \frac{1}{8}x^5$.

4.6. $1 - \frac{1}{2}x^2 - \frac{1}{6}x^3 + \frac{1}{40}x^5$.

Lecture 5
Solution at a Singular Point

In this lecture, through a simple example, first we shall show that at a regular singular point the power series used earlier at an ordinary point does not provide a solution, and hence we need to modify it. This modification is called *the method of Frobenius* after George Frobenius (1849–1917). Then we shall state and prove a general result which provides three possible different forms of the power series solution. Once the particular form of the solution is known its construction is almost routine. In fact, in the next lecture we shall illustrate this result through several examples; this includes a discussion of Laguerre's equation (3.18).

We recall that a second-order DE with a regular singular point x_0 is of the form (3.10), where the functions $p(x)$ and $q(x)$ are analytic at $x = x_0$. Further, in view of Problem 3.6, we can assume that $x_0 = 0$, so that equation (3.10) reduces to

$$y'' + \frac{p(x)}{x}y' + \frac{q(x)}{x^2}y = 0. \qquad (5.1)$$

In comparison with at an ordinary point, the construction of a series solution at a singular point is difficult. To understand the problem we consider Euler's equation

$$2x^2 y'' + xy' - y = 0; \qquad (5.2)$$

which has a regular singular point at $x = 0$; and its general solution

$$y(x) = c_1 x^{-1/2} + c_2 x \qquad (5.3)$$

exists in the interval $J = (0, \infty)$. Obviously, no solution of (5.2) can be represented by a power series with $x = 0$ as its point of expansion in any interval of the form $(0, a)$, $a > 0$. For if $y(x) = \sum_{m=0}^{\infty} c_m x^m$, $0 < x < a$ is a solution of (5.2), then $y(x)$ and all its derivatives possess finite right limits at $x = 0$, whereas no function of the form (5.3) has this property. Hence, at a regular singular point, solutions of (5.1) need not be analytic (in some instances solutions may be analytic, e.g., $y(x) = c_1 x + c_2 x^2$ is the general solution of $x^2 y'' - 2xy' + 2y = 0$). However, we shall see that every such DE does possess at least one solution of the form

$$y(x) = x^r \sum_{m=0}^{\infty} c_m x^m, \quad c_0 \neq 0. \qquad (5.4)$$

R.P. Agarwal, D. O'Regan, *Ordinary and Partial Differential Equations*,
Universitext, DOI 10.1007/978-0-387-79146-3_5,
© Springer Science+Business Media, LLC 2009

Since $p(x)$ and $q(x)$ are analytic at $x = 0$, these functions can be expressed as power series in x; i.e.,

$$p(x) = \sum_{m=0}^{\infty} p_m x^m \quad \text{and} \quad q(x) = \sum_{m=0}^{\infty} q_m x^m. \qquad (5.5)$$

Substituting (5.4) and (5.5) in (5.1), we obtain

$$x^{r-2} \sum_{m=0}^{\infty} (m+r)(m+r-1)c_m x^m + \frac{1}{x} \left(\sum_{m=0}^{\infty} p_m x^m \right) \left(x^{r-1} \sum_{m=0}^{\infty} (m+r)c_m x^m \right)$$
$$+ \frac{1}{x^2} \left(\sum_{m=0}^{\infty} q_m x^m \right) \left(x^r \sum_{m=0}^{\infty} c_m x^m \right) = 0,$$

which is the same as

$$\sum_{m=0}^{\infty} \left\{ (m+r)(m+r-1)c_m + \sum_{k=0}^{m} [(k+r)p_{m-k} + q_{m-k}]c_k \right\} x^{m+r-2} = 0. \qquad (5.6)$$

In (5.6) the coefficient of x^{r-2} does not lead to a recurrence relation, but gives

$$c_0 F(r) = c_0[r(r-1) + p_0 r + q_0] = 0. \qquad (5.7)$$

The other terms lead to the recurrence relation

$$(m+r)(m+r-1)c_m + \sum_{k=0}^{m} [(k+r)p_{m-k} + q_{m-k}]c_k = 0, \quad m = 1, 2, \cdots$$

which can be written as

$$\begin{aligned} F(r+m)c_m &= [(m+r)(m+r-1) + (m+r)p_0 + q_0]c_m \\ &= -\sum_{k=0}^{m-1} [(k+r)p_{m-k} + q_{m-k}]c_k, \quad m = 1, 2, \cdots. \end{aligned} \qquad (5.8)$$

Since $c_0 \neq 0$, the possible values of r are those which are the roots of the *indicial equation* $F(r) = 0$. The roots r_1 and r_2 are called the *exponents* of the regular singular point $x = 0$. Once r is fixed the relation (5.8) determines c_m as successive multiples of c_0. Thus, for two exponents r_1 and r_2 we can construct two solutions of the DE (5.1). However, if $r_1 = r_2$, then this method gives only one formal solution. Further, if at any stage $F(r+m)$ vanishes then this method obviously breaks down. A simple calculation shows that

$$F(r+m) = F(r) + m(2r + p_0 + m - 1) = 0. \qquad (5.9)$$

But from (5.7), we have $r_1 + r_2 = 1 - p_0$ and hence if $r = r_1$ or r_2, then (5.9) implies that $m = \pm(r_2 - r_1)$. Therefore, $F(r + m)$ vanishes if and only if the exponents differ by an integer, and r is chosen to be the smaller exponent. Thus, if r is taken to be the larger exponent, we can construct one formal solution.

In conclusion, the DE (5.1) always has at least one solution of the form (5.4), and the coefficients c_m, $m \geq 1$ can be obtained by substituting it in the equation directly. Further, to find the second solution either the method provided in Lecture 2, or the method of *Frobenius*, can be employed. In the following result we summarize the conclusions of Frobenius method.

Theorem 5.1. Let the functions $p(x)$ and $q(x)$ be analytic at $x = 0$, and hence these can be expressed as power series given in (5.5) for $|x| < \mu$. Further, let r_1 and r_2 be the roots of the indicial equation $F(r) = r(r - 1) + p_0 r + q_0 = 0$. Then,

(i). if $\mathrm{Re}(r_1) \geq \mathrm{Re}(r_2)$ and $r_1 - r_2$ is not a nonnegative integer, then the two linearly independent solutions of the DE (5.1) are

$$y_1(x) = |x|^{r_1} \sum_{m=0}^{\infty} c_m x^m, \tag{5.10}$$

and

$$y_2(x) = |x|^{r_2} \sum_{m=0}^{\infty} \bar{c}_m x^m; \tag{5.11}$$

(ii). if the roots of the indicial equation are equal, i.e., $r_2 = r_1$ then the two linearly independent solutions of the DE (5.1) are (5.10) and

$$y_2(x) = y_1(x) \ln |x| + |x|^{r_1} \sum_{m=1}^{\infty} d_m x^m; \tag{5.12}$$

(iii). if the roots of the indicial equation are such that $r_1 - r_2 = n$ (a positive integer) then the two linearly independent solutions of the DE (5.1) are (5.10) and

$$y_2(x) = c y_1(x) \ln |x| + |x|^{r_2} \sum_{m=0}^{\infty} e_m x^m, \tag{5.13}$$

where the coefficients c_m, $\bar{c}_m$, d_m, e_m and the constant c can be determined by substituting the form of the series for $y(x)$ in the equation (5.1). The constant c may turn out to be zero, in which case there is no logarithmic term in the solution (5.13). Each of the solutions given in (5.10) – (5.13) converges at least for $0 < |x| < \mu$.

Proof. (i) Since r_1 and r_2 are the roots of the indicial equation $F(r) = 0$, we have from (5.7) that

$$F(r) = r(r - 1) + p_0 r + q_0 = (r - r_1)(r - r_2)$$

and from (5.9) that $F(r_1 + m) = m(m + r_1 - r_2)$, and hence

$$|F(r_1 + m)| \geq m(m - |r_1 - r_2|). \tag{5.14}$$

Also, as in the proof of Theorem 4.1, for any $|x| = \mu_0 < \mu$ there exists a constant $M > 0$ such that $|p_j|\mu_0^j \leq M$, and $|q_j|\mu_0^j \leq M$, $j = 0, 1, \cdots$. Thus, on using these inequalities, from (5.14) and (5.8) it follows that

$$m(m - |r_1 - r_2|)|c_m| \leq M \sum_{k=0}^{m-1} (k + |r_1| + 1)\mu_0^{-m+k}|c_k|, \quad m = 1, 2, \cdots.$$

Now we choose an integer n such that $n - 1 \leq |r_1 - r_2| < n$, and define the positive constants C_j as follows:

$$C_j = |c_j|, \quad j = 0, 1, \cdots, n - 1$$

$$j(j - |r_1 - r_2|)C_j = M \sum_{k=0}^{j-1} (k + |r_1| + 1)\mu_0^{-j+k} C_k, \quad j = n, n + 1, \cdots.$$

$$\tag{5.15}$$

By an easy induction argument it is clear that $|c_m| \leq C_m$, $m = 0, 1, \cdots$.

Now the result of combining (5.15) with the equations obtained by replacing j by m and $m - 1$ leads to

$$\frac{C_m}{C_{m-1}} = \frac{(m - 1)(m - 1 - |r_1 - r_2|) + M(m + |r_1|)}{\mu_0 m(m - |r_1 - r_2|)}$$

and hence

$$\lim_{m \to \infty} \left| \frac{C_m x^m}{C_{m-1} x^{m-1}} \right| = \frac{|x|}{\mu_0}.$$

Thus, the ratio test shows that the series $\sum_{m=0}^{\infty} C_m x^m$ converges for $|x| < \mu_0$, and now by the comparison test $\sum_{m=0}^{\infty} c_m x^m$ converges absolutely in the interval $|x| < \mu_0$. However, since μ_0 is arbitrary, the series $\sum_{m=0}^{\infty} c_m x^m$ converges absolutely for $|x| < \mu$. Finally, the presence of the factor $|x|^{r_1}$ may introduce a singular point at the origin. Thus, we can at least say that $|x|^{r_1} \sum_{m=0}^{\infty} c_m x^m$ is a solution of the DE (5.1) and it is analytic for $0 < |x| < \mu$.

If we replace r_1 by r_2 in the above considerations, then it follows that $|x|^{r_2} \sum_{m=0}^{\infty} \bar{c}_m x^m$ is the second solution of (5.1) which is also analytic for $0 < |x| < \mu$.

(ii) Since the roots of the indicial equation $F(r) = 0$ are repeated, i.e., $r_1 = r_2$, we have $F(r_1) = (\partial F / \partial r)_{r=r_1} = 0$, and there exists a solution $y_1(x) = x^{r_1} \sum_{m=0}^{\infty} c_m x^m$ in the interval $0 < x < \mu$.

Now in (5.6) we assume that r is not a solution of the indicial equation, but the coefficients c_m satisfy the recurrence relation (5.8). Thus, if $\mathcal{L}_2[y]$ represents the left side of (5.1), then

$$\mathcal{L}_2[y(x)] = c_0 x^{r-2} F(r), \qquad (5.16)$$

where $y(x) = x^r \sum_{m=0}^{\infty} c_m x^m$.

From (5.16), it follows that

$$\frac{\partial}{\partial r}\mathcal{L}_2[y(x)] = \mathcal{L}_2\left[\frac{\partial y(x)}{\partial r}\right] = c_0 x^{r-2}\left[\frac{\partial F(r)}{\partial r} + F(r)\ln x\right]$$

and hence $\mathcal{L}_2\left[(\partial y(x)/\partial r)_{r=r_1}\right] = 0$, i.e., $(\partial y(x)/\partial r)_{r=r_1}$ is the second formal solution. Since

$$\frac{\partial y(x)}{\partial r} = x^r \ln x \sum_{m=0}^{\infty} c_m x^m + x^r \sum_{m=0}^{\infty} \frac{\partial c_m}{\partial r} x^m,$$

we find

$$y_2(x) = \left(\frac{\partial y(x)}{\partial r}\right)_{r=r_1} = y_1(x)\ln x + x^{r_1}\sum_{m=0}^{\infty} d_m x^m,$$

where

$$d_m = \left(\frac{\partial c_m}{\partial r}\right)_{r=r_1}, \qquad m = 0, 1, \cdots. \qquad (5.17)$$

Clearly, c_0 does not depend on r, and hence d_0 is zero.

Finally, we note that the case $-\mu < x < 0$ can be considered similarly. Further, since $\sum_{m=0}^{\infty} c_m x^m$ is uniformly and absolutely convergent for $|x| \leq \mu_1 < \mu$, it follows that $\sum_{m=1}^{\infty} d_m x^m$ is uniformly and absolutely convergent for $|x| \leq \mu_1 \leq \mu$; this also justifies our assumption that differentiation with respect to r can be performed term by term. Consequently, the solution $y_2(x)$ is analytic for $0 < |x| < \mu$.

(iii). Since the roots r_1 and r_2 of the indicial equation $F(r) = 0$ are such that $r_1 - r_2 = n$ (a positive integer), it is immediate that the solution $y_1(x)$ corresponding to the exponent r_1 can be given by (5.10). Further, $y_1(x)$ is indeed analytic for $0 < |x| < \mu$.

Corresponding to r_2 we can obtain $c_m(r_2)$ for $m = 1, 2, \cdots, n-1$ as a linear multiple of c_0 from the recurrence relation (5.8). However, since in (5.8) the coefficient of $c_n(r_2)$ is $F(r_2 + n) = F(r_1) = 0$ we cannot obtain a finite value of $c_n(r_2)$. To obviate this difficulty we choose $c_0(r) = r - r_2$, so that $c_m(r_2) = 0$ for $m = 0, 1, \cdots, n-1$ and $c_n(r_2)$ is indeterminate. Let us choose an arbitrary value of $c_n(r_2)$. Repeated application of (5.8) now yields $c_{n+m}(r_2)$ as linear multiples of $c_n(r_2)$ for positive integers m. This process produces the solution

$$x^{r_2} \sum_{m=n}^{\infty} c_m(r_2)x^m = x^{r_1-n} \sum_{m=n}^{\infty} c_m(r_2)x^m = x^{r_1} \sum_{m=n}^{\infty} c_m(r_2)x^{m-n}$$

$$= x^{r_1} \sum_{\tau=0}^{\infty} c_\tau^*(r_2)x^\tau,$$

(5.18)

where $c_\tau^*(r_2) = c_{n+\tau}(r_2)$, $\tau = 0, 1, \cdots$. However, since the successive coefficients are calculated from (5.8), this solution is a constant multiple of the solution $y_1(x)$.

Once again, as in Part (ii), we take r not to be the solution of the indicial equation but the coefficients c_m satisfy the recurrence relation (5.8), so that

$$\mathcal{L}_2[y(x)] = c_0 x^{r-2} F(r) = c_0 x^{r-2}(r - r_1)(r - r_2) = x^{r-2}(r - r_1)(r - r_2)^2.$$

Now on account of the repeated factor, we have $\mathcal{L}_2\left[(\partial y(x)/\partial r)_{r=r_2}\right] = 0$, and hence $(\partial y(x)/\partial r)_{r=r_2}$ is the second formal solution, i.e.,

$$y_2(x) = x^{r_2} \ln|x| \sum_{m=0}^{\infty} c_m(r_2)x^m + x^{r_2} \sum_{m=0}^{\infty} \left(\frac{\partial c_m}{\partial r}\right)_{r=r_2} x^m,$$

which from (5.18) is the same as (5.13), where

$$c = \lim_{r \to r_2}(r - r_2)c_n(r) \quad \text{and} \quad e_m = \left(\frac{\partial c_m}{\partial r}\right)_{r=r_2}, \quad m = 0, 1, \cdots. \quad (5.19)$$

Clearly, this solution $y_2(x)$ is also analytic for $0 < |x| < \mu$. $\blacksquare$

Lecture 6
Solution at a Singular Point (Cont'd.)

In this lecture, we shall illustrate Theorem 5.1 through several examples. We begin with Laguerre's equation (3.18) which shows how easily Theorem 5.1(i) is applied in practice.

Laguerre's equation. In the DE (3.18), a and b are real constants (parameters). It arises in quantum mechanics. Clearly, in this equation $p(x) = (a + 1 - x)$ and $q(x) = bx$ are analytic for all x, and hence the point $x = 0$ is a regular singular point. Since $p_0 = a + 1$, $q_0 = 0$ the indicial equation is $F(r) = r(r - 1) + (a + 1)r = 0$, and therefore the exponents are $r_1 = 0$ and $r_2 = -a$. Further, since $p_1 = -1$, $q_1 = b$ and $p_m = q_m = 0$, $m \geq 2$ the recurrence relation (5.8) for $r_1 = 0$ and $r_2 = -a$, respectively, reduces to

$$m(m + a)c_m = (m - 1 - b)c_{m-1}$$

and

$$m(m - a)c_m = (m - 1 - a - b)c_{m-1}.$$

Thus, if a is not zero or an integer we easily obtain the solutions

$$
\begin{aligned}
y_1(x) &= 1 - \frac{b}{a+1}x + \frac{b(b-1)}{2!\,(a+1)(a+2)}x^2 - \cdots \\
&= \sum_{m=0}^{\infty} \frac{(-1)^m\,\Gamma(a+1)\,\Gamma(b+1)}{m!\,\Gamma(m+a+1)\,\Gamma(b+1-m)}x^m
\end{aligned}
\tag{6.1}
$$

and

$$
\begin{aligned}
y_2(x) &= |x|^{-a}\left(1 - \frac{a+b}{1-a}x + \frac{1}{2!}\frac{(a+b)(a+b-1)}{(1-a)(2-a)}x^2 - \cdots\right) \\
&= |x|^{-a}\sum_{m=0}^{\infty}\frac{(-1)^m\,\Gamma(a+b+1)\,\Gamma(1-a)}{m!\,\Gamma(a+b+1-m)\,\Gamma(m+1-a)}x^m.
\end{aligned}
\tag{6.2}
$$

Clearly, in view of Theorem 5.1(i) both of these solutions converge at least for $0 < |x| < \infty$. Further, the general solution of (3.18) appears as $y(x) = Ay_1(x) + By_2(x)$, where A and B are arbitrary constants.

R.P. Agarwal, D. O'Regan, *Ordinary and Partial Differential Equations*,
Universitext, DOI 10.1007/978-0-387-79146-3_6,
© Springer Science+Business Media, LLC 2009

The following example also dwells upon the importance of Theorem 5.1(i).

Example 6.1. In the DE

$$x^2 y'' + x\left(x - \frac{1}{2}\right) y' + \frac{1}{2} y = 0 \tag{6.3}$$

$p(x) = x - (1/2)$ and $q(x) = 1/2$ are analytic for all x, and hence the point $x = 0$ is a regular singular point. Since $p_0 = -1/2$ and $q_0 = 1/2$ the indicial equation is

$$F(r) = r(r-1) - \frac{1}{2} r + \frac{1}{2} = (r-1)\left(r - \frac{1}{2}\right) = 0$$

and therefore the exponents are $r_1 = 1$ and $r_2 = 1/2$. Thus, Theorem 5.1(i) is applicable and we can construct two linearly independent solutions in any interval not containing the origin.

The recurrence relation (5.8) for $r_1 = 1$ reduces to

$$m\left(m + \frac{1}{2}\right) c_m = -m c_{m-1},$$

which is the same as

$$c_m = -\frac{2}{2m+1} c_{m-1}$$

and gives

$$c_m = (-1)^m \frac{2^m}{(2m+1)(2m-1)\cdots 3} c_0, \quad m = 1, 2, \cdots.$$

Similarly, the recurrence relation (5.8) for $r_2 = 1/2$ is simplified to $c_m = -c_{m-1}/m$, which gives

$$c_m = (-1)^m \frac{1}{m!} c_0, \quad m = 1, 2, \cdots.$$

Thus, the linearly independent solutions of the DE (6.3) are

$$y_1(x) = |x| \sum_{m=0}^{\infty} (-1)^m \frac{(2x)^m}{(2m+1)(2m-1)\cdots 3}$$

and

$$y_2(x) = |x|^{1/2} \sum_{m=0}^{\infty} (-1)^m \frac{x^m}{m!}.$$

Further, the general solution of (6.3) in any interval not containing the origin is $y(x) = Ay_1(x) + By_2(x)$, where A and B are arbitrary constants.

Our next example uses Theorem 5.1(ii).

Example 6.2. In the DE

$$x(1-x)y'' + (1-x)y' - y = 0 \tag{6.4}$$

$p(x) = 1$ and $q(x) = -x/(1-x)$ are analytic for all $|x| < 1$, and hence the point $x = 0$ is a regular singular point. Since $p_0 = 1$ and $q_0 = 0$, the indicial equation is $F(r) = r(r-1) + r = 0$, and therefore the exponents are $r_1 = r_2 = 0$. Substituting directly $y(x) = x^r \sum_{m=0}^{\infty} c_m x^m$ in the DE (6.4), we obtain

$$r^2 c_0 x^{r-1} + \sum_{m=1}^{\infty} \left[(m+r)^2 c_m - (m+r-1)^2 c_{m-1} - c_{m-1} \right] x^{m+r-1} = 0.$$

Thus, the recurrence relation is

$$c_m = \frac{(m+r-1)^2 + 1}{(m+r)^2} c_{m-1}, \quad m = 1, 2, \cdots.$$

Now a simple calculation gives

$$c_m = \frac{(r^2+1)((r+1)^2+1) \cdots ((r+m-1)^2+1)}{(r+1)^2(r+2)^2 \cdots (r+m)^2} c_0, \quad m = 1, 2, \cdots \tag{6.5}$$

and hence the first solution corresponding to $r = 0$ is

$$y_1(x) = 1 + \sum_{m=1}^{\infty} \frac{1.2.5 \cdots ((m-1)^2+1)}{(m!)^2} x^m.$$

To find the second solution we logarithmically differentiate (6.5) with respect to r, to find

$$\frac{c_m'}{c_m} = \sum_{k=1}^{m} \frac{2(r+k-1)}{(r+k-1)^2+1} - \sum_{k=1}^{m} \frac{2}{(r+k)};$$

and hence taking $r = 0$, we obtain

$$d_m = c_m' \big|_{r=0} = 2c_m \big|_{r=0} \sum_{k=1}^{m} \frac{k-2}{k((k-1)^2+1)}.$$

Thus, the second solution can be written as

$$y_2(x) = y_1(x) \ln|x| + 2 \sum_{m=1}^{\infty} \frac{1.2.5 \cdots ((m-1)^2+1)}{(m!)^2} \left(\sum_{k=1}^{m} \frac{k-2}{k((k-1)^2+1)} \right) x^m.$$

Clearly, in view of Theorem 5.1(ii) both of these solutions converge at least for $0 < |x| < 1$. Further, the general solution of (6.4) appears as $y(x) = Ay_1(x) + By_2(x)$, where A and B are arbitrary constants.

Our next two examples explain the importance of Theorem 5.1(iii).

Example 6.3. In the DE

$$xy'' + 2y' - y = 0 \qquad (6.6)$$

both the functions $p(x) = 2$, $q(x) = -x$ are analytic for $|x| < \infty$, and hence the origin is a regular singular point. Since $p_0 = 2$ and $q_0 = 0$, the indicial equation is $F(r) = r(r-1) + 2r = r^2 + r = 0$, and therefore the exponents are $r_1 = 0$, $r_2 = -1$. Further, we note that the recurrence relation (5.8) for the equation (6.6) reduces to

$$(m+r)(m+r+1)c_m = c_{m-1}, \quad m = 1, 2, \cdots$$

which easily gives

$$c_m = \frac{1}{(r+1)(r+2)^2(r+3)^2 \cdots (r+m)^2(r+m+1)}c_0, \quad m = 1, 2, \cdots. \qquad (6.7)$$

For the exponent $r_1 = 0$, (6.7) reduces to

$$c_m = \frac{1}{m! \, (m+1)!}c_0, \quad m = 1, 2, \cdots;$$

therefore, the first solution $y_1(x)$ is given by

$$y_1(x) = \sum_{m=0}^{\infty} \frac{1}{m! \, (m+1)!}x^m.$$

Now to find the second solution we let $c_0 = r - r_2 = (r+1)$, so that

$$c = \lim_{r \to -1}(r+1)c_1(r) = \lim_{r \to -1} \frac{(r+1)}{(r+1)(r+2)} = 1,$$

and (6.7) is the same as

$$c_m = \frac{1}{(r+2)^2 \cdots (r+m)^2(r+m+1)}, \quad m = 1, 2, \cdots. \qquad (6.8)$$

Now a logarithmic differentiation of (6.8) with respect to r gives

$$\frac{c_m'}{c_m} = -2\sum_{k=2}^{m} \frac{1}{(r+k)} - \frac{1}{(r+m+1)}, \quad m = 1, 2, \cdots$$

and hence

$$e_0 = c_0'(-1) \quad = \quad 1$$

$$e_m = c_m'(-1) \quad = \quad \frac{1}{1^2 \cdot 2^2 \cdots (m-1)^2 m} \left[-2 \sum_{k=1}^{m-1} \frac{1}{k} - \frac{1}{m} \right]$$

$$= \quad -\frac{1}{m! \, (m-1)!} \left[2 \sum_{k=1}^{m-1} \frac{1}{k} + \frac{1}{m} \right], \quad m = 1, 2, \cdots.$$

Thus, the second solution $y_2(x)$ appears as

$$y_2(x) = y_1(x) \ln |x| + |x|^{-1} \left[1 - \sum_{m=1}^{\infty} \frac{1}{m! \, (m-1)!} \left(2 \sum_{k=1}^{m-1} \frac{1}{k} + \frac{1}{m} \right) x^m \right].$$

Clearly, in view of Theorem 5.1(iii) both of these solutions converge at least for $0 < |x| < \infty$. Moreover, the general solution of (6.6) can be written as $y(x) = Ay_1(x) + By_2(x)$, where A and B are arbitrary constants.

Example 6.4. In the DE

$$xy'' - y' + 4x^3 y = 0 \tag{6.9}$$

both the functions $p(x) = -1$, $q(x) = 4x^4$ are analytic for $|x| < \infty$, and hence the origin is a regular singular point. Since $p_0 = -1$ and $q_0 = 0$, the indicial equation is $F(r) = r(r-1) - r = r^2 - 2r = 0$, and therefore the exponents are $r_1 = 2$, $r_2 = 0$. Thus, two linearly independent solutions of (6.9) are of the form (5.10) and (5.13). A direct substitution of these in the equation (6.9) computes the solutions explicitly as

$$y_1(x) = x^2 \sum_{m=0}^{\infty} \frac{(-1)^m}{(2m+1)!} x^{4m}$$

and

$$y_2(x) = \sum_{m=0}^{\infty} \frac{(-1)^m}{(2m)!} x^{4m}.$$

Note that for the equation (6.9) in (5.13) the constant $c = 0$. Further, in view of Theorem 5.1(iii) both of these solutions converge at least for $0 < |x| < \infty$. Again, the general solution of (6.9) appears as $y(x) = Ay_1(x) + By_2(x)$, where A and B are arbitrary constants.

Problems

6.1. Compute the indicial equation and their roots for the following DEs:

(i) $2xy'' + y' + xy = 0$

(ii) $x^2y'' + xy' + (x^2 - 1/9)y = 0$

(iii) $x^2y'' + (x + x^2)y' - y = 0$

(iv) $x^2y'' + xy' + (x^2 - 1/4)y = 0$

(v) $x(x - 1)y'' + (2x - 1)y' - 2y = 0$

(vi) $x^2y'' + 3\sin xy' - 2y = 0$

(vii) $x^2y'' + (1/2)(x + \sin x)y' + y = 0$

(viii) $x^2y'' + xy' + (1 - x)y = 0.$

6.2. Verify that each of the given DEs has a regular singular point at the indicated point $x = x_0$, and express their solutions in terms of power series valid for $x > x_0$:

(i) $4xy'' + 2y' + y = 0, \quad x = 0$

(ii) $9x^2y'' + 9xy' + (9x^2 - 1)y = 0, \quad x = 0$

(iii) $2x^2y'' + xy' - (x + 1)y = 0, \quad x = 0$

(iv) $(1 - x^2)y'' + y' + 2y = 0, \quad x = -1$

(v) $x^2y'' + (x^2 - 7/36)y = 0, \quad x = 0$

(vi) $x^2y'' + (x^2 - x)y' + 2y = 0, \quad x = 0$

(vii) $x^2y'' + (x^2 - x)y' + y = 0, \quad x = 0$

(viii) $x(1 - x)y'' + (1 - 5x)y' - 4y = 0, \quad x = 0$

(ix) $(x^2 + x^3)y'' - (x + x^2)y' + y = 0, \quad x = 0$

(x) $x^2y'' + 2xy' + xy = 0, \quad x = 0$

(xi) $x^2y'' + 4xy' + (2 + x)y = 0, \quad x = 0$

(xii) $x(1 - x)y'' - 3xy' - y = 0, \quad x = 0$

(xiii) $x^2y'' - (x + 2)y = 0, \quad x = 0$

(xiv) $x(1 + x)y'' + (x + 5)y' - 4y = 0, \quad x = 0$

(xv) $(x - x^2)y'' - 3y' + 2y = 0, \quad x = 0.$

6.3 A supply of hot air can be obtained by passing the air through a heated cylindrical tube. It can be shown that the temperature T of the air in the tube satisfies the differential equation

$$\frac{d^2T}{dx^2} - \frac{upC}{kA}\frac{dT}{dx} + \frac{2\pi rh}{kA}(T_w - T) = 0, \tag{6.10}$$

where x = distance from intake end of the tube, u = flow rate of air, p = density of air, C = heat capacity of air, k = thermal conductivity, A = cross–sectional area of the tube, r = radius of the tube, h = heat transfer coefficient of air (nonconstant), T_w = temperature of the tube (see Jenson and Jefferys, 1977). For the parameters they have taken, the differential equation (6.10) becomes

$$\frac{d^2T}{dx^2} - 26200\frac{dT}{dx} - 11430x^{-1/2}(T_w - T) = 0. \tag{6.11}$$

(i) Show that the substitution $y = T_w - T$, $x = z^2$ transforms (6.11) into

$$z\frac{d^2y}{dz^2} - (1 + 52400z^2)\frac{dy}{dz} - 45720z^2y = 0, \qquad (6.12)$$

for which $z = 0$ is a regular singular point with exponents 0 and 2.

(ii) Find first few terms of the series solution of (6.12) at $z = 0$ for the exponent 2.

6.4 In building design it is sometimes useful to use supporting columns that are special geometrical designs. In studying the buckling of columns of varying cross sections, we obtain the following differential equation:

$$x^n\frac{d^2y}{dx^2} + k^2y = 0,$$

where $k > 0$ and n is a positive integer. In particular, if $n = 1$, the column is rectangular with one dimension constant, whereas if $n = 4$, the column is a truncated pyramid or cone. Show that for the case $n = 1$, the point $x = 0$ is regular singular with exponents 0 and 1. Also, find the series solution at $x = 0$ for the exponent 1.

6.3 A large-diameter pipe such as the 30-ft-diameter pipe used in the construction of Hoover Dam is strengthened by a device called a *stiffener ring*. To cut down the stress on the stiffener ring, a fillet insert device is used. In determining the radial displacement of the fillet insert due to internal water pressure, one encounters the fourth order equation

$$x^2y^{iv} + 6xy''' + 6y'' + y = 0, \quad x > 0. \qquad (6.13)$$

Here y is proportional to the radial displacement and x is proportional to the distance measured along an inside element of the pipe shell from some fixed point. Find series solution of (6.13) at $x = 0$ for which the limit as $x \to 0$ exists.

6.6 Consider the *Lane–Emden equation*

$$xy'' + 2y' + xy^n = 0 \qquad (6.14)$$

with the initial conditions

$$y(0) = 1, \quad y'(0) = 0. \qquad (6.15)$$

Astrophysicists and astronomers use equation (6.14) to approximate the density and internal temperatures of certain stars and nebula. Show that the series solution of (6.14), (6.15) can be written as

$$y(x) = 1 - \frac{x^2}{3!} + n\frac{x^4}{5!} + (5n - 8n^2)\frac{x^6}{3\cdot7!} + (70n - 183n^2 + 122n^3)\frac{x^8}{9\cdot9!}$$

$$+ (3150n - 1080n^2 + 12642n^3 - 5032n^4)\frac{x^{10}}{45\cdot11!} + \cdots.$$

Thus, in particular deduce that

$$\text{for } n = 0, \quad y(x) = 1 - \frac{x^2}{6}$$

$$\text{for } n = 1, \quad y(x) = \frac{\sin x}{x}$$

$$\text{for } n = 5, \quad y(x) = \left(1 + \frac{x^2}{3}\right)^{-1/2}.$$

6.7 The *Lane–Emden equation*

$$xy'' + 2y' + xe^y = 0 \tag{6.16}$$

appears in a study of isothermal gas spheres. Show that the series solution of (6.16) with the initial conditions $y(0) = y'(0) = 0$ can be written as

$$y(x) = -\frac{1}{6}x^2 + \frac{1}{5 \cdot 4!}x^4 - \frac{8}{21 \cdot 6!}x^6 + \frac{122}{81 \cdot 8!}x^8 - \frac{61 \cdot 67}{495 \cdot 10!}x^{10} + \cdots.$$

6.8 The *Lane–Emden equation*

$$xy'' + 2y' + xe^{-y} = 0 \tag{6.17}$$

appears in the theory of thermionic currents when one seeks to determine the density and electric force of an electron gas in the neighborhood of a hot body in thermal equilibrium. Find first few terms of the series expansion of the solution of (6.17) satisfying the initial conditions $y(0) = y'(0) = 0$.

6.9 The *White-Dwarf* equation,

$$xy'' + 2y' + x(y^2 - C)^{3/2} = 0, \tag{6.18}$$

was introduced by S. Chandrasekhar in his study of gravitational potential of the degenerate (white-dwarf) stars. This equation for $C = 0$ is the same as (6.14) with $n = 3$. Show that the series solution of (6.18) with the initial conditions (6.15) in terms of $q^2 = 1 - C$ can be written as

$$\begin{aligned}
y(x) &= 1 - \frac{q^3}{6}x^2 + \frac{q^4}{40}x^4 - \frac{q^5}{7!}(5q^2 + 14)x^6 + \frac{q^6}{3 \cdot 9!}(339q^2 + 280)x^8 \\
&\quad + \frac{q^7}{5 \cdot 11!}(1425q^4 + 11436q^2 + 4256)x^{10} + \cdots.
\end{aligned}$$

6.10 Show that for the DE

$$x^3y'' + x(x+1)y' - y = 0$$

$x_0 = 0$ is an irregular singular point. Find its solution in the form $y = \sum_{m=0}^{\infty} c_m x^{-m}$.

Answers or Hints

6.1. (i) $r(r-1) + \frac{1}{2}r = 0$, $r_1 = 0, r_2 = \frac{1}{2}$ (ii) $r(r-1) + r - \frac{1}{9} = 0$, $r_1 = \frac{1}{3}, r_2 = -\frac{1}{3}$ (iii) $r(r-1) + r - 1 = 0$, $r_1 = 1, r_2 = -1$ (iv) $r(r-1) + r - \frac{1}{4} = 0$, $r_1 = \frac{1}{2}, r_2 = -\frac{1}{2}$ (v) $r(r-1) + r = 0$, $r_1 = r_2 = 0$ (vi) $r(r-1) + 3r - 2 = 0$, $r_1 = -1 + \sqrt{3}, r_2 = -1 - \sqrt{3}$ (vii) $r(r-1) + r + 1 = 0$, $r_1 = i, r_2 = -i$ (viii) $r(r-1) + r + 1 = 0$, $r_1 = i, r_2 = -i$.

6.2. (i) $y_1(x) = \sum_{m=0}^{\infty} \frac{(-1)^m (\sqrt{x})^{2m}}{(2m)!}$, $y_2(x) = x^{1/2} \sum_{m=0}^{\infty} \frac{(-1)^m (\sqrt{x})^{2m}}{(2m+1)!}$

(ii) $y_1(x) = x^{1/3} \left[1 + \sum_{m=1}^{\infty} \frac{(-1)^m 3^m x^{2m}}{2 \cdot 4 \cdot 6 \cdots (2m)\, 8 \cdot 14 \cdot 20 \cdots (6m+2)} \right]$,

$y_2(x) = x^{-1/3} \left[1 + \sum_{m=1}^{\infty} \frac{(-1)^m 3^m x^{2m}}{2 \cdot 4 \cdot 6 \cdots (2m)\, 4 \cdot 10 \cdot 16 \cdots (6m-2)} \right]$

(iii) $y_1(x) = x \left[1 + \frac{1}{5}x + \frac{1}{70}x^2 + \cdots \right]$,

$y_2(x) = x^{-1/2} \left[1 - x - \frac{1}{2}x^2 - \frac{1}{18}x^3 - \cdots \right]$

(iv) $y_1(x) = 1 - 2(x+1) + \frac{2}{3}(x+1)^2$,

$y_2(x) = (x+1)^{1/2} \left[1 - \frac{3}{4}(x+1) + \sum_{m=2}^{\infty} \frac{3(2m-4)!}{2^{3m-2} m!\, (m-2)!} (x+1)^m \right]$

(v) $y_1(x) = x^{7/6} \left[1 + \sum_{m=1}^{\infty} \frac{(-1)^m 3^m x^{2m}}{2^{2m} m!\, 5 \cdot 8 \cdots (3m+2)} \right]$,

$y_2(x) = x^{-1/6} \left[1 + \sum_{m=1}^{\infty} \frac{(-1)^m 3^m}{2^{2m} m!\, 1 \cdot 4 \cdots (3m-2)} x^{2m} \right]$

(vi) $y_1(x) = x[\phi(x) \cos(\ln x) - \psi(x) \sin(\ln x)]$,

$y_2(x) = x[\psi(x) \cos(\ln x) + \phi(x) \sin(\ln x)]$, where

$\phi(x) = 1 - \frac{3}{5}x + \frac{1}{5}x^2 \cdots$, $\psi(x) = \frac{1}{5}x - \frac{3}{20}x^2 \cdots$

(vii) $y_1(x) = x \sum_{m=0}^{\infty} \frac{(-1)^m}{m!} x^m$,

$y_2(x) = y_1(x) \ln x + x \left[x - \frac{1}{2!}\left(1 + \frac{1}{2}\right) x^2 + \frac{1}{3!}\left(1 + \frac{1}{2} + \frac{1}{3}\right) x^3 - \cdots \right]$

(viii) $y_1(x) = \sum_{m=0}^{\infty} (1+m)^2 x^m$, $y_2(x) = y_1(x) \ln x - 2 \sum_{m=1}^{\infty} m(m+1) x^m$

(ix) $y_1(x) = x(1+x)$, $y_2(x) = y_1(x) \ln x + x \left[-2x - \sum_{m=2}^{\infty} \frac{(-1)^m}{m(m-1)} x^m \right]$

(x) $y_1(x) = \sum_{m=0}^{\infty} \frac{(-1)^m}{m!\, (m+1)!} x^m$,

$y_2(x) = -y_1(x) \ln x + x^{-1} \left[1 - \sum_{m=1}^{\infty} \frac{(-1)^m}{m!\, (m-1)!} \left(2\sum_{k=1}^{m-1} \frac{1}{k} + \frac{1}{m} \right) x^m \right]$

(xi) $y_1(x) = x^{-1} \sum_{m=0}^{\infty} \frac{(-1)^m}{m!\, (m+1)!} x^m$, $y_2(x) = -y_1(x) \ln x$

$+ x^{-2} \left[1 - \sum_{m=1}^{\infty} \frac{(-1)^m}{m!\, (m-1)!} \left(2\sum_{k=1}^{m-1} \frac{1}{k} + \frac{1}{m} \right) x^m \right]$

(xii) $y_1(x) = x(1-x)^{-2}$, $y_2(x) = y_1(x) \ln x + (1-x)^{-1}$

(xiii) $y_1(x) = x^2 \sum_{m=0}^{\infty} \frac{3!}{m!\, (m+3)!} x^m$,

$y_2(x) = \frac{1}{12} y_1(x) \ln x + x^{-1} \left(1 - \frac{1}{2}x + \frac{1}{4}x^2 - \frac{1}{36}x^3 + \cdots \right)$

(xiv) $y_1(x) = 1 + \frac{4}{5}x + \frac{1}{5}x^2$, $y_2(x) = x^{-4}(1 + 4x + 5x^2)$

(xv) $y_1(x) = \sum_{m=4}^{\infty} (m-3) x^m$, $y_2(x) = 1 + \frac{2}{3}x + \frac{1}{3}x^2$.

6.3. $c_0 \left[z^2 + 13100 z^4 + 3048 z^5 + \frac{343220000}{3} z^6 + \cdots \right]$ (although coefficients c_n are large, the series converges).

6.4. $c_0 \left(x - \frac{k^2}{2!} x^2 + \frac{k^4}{2!3!} x^3 - \frac{k^6}{3!4!} x^4 + \frac{k^8}{4!5!} x^5 - \cdots \right).$

6.5. $c_0 \sum_{m=0}^{\infty} (-1)^m \frac{x^{2m}}{((2m!))^2 (2m+1)} + c_1 \sum_{m=0}^{\infty} (-1)^m \frac{x^{2m+1}}{((2m+1)!)^2 (m+1)}.$

6.8. If $y(x)$ is a solution of (6.16), then $-y(ix)$ a solution of (6.17), $y(x) = -\frac{1}{6} x^2 - \frac{1}{5 \cdot 4!} x^4 - \frac{8}{21 \cdot 6!} x^6 - \frac{122}{81 \cdot 8!} x^8 - \frac{61 \cdot 67}{495 \cdot 10!} x^{10} - \cdots.$

6.10. $y = c_0 \sum_{m=0}^{\infty} \frac{x^{-m}}{m!} = c_0 e^{1/x}.$

Lecture 7
Legendre Polynomials and Functions

The Legendre DE (3.19), where a is a real constant (parameter), arises in problems such as the flow of an ideal fluid past a sphere, the determination of the electric field due to a charged sphere, and the determination of the temperature distribution in a sphere given its surface temperature. In this lecture we shall show that if the parameter a is a nonnegative integer n, then one of the solutions of (3.19) reduces to a polynomial of degree exactly n. These polynomial solutions are known as Legendre polynomials. We shall obtain explicit representations of these polynomials and discuss their various properties.

Since the functions

$$p_1(x) = -\frac{2x}{1-x^2} \quad \text{and} \quad p_2(x) = \frac{a(a+1)}{1-x^2}$$

are analytic for $|x| < 1$, $x = x_0 = 0$ is an ordinary point for (3.19). Thus, Theorem 4.1 ensures that its series solution $y(x) = \sum_{m=0}^{\infty} c_m x^m$ converges for $|x| < 1$. To find this solution, we substitute it directly in (3.19), to obtain

$$(1-x^2) \sum_{m=0}^{\infty} (m+1)(m+2)c_{m+2}x^m \ -2x \sum_{m=0}^{\infty} (m+1)c_{m+1}x^m$$

$$+a(a+1) \sum_{m=0}^{\infty} c_m x^m = 0,$$

which is the same as

$$\sum_{m=0}^{\infty} \left[(m+1)(m+2)c_{m+2} - \{(m-1)m + 2m - a(a+1)\} c_m \right] x^m = 0,$$

or

$$\sum_{m=0}^{\infty} \left[(m+1)(m+2)c_{m+2} + (a+m+1)(a-m)c_m \right] x^m = 0.$$

But this is possible if and only if

$$(m+1)(m+2)c_{m+2} + (a+m+1)(a-m)c_m = 0, \quad m = 0, 1, \cdots$$

R.P. Agarwal, D. O'Regan, *Ordinary and Partial Differential Equations*,
Universitext, DOI 10.1007/978-0-387-79146-3_7,
© Springer Science+Business Media, LLC 2009

or

$$c_{m+2} = -\frac{(a+m+1)(a-m)}{(m+1)(m+2)}c_m, \quad m = 0, 1, \cdots \tag{7.1}$$

which is the required *recurrence relation*.

Now a little computation gives

$$c_{2m} = \frac{(-1)^m(a+2m-1)(a+2m-3)\cdots(a+1)a(a-2)\cdots(a-2m+2)}{(2m)!}c_0$$

$$= (-1)^m \frac{\Gamma\left(\frac{1}{2}a+1\right)\Gamma\left(\frac{1}{2}a+m+\frac{1}{2}\right)2^{2m}}{\Gamma\left(\frac{1}{2}a+\frac{1}{2}\right)\Gamma\left(\frac{1}{2}a-m+1\right)(2m)!}c_0, \quad m = 1, 2, \cdots$$

$$\tag{7.2}$$

and

$$c_{2m+1} = \frac{(-1)^m(a+2m)(a+2m-1)\cdots(a+2)(a-1)(a-3)\cdots(a-2m+1)}{(2m+1)!}c_1$$

$$= (-1)^m \frac{\Gamma\left(\frac{1}{2}a+\frac{1}{2}\right)\Gamma\left(\frac{1}{2}a+m+1\right)2^{2m+1}}{2\,\Gamma\left(\frac{1}{2}a+1\right)\Gamma\left(\frac{1}{2}a-m+\frac{1}{2}\right)(2m+1)!}c_1, \quad m = 1, 2, \cdots.$$

$$\tag{7.3}$$

Thus, the series solution of (3.19) can be written as

$$y(x) = c_0\left[1 - \frac{(a+1)a}{2!}x^2 + \frac{(a+3)(a+1)a(a-2)}{4!}x^4 - \cdots\right]$$

$$+c_1\left[x - \frac{(a+2)(a-1)}{3!}x^3 + \frac{(a+4)(a+2)(a-1)(a-3)}{5!}x^5 - \cdots\right]$$

$$= c_0y_1(x) + c_1y_2(x).$$

$$\tag{7.4}$$

It is clear that $y_1(x)$ and $y_2(x)$ are linearly independent solutions of Legendre's equation.

If in the DE (3.19), a is an even integer $2n$, then from (7.2) it is clear that $c_{2n+2} = c_{2n+4} = \cdots = 0$; i.e., $y_1(x)$ reduces to a polynomial of degree $2n$ involving only even powers of x. Similarly, if $a = 2n + 1$, then $y_2(x)$ reduces to a polynomial of degree $2n + 1$, involving only odd powers of x. Since $y_1(x)$ and $y_2(x)$ are themselves solutions of (3.19) we conclude that Legendre's DE has a polynomial solution for each nonnegative integer value of the parameter a. The interest is now to obtain these polynomials in descending powers of x. For this we note that the recurrence relation (7.1) can be written as

$$c_s = -\frac{(s+1)(s+2)}{(n-s)(n+s+1)}c_{s+2}, \quad s \le n - 2, \tag{7.5}$$

where we have taken s as the index, and a as an integer n. With the help of (7.5) we can express all nonvanishing coefficients in terms of the coefficient

c_n of the highest power of x. It is customary to choose

$$c_n = \frac{(2n)!}{2^n (n!)^2} = \frac{1.3.5. \cdots (2n-1)}{n!} \qquad (7.6)$$

so that the polynomial solution of (3.19) will have the value 1 at $x = 1$.

From (7.5) and (7.6) it is easy to obtain

$$c_{n-2m} = (-1)^m \frac{(2n-2m)!}{2^n \, m! \, (n-m)! \, (n-2m)!} \qquad (7.7)$$

as long as $n - 2m \geq 0$.

The resulting solution of (3.19) is called the *Legendre polynomial of degree* n and is denoted as $P_n(x)$. From (7.7) this solution can be written as

$$P_n(x) = \sum_{m=0}^{[\frac{n}{2}]} (-1)^m \frac{(2n-2m)!}{2^n \, m! \, (n-m)! \, (n-2m)!} x^{n-2m}. \qquad (7.8)$$

From (7.8), we easily obtain

$$P_0(x) = 1, \quad P_1(x) = x, \quad P_2(x) = \frac{1}{2}(3x^2 - 1), \quad P_3(x) = \frac{1}{2}(5x^3 - 3x),$$

$$P_4(x) = \frac{1}{8}(35x^4 - 30x^2 + 3), \quad P_5(x) = \frac{1}{8}(63x^5 - 70x^3 + 15x).$$

The other nonpolynomial solution of (3.19) is usually denoted as $Q_n(x)$ (see Problem 7.1). Note that for each n, $Q_n(x)$ is unbounded at $x = \pm 1$.

Legendre polynomials $P_n(x)$ can be represented in a very compact form. This we shall show in the following theorem.

Theorem 7.1 (Rodrigues' Formula).

$$P_n(x) = \frac{1}{2^n \, n!} \frac{d^n}{dx^n} (x^2 - 1)^n. \qquad (7.9)$$

Proof. Let $v = (x^2 - 1)^n$, then

$$(x^2 - 1) \frac{dv}{dx} = 2nxv. \qquad (7.10)$$

Differentiating (7.10), $(n + 1)$ times by Leibniz's rule, we obtain

$$(x^2-1)\frac{d^{n+2}v}{dx^{n+2}} + 2(n+1)x\frac{d^{n+1}v}{dx^{n+1}} + n(n+1)\frac{d^n v}{dx^n} = 2n\left\{x\frac{d^{n+1}v}{dx^{n+1}} + (n+1)\frac{d^n v}{dx^n}\right\},$$

which is the same as

$$(1 - x^2)\frac{d^{n+2}v}{dx^{n+2}} - 2x\frac{d^{n+1}v}{dx^{n+1}} + n(n+1)\frac{d^n v}{dx^n} = 0. \qquad (7.11)$$

If we substitute $z = d^n v/dx^n$, then (7.11) becomes

$$(1 - x^2)\frac{d^2 z}{dx^2} - 2x\frac{dz}{dx} + n(n+1)z = 0,$$

which is the same as (3.19) with $a = n$. Thus, it is necessary that

$$z = \frac{d^n v}{dx^n} = cP_n(x),$$

where c is a constant. Since $P_n(1) = 1$, we have

$$
\begin{aligned}
c &= \left(\frac{d^n v}{dx^n}\right)_{x=1} = \frac{d^n}{dx^n}(x^2 - 1)^n \Big|_{x=1} = \frac{d^n}{dx^n}(x-1)^n(x+1)^n \Big|_{x=1} \\
&= \sum_{k=0}^{n} \binom{n}{k}\frac{n!}{(n-k)!}(x-1)^{n-k}\frac{n!}{k!}(x+1)^k \Big|_{x=1} = 2^n\, n!.
\end{aligned}
$$

Thus, it follows that

$$P_n(x) = \frac{1}{c}\frac{d^n v}{dx^n} = \frac{1}{2^n\, n!}\frac{d^n}{dx^n}(x^2 - 1)^n. \qquad \blacksquare$$

Let $\{f_n(x)\}$ be a sequence of functions in some interval J. A function $F(x,t)$ is said to be a *generating function* of $\{f_n(x)\}$ if

$$F(x,t) = \sum_{n=0}^{\infty} f_n(x)t^n.$$

The following result provides the generating function for the sequence of Legendre polynomials $\{P_n(x)\}$.

Theorem 7.2 (Generating Function).

$$(1 - 2xt + t^2)^{-1/2} = \sum_{n=0}^{\infty} P_n(x)t^n. \qquad (7.12)$$

Proof. If $|x| \leq r$ where r is arbitrary, and $|t| < (1 + r^2)^{1/2} - r$, then it follows that

$$
\begin{aligned}
|2xt - t^2| &\leq 2|x||t| + |t^2| \\
&< 2r(1 + r^2)^{1/2} - 2r^2 + 1 + r^2 + r^2 - 2r(1 + r^2)^{1/2} = 1
\end{aligned}
$$

and hence we can expand $(1 - 2xt + t^2)^{-1/2}$ binomially, to obtain

$$[1 - t(2x - t)]^{-1/2} = 1 + \frac{1}{2}t(2x - t) + \frac{1}{2}\frac{3}{4}t^2(2x - t)^2 + \cdots$$
$$+ \frac{1.3 \cdots (2n - 1)}{2.4 \cdots (2n)} t^n (2x - t)^n + \cdots.$$

The coefficient of t^n in this expansion is

$$\frac{1.3. \cdots (2n - 1)}{2.4. \cdots (2n)}(2x)^n - \frac{1.3. \cdots (2n - 3)}{2.4 \cdots (2n - 2)} \frac{(n - 1)}{1!}(2x)^{n-2}$$
$$+ \frac{1.3 \cdots (2n - 5)}{2.4 \cdots (2n - 4)} \frac{(n - 2)(n - 3)}{2!}(2x)^{n-4} - \cdots$$
$$= \frac{1.3 \cdots (2n-1)}{n!}\left[x^n - \frac{n(n-1)}{(2n-1)!1.2}x^{n-2} + \frac{n(n-1)(n-2)(n-3)}{(2n-1)(2n-3)\,2.4}x^{n-4} - \cdots \right]$$
$$= P_n(x). \quad \blacksquare$$

Now as an application of (7.12) we shall prove the following recurrence relation.

Theorem 7.3 (Recurrence Relation).

$$(n + 1)P_{n+1}(x) = (2n + 1)xP_n(x) - nP_{n-1}(x), \quad n = 1, 2, \cdots. \quad (7.13)$$

Proof. Differentiating (7.12) with respect to t, we get

$$(x - t)(1 - 2xt + t^2)^{-3/2} = \sum_{n=1}^{\infty} nP_n(x)t^{n-1}$$

and hence

$$(x - t)(1 - 2xt + t^2)^{-1/2} = (1 - 2xt + t^2) \sum_{n=1}^{\infty} nP_n(x)t^{n-1},$$

which is the same as

$$(x - t) \sum_{n=0}^{\infty} P_n(x)t^n = (1 - 2xt + t^2) \sum_{n=1}^{\infty} nP_n(x)t^{n-1}.$$

Equating the coefficients of t^n, we get the relation (7.13). $\quad \blacksquare$

Since $P_0(x) = 1$ and $P_1(x) = x$ the relation (7.13) can be used to compute Legendre polynomials of higher degrees. Several other recurrence relations are given in Problem 7.8.

Next we shall consider (3.19) in the neighborhood of the regular singularity at $x = 1$. The transformation $x = 1 - 2t$ replaces this singularity to the origin, and then the DE (3.19) becomes

$$t(1-t)\frac{d^2y}{dt^2} + (1-2t)\frac{dy}{dt} + a(a+1)y = 0. \qquad (7.14)$$

Considering the solution of (7.14) in the form $y(t) = t^r \sum_{m=0}^{\infty} c_m t^m$, we obtain the indicial equation $r^2 = 0$, i.e., both the exponents are zero. Further, it is easy to find the recurrence relation

$$(m+r)^2 c_m = (m+r+a)(m+r-a-1)c_{m-1}, \quad m = 1, 2, \cdots. \qquad (7.15)$$

For $r = 0$, (7.15) when a not an integer and $c_0 = 1$, gives

$$c_m = \frac{\Gamma(m+a+1)\,\Gamma(m-a)}{(m!)^2\,\Gamma(a+1)\,\Gamma(-a)}, \quad m = 0, 1, \cdots.$$

Thus, in the neighborhood of $x = 1$ the first solution of (3.19) can be written as

$$y_1(x) = \frac{1}{\Gamma(a+1)\,\Gamma(-a)} \sum_{m=0}^{\infty} \frac{\Gamma(a+m+1)\,\Gamma(m-a)}{(m!)^2}\left(\frac{1-x}{2}\right)^m. \qquad (7.16)$$

Since a is not an integer, and

$$\lim_{m\to\infty}\left|\frac{c_m}{c_{m-1}}\right| = \lim_{m\to\infty}\left|\left(1 - \frac{a+1}{m}\right)\left(1 + \frac{a}{m}\right)\right| = 1,$$

it follows that the series solution (7.16) converges for $|x-1| < 2$ and diverges for $|x-1| > 2$.

When a is a positive integer n, then $c_m = 0$ for all $m \geq n+1$, and

$$c_m = (-1)^m \frac{(n+m)!}{(n-m)!\,(m!)^2}c_0, \quad 0 \leq m \leq n$$

and hence (7.16) reduces to a polynomial solution of degree n,

$$\sum_{m=0}^{n} \frac{(-1)^m(n+m)!}{(n-m)!\,(m!)^2}\left(\frac{1-x}{2}\right)^m. \qquad (7.17)$$

To obtain the second solution of (7.14), we return to the recurrence relation (7.15), whose solution with $c_0 = 1$ is

$$c_m = \prod_{k=1}^{m}(k+r+a) \prod_{k=1}^{m}(k+r-a-1) \bigg/ \prod_{k=1}^{m}(k+r)^2.$$

Now a logarithmic differentiation with respect to r yields

$$\frac{c_m'}{c_m} = \sum_{k=1}^{m} \left(\frac{1}{k+r+a} + \frac{1}{k+r-a-1} - \frac{2}{k+r} \right);$$

and hence for $r = 0$, we get

$$d_m = c_m'(0) = c_m(0) \sum_{k=1}^{m} \left(\frac{1}{k+a} + \frac{1}{k-a-1} - \frac{2}{k} \right), \quad m = 1, 2, \cdots.$$

Thus, from Theorem 5.1(ii) the second solution is given by

$$y_2(x) = y_1(x) \ln \left| \frac{1-x}{2} \right| + \frac{1}{\Gamma(a+1)\Gamma(-a)} \sum_{m=1}^{\infty} \frac{\Gamma(a+m+1)\Gamma(m-a)}{(m!)^2}$$
$$\times \left\{ \sum_{k=1}^{m} \left(\frac{1}{k+a} + \frac{1}{k-a-1} - \frac{2}{k} \right) \right\} \left(\frac{1-x}{2} \right)^m.$$

$$(7.18)$$

The general solution of (3.19) in the neighborhood of $x = 1$ can be easily obtained from (7.16) and (7.18).

Problems

7.1.   For $a = n$, one solution of the DE (3.19) is $P_n(x)$ given in (7.8). Show that the second solution is

$$Q_n(x) = P_n(x) \int^x \frac{dt}{(1-t^2)[P_n(t)]^2}$$

and hence deduce that

$$Q_0(x) = \frac{1}{2} \ln \frac{1+x}{1-x}, \quad Q_1(x) = \frac{x}{2} \ln \frac{1+x}{1-x} - 1$$

and

$$Q_2(x) = \frac{1}{4}(3x^2 - 1) \ln \frac{1+x}{1-x} - \frac{3}{2}x.$$

7.2.   Use Rodrigues' formula (7.9) to show that

$$P_n(0) = \begin{cases} 0, & \text{if } n \text{ is odd} \\ (-1)^{n/2} \dfrac{1.3. \cdots (n-1)}{2.4 \cdots n}, & \text{if } n \text{ is even.} \end{cases}$$

7.3.   Use Rodrigues' formula (7.9) to show that all the roots of $P_n(x)$ lie in the interval $(-1, 1)$ and are distinct.

7.4.   Prove the following relations:

(i) $\dfrac{1 - t^2}{(1 - 2xt + t^2)^{3/2}} = \sum\limits_{n=0}^{\infty} (2n + 1) P_n(x) t^n$

(ii) $\dfrac{1 + t}{t(1 - 2xt + t^2)^{1/2}} - \dfrac{1}{t} = \sum\limits_{n=0}^{\infty} \{ P_n(x) + P_{n+1}(x) \} t^n.$

7.5.   Show that if $x = \cos\theta = (e^{i\theta} + e^{-i\theta})/2$, then the Legendre DE (3.19) can be written as

$$\frac{d}{d\theta}\left(\sin\theta \frac{dy}{d\theta} \right) + n(n+1)\sin\theta y = 0. \tag{7.19}$$

Further, with this substitution, the relation (7.12) is the same as

$$(1 - te^{i\theta})^{-1/2}(1 - te^{-i\theta})^{-1/2} = \sum\limits_{n=0}^{\infty} P_n(\cos\theta) t^n;$$

and hence

$$y = P_n(\cos\theta) = \frac{1.3\cdots(2n-1)}{2.4\cdots 2n}\left[2\cos n\theta + \frac{1.n}{1.(2n-1)}2\cos(n-2)\theta \right.$$
$$\left. + \frac{n(n-1)}{(2n-1)(2n-3)}\frac{1.3}{1.2}2\cos(n-4)\theta + \cdots + T_n \right]$$

is a solution of (7.19), where the final term T_n is $\cos\theta$ if n is odd, and half the constant term indicated if n is even. These functions are called the *Legendre coefficients.* Deduce that the Legendre coefficients are uniformly bounded, in fact $|P_n(\cos\theta)| \le 1$, $n = 0, 1, 2, \cdots$ for all real values of θ.

7.6.   Prove *Laplace's first integral representation*

$$P_n(x) = \frac{1}{\pi}\int_0^\pi [x \pm \sqrt{x^2 - 1}\cos\phi]^n d\phi.$$

7.7.   Prove *Laplace's second integral representation*

$$P_n(x) = \frac{1}{\pi}\int_0^\pi \frac{d\phi}{[x \pm \sqrt{x^2 - 1}\cos\phi]^{n+1}}.$$

7.8.   Prove the following recurrence relations:

(i) $nP_n(x) = xP_n'(x) - P_{n-1}'(x)$

(ii) $(2n + 1)P_n(x) = P_{n+1}'(x) - P_{n-1}'(x)$

(iii) $(n + 1)P_n(x) = P_{n+1}'(x) - xP_n'(x)$

(iv) $(1 - x^2)P_n'(x) = n(P_{n-1}(x) - xP_n(x))$

(v) $(1 - x^2)P_n'(x) = (n+1)(xP_n(x) - P_{n+1}(x))$

(vi) $(2n+1)(1 - x^2)P_n'(x) = n(n+1)(P_{n-1}(x) - P_{n+1}(x))$.

7.9. Show that

(i) $P_n(1) = 1$

(ii) $P_n(-x) = (-1)^n P_n(x)$, and hence $P_n(-1) = (-1)^n$.

7.10. Show that in the neighborhood of the point $x = 1$ the DE (3.19) for $a = -n$ has a polynomial solution of degree $n - 1$. Find this solution.

7.11. Show that

(i) Christoffel's expansion

$$P_n'(x) = (2n - 1)P_{n-1}(x) + (2n - 5)P_{n-3}(x) + (2n - 9)P_{n-5}(x) + \cdots,$$

where the last term is $3P_1(x)$ or $P_0(x)$ according to whether n is even or odd.

(ii) Christoffel's summation formula

$$\sum_{r=0}^{n}(2r + 1)P_r(x)P_r(y) = (n+1)\frac{P_{n+1}(x)P_n(y) - P_{n+1}(y)P_n(x)}{x - y}.$$

7.12. The DE

$$(1 - x^2)y'' - 2xy' + \left[n(n+1) - \frac{m^2}{1 - x^2}\right]y = 0 \qquad (7.20)$$

is called *Legendre's associated DE*. If $m = 0$, it reduces to (3.19). Show that when m and n are nonnegative integers, the general solution of (7.20) can be written as

$$y(x) = AP_n^m(x) + BQ_n^m(x),$$

where $P_n^m(x)$ and $Q_n^m(x)$ are called *associated Legendre's functions of the first and second kinds*, respectively, and in terms of $P_n(x)$ and $Q_n(x)$ are given by

$$P_n^m(x) = (1 - x^2)^{m/2}\frac{d^m}{dx^m}P_n(x) \qquad (7.21)$$

$$Q_n^m(x) = (1 - x^2)^{m/2}\frac{d^m}{dx^m}Q_n(x). \qquad (7.22)$$

Note that if $m > n$, $P_n^m(x) = 0$. The functions $Q_n^m(x)$ are unbounded for $x = \pm 1$.

Further, show that

$$\int_0^{\pi}[P_n^m(\cos\phi)]^2\sin\phi\,d\phi = \frac{2}{2n+1}\frac{(n+m)!}{(n-m)!}.$$

Answers or Hints

7.1. Use (2.5) and compute necessary partial fractions.

7.2. First use Leibniz's rule to obtain $P_n(0) = \frac{1}{2^n} \sum_{k=0}^{n} (-1)^k \left[\binom{n}{k} \right]^2$ and then equate the coefficients of x^n in the relation $(1+x)^n (1-x)^n = (1-x^2)^n$.

7.3. Use Rolle's theorem repeatedly.

7.4. (i) Differentiate (7.12) with respect to t (ii) Begin with the right-hand side.

7.5. Equate the coefficients of t^n.

7.6. Use $\int_0^\pi \frac{d\phi}{a \pm b \cos \phi} = \frac{\pi}{\sqrt{a^2 - b^2}}$ $(a^2 > b^2)$ with $a = 1 - tx$, $b = t\sqrt{(x^2 - 1)}$ and equate the coefficients of t^n.

7.7. Use the same formula as in Problem 7.6 with $a = xt - 1$, $b = t\sqrt{(x^2 - 1)}$ and equate the coefficients of t^{-n-1}.

7.8. (i) Differentiate (7.12) with respect to x and t and equate, to obtain the relation $t \sum_{n=0}^{\infty} n t^{n-1} P_n(x) = (x-t) \sum_{n=0}^{\infty} t^n P_n'(x)$. Finally equate the coefficients of t^n (ii) Differentiate (7.13) and use (i) (iii) Use (i) and (ii) (iv) Use (iii) and (i) (v) Use (7.13) and (iv) (vi) Use (v) in (iv).

7.9. (i) Use (7.12) (ii) In (7.12) first replace x by $-x$ and then t by $-t$ and then equate.

7.10. When $a = -n$, the recurrence relation (7.15) reduces to $m^2 c_m = (m - n)(m + n - 1) c_{m-1}$, $m \geq 1$, which can be solved to obtain

$$c_m = \begin{cases} (-1)^m \frac{(n+m-1)!}{(n-m-1)! \, (m!)^2} c_0, & 1 \leq m \leq n - 1 \\ 0, & m \geq n. \end{cases}$$

Therefore, the polynomial solution of (3.19) is

$$\sum_{m=0}^{n-1} \frac{(-1)^m (n+m-1)!}{(n-m-1)! \, (m!)^2} \left(\frac{1-x}{2} \right)^2 .$$

7.11. (i) Use Problem 7.8(ii) (ii) Use (7.13).

7.12. Verify directly.

Lecture 8
Chebyshev, Hermite, and Laguerre Polynomials

In this lecture we shall show that if the parameter a in (3.15) and (3.16), whereas b in (3.18) is a nonnegative integer n, then one of the solutions of each of these equations reduces to a polynomial of degree n. These polynomial solutions are, respectively, known as Chebyshev, Hermite, and Laguerre polynomials. We shall obtain explicit representations of these polynomials and discuss their various properties.

Chebyshev polynomials. From the recurrence relation (4.15) it is clear that in the case $a = n$ one of the solutions of the Chebyshev DE (3.15) reduces to a polynomial of degree n. Here we shall provide an easier method to construct these polynomials. For this, in the Chebyshev DE of order n,

$$(1 - x^2)y'' - xy' + n^2 y = 0 \tag{8.1}$$

we use the substitution $x = \cos\theta$, to obtain

$$(1 - \cos^2\theta)\frac{1}{\sin\theta}\left(\frac{1}{\sin\theta}\frac{d^2 y}{d\theta^2} - \frac{\cos\theta}{\sin^2\theta}\frac{dy}{d\theta}\right) + \cos\theta\frac{1}{\sin\theta}\frac{dy}{d\theta} + n^2 y = 0,$$

which is the same as

$$\frac{d^2 y}{d\theta^2} + n^2 y = 0.$$

Since for this DE $\sin n\theta$ and $\cos n\theta$ are the solutions, the solutions of (8.1) are $\sin(n\cos^{-1}x)$ and $\cos(n\cos^{-1}x)$. The solution $T_n(x) = \cos(n\cos^{-1}x)$ is a polynomial in x of degree n, and is called the *Chebyshev polynomial of the first kind*. To find its explicit representation, we need the following recurrence relation.

Theorem 8.1 (Recurrence Relation).

$$T_{n+1}(x) = 2xT_n(x) - T_{n-1}(x), \quad n \geq 1. \tag{8.2}$$

Proof. The proof is immediate from the identity

$$\cos((n+1)\cos^{-1}x) + \cos((n-1)\cos^{-1}x) = 2\cos(n\cos^{-1}x)\cos(\cos^{-1}x). \quad \blacksquare$$

R.P. Agarwal, D. O'Regan, *Ordinary and Partial Differential Equations*,
Universitext, DOI 10.1007/978-0-387-79146-3_8,
© Springer Science+Business Media, LLC 2009

Since $T_0(x) = \cos(0) = 1$, $T_1(x) = \cos(\cos^{-1} x) = x$ from the relation (8.2), it immediately follows that

$$T_2(x) = 2x^2 - 1, \quad T_3(x) = 4x^3 - 3x, \quad T_4(x) = 8x^4 - 8x^2 + 1.$$

Now we shall show that

$$T_n(x) = \cos(n \cos^{-1} x) = \frac{n}{2} \sum_{m=0}^{\left[\frac{n}{2}\right]} (-1)^m \frac{(n-m-1)!}{m!\,(n-2m)!} (2x)^{n-2m}, \quad n \geq 1.$$

$$(8.3)$$

For this, in view of (8.2), we have

$$
\begin{aligned}
T_{n+1}(x) &= \frac{n}{2} \sum_{m=0}^{\left[\frac{n}{2}\right]} (-1)^m \frac{(n-m-1)!}{m!\,(n-2m)!} (2x)^{n+1-2m} \\
&\quad - \frac{(n-1)}{2} \sum_{m=0}^{\left[\frac{n-1}{2}\right]} \frac{(-1)^m (n-1-m-1)!}{m!\,(n-1-2m)!} (2x)^{n-1-2m} \\
&= \frac{n}{2} \left[\frac{1}{n} (2x)^{n+1} + \sum_{m=1}^{\left[\frac{n}{2}\right]} \frac{(-1)^m (n-m-1)!}{m!\,(n-2m)!} (2x)^{n+1-2m} \right] \\
&\quad + \frac{(n-1)}{2} \sum_{m=1}^{\left[\frac{n-1}{2}\right]+1} \frac{(-1)^m (n-m-1)!}{(m-1)!\,(n-2m+1)!} (2x)^{n+1-2m}.
\end{aligned}
$$

Thus, if $n = 2k$, then $\left[\frac{n}{2}\right] = \left[\frac{n-1}{2}\right] + 1 = k = \left[\frac{n+1}{2}\right]$, and we get

$$
\begin{aligned}
T_{n+1}(x) &= \frac{1}{2} (2x)^{n+1} + \sum_{m=1}^{\left[\frac{n+1}{2}\right]} \frac{1}{2} (-1)^m \frac{(n-m-1)!(n-m)(n+1)}{m!\,(n-2m+1)!} (2x)^{n+1-2m} \\
&= \frac{(n+1)}{2} \sum_{m=0}^{\left[\frac{n+1}{2}\right]} (-1)^m \frac{(n-m)!}{m!\,(n+1-2m)!} (2x)^{n+1-2m}.
\end{aligned}
$$

Similarly, if $n = 2k+1$, then $\left[\frac{n}{2}\right] = k$, $\left[\frac{n-1}{2}\right] + 1 = k+1$, and we find

$$
\begin{aligned}
T_{n+1}(x) &= \frac{1}{2} (2x)^{n+1} + \sum_{m=1}^{\left[\frac{n}{2}\right]} \frac{1}{2} (-1)^m \frac{(n-m-1)!(n-m)(n+1)}{m!\,(n-2m+1)!} (2x)^{n+1-2m} \\
&\qquad\qquad\qquad\qquad\qquad\qquad\qquad\qquad + (-1)^{\left[\frac{n+1}{2}\right]} \\
&= \frac{(n+1)}{2} \sum_{m=0}^{\left[\frac{n+1}{2}\right]} (-1)^m \frac{(n-m)!}{m!\,(n+1-2m)!} (2x)^{n+1-2m}.
\end{aligned}
$$

Next for Chebyshev polynomials of the first kind we shall prove the following results.

Theorem 8.2 (Rodrigues' Formula).

$$T_n(x) = \frac{(-2)^n \, n!}{(2n)!}(1 - x^2)^{1/2}\frac{d^n}{dx^n}(1 - x^2)^{n-1/2}. \tag{8.4}$$

Proof. Set $v = (1 - x^2)^{n-1/2}$, to obtain $(1 - x^2)v' + (2n - 1)vx = 0$. Differentiating this relation $(n + 1)$ times, we obtain

$$(1 - x^2)D^{n+2}v - 3xD^{n+1}v + (n^2 - 1)D^n v = 0, \quad D = \frac{d}{dx}.$$

Let $w = (1 - x^2)^{1/2}D^n v$, to get

$$(1 - x^2)w'' - xw' + n^2 w$$
$$= (1 - x^2)^{1/2}\left[(1 - x^2)D^{n+2}v - 3xD^{n+1}v + (n^2 - 1)D^n v\right] = 0.$$

Thus, both $w(x)$ and $T_n(x)$ are the polynomial solutions of the Chebyshev DE (8.1), and hence $T_n(x) = cw(x)$, where the constant c is yet to be determined.

Since in view of Leibniz's rule

$$(1 - x^2)^{1/2}D^n(1 - x^2)^{n-1/2}$$

$$= (1 - x^2)^{1/2}\sum_{j=0}^{n}\binom{n}{j}D^{n-j}(1 + x)^{n-1/2}D^j(1 - x)^{n-1/2}$$

$$= n! \, (-1)^n \sum_{j=0}^{n}\binom{n - 1/2}{j}\binom{n - 1/2}{n - j}(x - 1)^{n-j}(x + 1)^j,$$

the coefficient of x^n in the right side of the above relation is

$$n! \, (-1)^n \sum_{j=0}^{n}\binom{n - 1/2}{j}\binom{n - 1/2}{n - j} = (-1)^n n!\binom{2n - 1}{n},$$

where the last identity follows from Vandermonde's equality. Thus, on comparing the coefficients of x^n in both sides of $T_n(x) = cw(x)$ we find $c = (-1)^n 2^n n!/(2n)!$. ■

Theorem 8.3 (Generating Function).

$$\frac{1 - t^2}{1 - 2tx + t^2} = T_0(x) + 2\sum_{n=1}^{\infty}T_n(x)t^n. \tag{8.5}$$

Proof. In the expansion

$$(1 - t^2)(1 - t(2x - t))^{-1} = (1 - t^2) \sum_{n=0}^{\infty} t^n (2x - t)^n$$

clearly the constant term is 1, and the coefficient of t^n is

$$(2x)^n + \sum_{m=1}^{\left[\frac{n}{2}\right]} (-1)^m \left\{ \binom{n-m}{m} + \binom{n-m-1}{m-1} \right\} (2x)^{n-2m}$$

$$= \sum_{m=0}^{\left[\frac{n}{2}\right]} \frac{n(-1)^m(n-m-1)!}{m!\,(n-2m)!} (2x)^{n-2m} = 2T_n(x). \quad \blacksquare$$

Hermite polynomials. For the Hermite DE of order n, i.e.,

$$y'' - 2xy' + 2ny = 0 \tag{8.6}$$

we follow exactly as for Legendre polynomials. Indeed, in the case $a = n$ the recurrence relation (4.13) can be written as

$$c_s = \frac{(s+2)(s+1)}{2(s-n)} c_{s+2}, \quad s \leq n - 2.$$

This relation with $c_n = 2^n$ gives

$$c_{n-2m} = \frac{(-1)^m \, n! \, 2^{n-2m}}{(n-2m)! \, m!}.$$

Thus, Hermite polynomials of degree n represented as $H_n(x)$ appear as

$$H_n(x) = \sum_{m=0}^{\left[\frac{n}{2}\right]} \frac{(-1)^m \, n!}{m!\,(n-2m)!} (2x)^{n-2m}. \tag{8.7}$$

From (8.7), we find

$$H_0(x) = 1, \quad H_1(x) = 2x, \quad H_2(x) = 4x^2 - 2, \quad H_3(x) = 8x^3 - 12x.$$

Now for Hermite polynomials we shall prove the following result.

Theorem 8.4 (Rodrigues' Formula).

$$H_n(x) = (-1)^n e^{x^2} \frac{d^n}{dx^n} e^{-x^2}. \tag{8.8}$$

Proof. We note that the DE (8.6) can be written as

$$\left(e^{-x^2}y'\right)' + 2ne^{-x^2}y = 0. \tag{8.9}$$

Let $y = (-1)^n e^{x^2} D^n e^{-x^2}$, $(D = d/dx)$ so that

$$y' = (-1)^n e^{x^2}\left(2xD^n e^{-x^2} + D^n\left(De^{-x^2}\right)\right),$$

and hence

$$\begin{aligned}
\left(e^{-x^2}y'\right)' &= (-1)^n\left[2D^n e^{-x^2} + 2xD^n(De^{-x^2}) + D^{n+1}(-2xe^{-x^2})\right] \\
&= 2n(-1)^{n+1}D^n e^{-x^2} = -2ne^{-x^2}y.
\end{aligned}$$

Thus, $(-1)^n e^{x^2} D^n e^{-x^2}$ and $H_n(x)$ both are solutions of the DE (8.9). But this is possible only if $H_n(x) = c(-1)^n e^{x^2} D^n e^{-x^2}$, where c is a constant. Finally, a comparison of the coefficients of x^n on both sides gives $c = 1$. ∎

Laguerre polynomials. For Laguerre's DE of order n, i.e.,

$$xy'' + (a+1-x)y' + ny = 0 \tag{8.10}$$

the solution $y_1(x)$ obtained in (6.1) reduces to a polynomial of degree n. This solution multiplied by the constant $\Gamma(n+a+1)/[n!\,\Gamma(a+1)]$ is called the *Laguerre polynomial* $L_n^{(a)}(x)$ and can be written as

$$L_n^{(a)}(x) = \Gamma(n+a+1)\sum_{m=0}^{n}\frac{(-1)^m x^m}{m!\,(n-m)!\,\Gamma(m+a+1)}. \tag{8.11}$$

In the particular case $a = 0$, Laguerre polynomial of degree n is simply represented as $L_n(x)$ and reduces to

$$L_n(x) = \sum_{m=0}^{n}\frac{1}{m!}\binom{n}{m}(-x)^m. \tag{8.12}$$

Problems

8.1. Verify that the nth degree *Chebyshev polynomial of the second kind* defined by the relation

$$U_n(x) = \frac{\sin((n+1)\cos^{-1}x)}{\sin(\cos^{-1}x)}$$

satisfies the DE

$$(1 - x^2)y'' - 3xy' + n(n+2)y = 0.$$

Show that

(i) $U_n(x) = \dfrac{1}{n+1}T'_{n+1}(x)$

(ii) $U_n(x) = \displaystyle\sum_{m=0}^{\left[\frac{n}{2}\right]}(-1)^m \dfrac{(n-m)!}{m!\,(n-2m)!}(2x)^{n-2m}$

(iii) $U_n(x) = \dfrac{(-2)^n\,(n+1)!}{(2n+1)!}(1-x^2)^{-1/2}\dfrac{d^n}{dx^n}(1-x^2)^{n+1/2}$

(Rodrigues' Formula)

(iv) $\dfrac{1}{1-2tx+t^2} = \displaystyle\sum_{n=0}^{\infty}U_n(x)t^n$ (Generating Function)

(v) $U_{n+1}(x) = 2xU_n(x) - U_{n-1}(x),\quad U_0(x) = 1,\quad U_1(x) = 2x$

(Recurrence Relation).

8.2.   For Hermite polynomials show that

(i) $e^{2xt-t^2} = \displaystyle\sum_{n=0}^{\infty}H_n(x)\dfrac{t^n}{n!}$ (Generating Function)

(ii) $H_{n+1}(x) = 2xH_n(x) - 2nH_{n-1}(x),\quad H_0(x) = 1,\quad H_1(x) = 2x$

(Recurrence Relation)

(iii) $H'_n(x) = 2nH_{n-1}(x).$

8.3.   For Laguerre polynomials show that

(i) $L_n^{(a)}(x) = \dfrac{e^x x^{-a}}{n!}\dfrac{d^n}{dx^n}\left(e^{-x}x^{n+a}\right)$ (Rodrigues' Formula)

(ii) $(1-t)^{-1-a}\exp\left(-\dfrac{xt}{(1-t)}\right) = \displaystyle\sum_{n=0}^{\infty}L_n^{(a)}(x)t^n$ (Generating Function)

(iii) $L_{n+1}^{(a)}(x) = \dfrac{2n+a+1-x}{n+1}L_n^{(a)}(x) - \dfrac{n+a}{n+1}L_{n-1}^{(a)}(x),$

$L_0^{(a)}(x) = 1,\quad L_1^{(a)}(x) = 1+a-x$ (Recurrence Relation).

Answers or Hints

8.1.   (i) Differentiate $T_{n+1}(x) = \cos((n+1)\cos^{-1}x)$ (ii) Use (8.3) and
(i) (iii) Use (8.4) and (i) (iv) Expand $(1-[2tx-t^2])^{-1}$ (v) Set $\theta = \cos^{-1}x$
in the relation $\sin(n+2)\theta + \sin n\theta = 2\sin(n+1)\theta\cos\theta$.

8.2. (i) Expand $e^{-t^2}e^{2xt}$ (ii) Differentiate both sides of (i) with respect to t and equate the coefficients of t^n (ii) Differentiate (8.8).

8.3. (i) Use Leibniz's rule (ii) Expand $(1-t)^{-1-a}\exp\left(-\frac{xt}{1-t}\right)$ (iii) Differentiate both sides of (ii) with respect to t and equate the coefficients of t^n.

Lecture 9
Bessel Functions

Bessel's differential equation (2.15) with $a = n$ first appeared in the works of Euler and Bernoulli whereas Bessel functions, also sometimes termed *cylindrical functions*, were introduced by Bessel, in 1824, in the discussion of a problem in dynamical astronomy. Since many problems of mathematical physics reduce to the Bessel equation, in this lecture we shall investigate it in somewhat more detail.

In the Bessel DE (2.15) the functions

$$xp_1(x) = x\left(\frac{x}{x^2}\right) = 1 \quad \text{and} \quad x^2 p_2(x) = x^2\left(\frac{x^2 - a^2}{x^2}\right) = x^2 - a^2$$

are analytic for all x, and hence the origin is a regular singular point.

Thus, in view of Theorem 5.1, we can assume that a solution of (2.15) can be written as $y(x) = x^r \sum_{m=0}^{\infty} c_m x^m$. Now a direct substitution of this in (2.15) leads to the equations

$$c_0(r + a)(r - a) = 0$$

$$c_1(1 + r + a)(1 + r - a) = 0$$

$$c_m(m + r + a)(m + r - a) = -c_{m-2}, \quad m = 2, 3, \cdots. \tag{9.1}$$

We assume that $c_0 \neq 0$, so that from the first equation $r_1 = a$ and $r_2 = -a$.

First, we shall consider the case when $r_1 - r_2 = 2a$ is not an integer. For this, we see that $c_1 = 0$ ($2a \neq \pm 1$), and for $r = a$ and $r = -a$, (9.1) reduces to

$$m(m + 2a)c_m = -c_{m-2}, \quad m = 2, 3, \cdots \tag{9.2}$$

and

$$m(m - 2a)c_m = -c_{m-2}, \quad m = 2, 3, \cdots \tag{9.3}$$

respectively. From these relations we easily obtain two linearly independent solutions $y_1(x)$ and $y_2(x)$ which appear as

$$y_1(x) = \left[1 - \frac{1}{2^2(1+a)1!}x^2 + \frac{1}{2^4(1+a)(2+a)2!}x^4 - \cdots\right]x^a c_0$$

and

$$y_2(x) = \left[1 - \frac{1}{2^2(1-a)1!}x^2 + \frac{1}{2^4(1-a)(2-a)2!}x^4 - \cdots\right]x^{-a} c_0^*.$$

R.P. Agarwal, D. O'Regan, *Ordinary and Partial Differential Equations*,
Universitext, DOI 10.1007/978-0-387-79146-3_9,
© Springer Science+Business Media, LLC 2009

In the above solutions we take the constants

$$c_0 = \frac{1}{2^a \Gamma(1+a)} \quad \text{and} \quad c_0^* = \frac{1}{2^{-a}\Gamma(1-a)},$$

to obtain

$$y_1(x) = \sum_{m=0}^{\infty} \frac{(-1)^m}{m!\, \Gamma(m+1+a)} \left(\frac{x}{2}\right)^{2m+a} \tag{9.4}$$

and

$$y_2(x) = \sum_{m=0}^{\infty} \frac{(-1)^m}{m!\, \Gamma(m+1-a)} \left(\frac{x}{2}\right)^{2m-a}. \tag{9.5}$$

These solutions are analytic for $|x| > 0$. The function $y_1(x)$ is called the *Bessel function* of order a of the *first kind* and is denoted by $J_a(x)$; $y_2(x)$ is the Bessel function of order $-a$ and is denoted by $J_{-a}(x)$. The general solution of the DE (2.15) is given by $y(x) = AJ_a(x) + BJ_{-a}(x)$, where A and B are arbitrary constants.

Now we shall consider the case when $r_1 - r_2 = 2a$ is a positive odd integer, i.e., $2a = 2n + 1$. Then, for $r = a$, $c_1 = 0$ and as before from (9.1) we get

$$J_{n+1/2}(x) = \sum_{m=0}^{\infty} \frac{(-1)^m}{m!\, \Gamma\left(m+n+\frac{3}{2}\right)} \left(\frac{x}{2}\right)^{2m+n+1/2}. \tag{9.6}$$

For $r = -a$ we have $-a = -n - 1/2$; then c_0 and c_{2n+1} are both arbitrary. Finding the coefficients c_{2m}, $m = 1, 2, \cdots$ and $c_{2n+2m+1}$, $m = 1, 2, \cdots$ from (9.1) in terms of c_0 and c_{2n+1} leads to the solution

$$
\begin{aligned}
y(x) &= c_0 x^{-n-1/2} \sum_{m=0}^{\infty} \frac{(-1)^m \Gamma\left(\frac{1}{2}-n\right)}{m!\, 2^{2m}\, \Gamma\left(\frac{1}{2}-n+m\right)} x^{2m} \\
&\quad + c_{2n+1} x^{-n-1/2} \sum_{m=0}^{\infty} \frac{(-1)^m \Gamma\left(n+\frac{3}{2}\right)}{m!\, 2^{2m}\, \Gamma\left(n+m+\frac{3}{2}\right)} x^{2n+1+2m} \\
&= AJ_{-n-1/2}(x) + BJ_{n+1/2}(x),
\end{aligned}
\tag{9.7}
$$

where $A = c_0 2^{-n-1/2} \Gamma\left(\frac{1}{2} - n\right)$ and $B = 2^{n+1/2}\Gamma\left(\frac{3}{2}+n\right) c_{2n+1}$. Since c_0 and c_{2n+1} are both arbitrary, A and B are arbitrary constants, and hence (9.7) represents the general solution of (2.15).

If we take $2a$ as a negative odd integer, i.e., $2a = -2n - 1$, then we get $r_1 = -n - 1/2$ and $r_2 = n + 1/2$, and hence for $r_2 = n + 1/2$ we have the same solution as (9.6), and for $r_1 = -n - 1/2$ the solution is the same as (9.7).

Next we consider the case when $a = 0$. Then the exponents are $r_1 = r_2 = 0$. The first solution $y_1(x)$ in this case is easily obtained and appears as

$$y_1(x) = J_0(x) = \sum_{m=0}^{\infty} \frac{(-1)^m}{(m!)^2} \left(\frac{x}{2}\right)^{2m}. \tag{9.8}$$

To obtain the second solution we return to the recurrence relation (9.1), which with $c_0 = 1$ gives

$$c_{2m} = \frac{(-1)^m}{(r+2)^2(r+4)^2\cdots(r+2m)^2}$$

and hence

$$\frac{c'_{2m}}{c_{2m}} = -2\sum_{k=1}^{m} \frac{1}{r+2k}.$$

Thus, we find

$$d_{2m} = c'_{2m}(0) = -c_{2m}(0)\sum_{k=1}^{m}\frac{1}{k}$$

and now Theorem 5.1(ii) gives the second solution $y_2(x)$, which is denoted as $J^0(x)$,

$$y_2(x) = J^0(x) = J_0(x)\ln|x| - \sum_{m=1}^{\infty}\frac{(-1)^m}{(m!)^2}\left(\sum_{k=1}^{m}\frac{1}{k}\right)\left(\frac{x}{2}\right)^{2m}. \tag{9.9}$$

Finally, the general solution in this case can be written as $y(x) = AJ_0(x) + BJ^0(x)$.

The remaining case, namely, when $r_1 - r_2$ is an even integer, i.e., when $2a = 2n$ is very important, which we shall consider now. For $r_1 = n$, there is no difficulty and the first solution $y_1(x)$ can be written as $y_1(x) = J_n(x)$. However, for the second solution $y_2(x)$ corresponding to the exponent $r_2 = -n$ we need some extra manipulation to obtain

$$y_2(x) = \frac{2}{\pi}\left[\left(\gamma + \ln\left|\frac{x}{2}\right|\right)J_n(x) - \frac{1}{2}\sum_{k=0}^{n-1}\frac{(n-k-1)!}{k!}\left(\frac{x}{2}\right)^{2k-n}\right.$$
$$\left. + \frac{1}{2}\sum_{m=0}^{\infty}(-1)^{m+1}\frac{\phi(m)+\phi(n+m)}{m!\,(n+m)!}\left(\frac{x}{2}\right)^{2m+n}\right], \tag{9.10}$$

where $\phi(0) = 0$, $\phi(m) = \sum_{k=1}^{m}(1/k)$ and γ is the *Euler constant* defined by

$$\gamma = \lim_{m\to\infty}\,(\phi(m) - \ln m) = 0.5772157\cdots.$$

This solution $y_2(x)$ is known as *Weber's Bessel function of the second kind* and is denoted by $Y_n(x)$. In the literature some authors also call $Y_n(x)$ the *Neumann function*.

Thus, the general solution in this case can be written as $y(x) = AJ_n(x) + BY_n(x)$.

The *Bessel functions of the third kind* also known as *Hankel functions* are the complex solutions of the DE (2.15) for $a = n$ and are defined by the relations

$$H_n^{(1)}(x) = J_n(x) + iY_n(x) \tag{9.11}$$

and

$$H_n^{(2)}(x) = J_n(x) - iY_n(x). \tag{9.12}$$

The *modified Bessel function of the first kind of order n* is defined as

$$I_n(x) = i^{-n}J_n(ix) = e^{-n\pi i/2}J_n(ix). \tag{9.13}$$

If n is an integer, $I_{-n}(x) = I_n(x)$; but if n is not an integer, $I_n(x)$ and $I_{-n}(x)$ are linearly independent.

The *modified Bessel function of the second kind of order n* is defined as

$$K_n(x) = \begin{cases} \dfrac{\pi}{2}\left[\dfrac{I_{-n}(x) - I_n(x)}{\sin n\pi}\right], & n \neq 0, 1, 2, \cdots \\ \lim\limits_{p \to n} \dfrac{\pi}{2}\left[\dfrac{I_{-p}(x) - I_p(x)}{\sin p\pi}\right], & n = 0, 1, 2, \cdots. \end{cases} \tag{9.14}$$

These functions are the solutions of the *modified Bessel DE*

$$x^2 y'' + xy' - (x^2 + n^2)y = 0. \tag{9.15}$$

The generating function for the sequence $\{J_n(x)\}_{n=-\infty}^{\infty}$ is given in the following theorem.

Theorem 9.1 (Generating Function).

$$\exp\left(\frac{1}{2}x\left(t - \frac{1}{t}\right)\right) = \sum_{n=-\infty}^{\infty} J_n(x)t^n. \tag{9.16}$$

Proof. We shall expand the left-hand side of (9.16) in powers of t and show that the coefficient of t^n is $J_n(x)$. In fact, we have

$$\begin{aligned} \exp\left(\frac{1}{2}x\left(t - \frac{1}{t}\right)\right) &= \exp\left(\frac{1}{2}xt\right)\exp\left(-\frac{x}{2t}\right) \\ &= \sum_{s=0}^{\infty}\frac{(xt)^s}{2^s\,s!}\sum_{r=0}^{\infty}\frac{(-x)^r}{2^r t^r r!} \\ &= \sum_{s=0}^{\infty}\sum_{r=0}^{\infty}(-1)^r\left(\frac{x}{2}\right)^{s+r}\frac{t^{s-r}}{s!\,r!}. \end{aligned}$$

The coefficient of t^n in the above expression is

$$\sum_{m=0}^{\infty} \frac{(-1)^m}{m!\,(n+m)!} \left(\frac{x}{2}\right)^{2m+n},$$

which, by definition, is the same as $J_n(x)$. ∎

If in (9.16) we replace t by $-1/\tau$, then it is the same as

$$\exp\left(\frac{1}{2}x\left(\tau - \frac{1}{\tau}\right)\right) = \sum_{n=-\infty}^{\infty} J_n(x)(-1)^n \tau^{-n} = \sum_{n=-\infty}^{\infty} J_n(x)\tau^n$$

and hence it is necessary that

$$J_n(x) = (-1)^n J_{-n}(x). \tag{9.17}$$

As in the case of Legendre polynomials, for Bessel functions $J_n(x)$ also several recurrence relations are known. In the following theorem we shall prove one and give a few others as problems.

Theorem 9.2 (Recurrence Relation).

$$xJ_n'(x) = nJ_n(x) - xJ_{n+1}(x). \tag{9.18}$$

Proof. From the definition of $J_n(x)$ in (9.4), we have

$$J_n(x) = \sum_{m=0}^{\infty} \frac{(-1)^m}{m!\,(m+n)!} \left(\frac{x}{2}\right)^{2m+n}$$

and hence

$$J_n'(x) = \sum_{m=0}^{\infty} \frac{(-1)^m(2m+n)}{m!\,(m+n)!} \frac{1}{2}\left(\frac{x}{2}\right)^{2m+n-1},$$

so

$$
\begin{aligned}
xJ_n'(x) &= n\sum_{m=0}^{\infty} \frac{(-1)^m}{m!(m+n)!} \left(\frac{x}{2}\right)^{2m+n} \\
&\quad + x\sum_{m=1}^{\infty} \frac{(-1)^m}{(m-1)!(m+n)!} \left(\frac{x}{2}\right)^{2m+n-1} \\
&= nJ_n(x) + x\sum_{m=0}^{\infty} \frac{(-1)^{m+1}}{m!\,(m+n+1)!} \left(\frac{x}{2}\right)^{2m+n+1} \\
&= nJ_n(x) - xJ_{n+1}(x). \qquad \blacksquare
\end{aligned}
$$

Finally, we note that when x is very small then from (9.5) with $a = n$, $y_2(x) \simeq$ constant $\times x^{-n}$; from (9.8) $J_0(x) \simeq$ constant; and from (9.9)

$J^0(x) \simeq \ln|x|$. Thus, when $a = n \ (\geq 0)$ the second solution of (2.15) for very small x can be approximated by

$$\text{constant} \times \begin{cases} \ln|x| & \text{if } n = 0 \\ x^{-n} & \text{if } n > 1. \end{cases}$$

Hence, the second solution tends to ∞ as $x \to 0$. Similarly, the functions $Y_n(x)$ and $K_n(x)$ are unbounded at $x = 0$.

We also note from the definition of $J_a(x)$ in (9.4) and the derivation of (9.18) that the recurrence relation (9.18) in fact holds for all real numbers n, i.e., n need not be an integer.

Problems

9.1. Prove the following relations:

(i) $xJ'_n(x) = -nJ_n(x) + xJ_{n-1}(x)$

(ii) $2J'_n(x) = J_{n-1}(x) - J_{n+1}(x)$

(iii) $2nJ_n(x) = x(J_{n-1}(x) + J_{n+1}(x))$

(iv) $\dfrac{d}{dx}(x^{-n}J_n(x)) = -x^{-n}J_{n+1}(x)$

(v) $\dfrac{d}{dx}(x^n J_n(x)) = x^n J_{n-1}(x)$

(vi) $\dfrac{d}{dx}(xJ_n(x)J_{n+1}(x)) = x(J_n^2(x) - J_{n+1}^2(x))$

(vii) $\dfrac{d}{dx}(J_n^2(x) + J_{n+1}^2(x)) = 2\left(\dfrac{n}{x}J_n^2(x) - \dfrac{n+1}{x}J_{n+1}^2(x)\right)$

(viii) $J_0^2(x) + 2J_1^2(x) + 2J_n^2(x) + \cdots = 1$, and hence $|J_0(x)| \leq 1$, $|J_n(x)| \leq 1/\sqrt{2}$.

9.2. From the definition of $J_a(x)$ show that

(i) $J_{1/2}(x) = \sqrt{\dfrac{2}{\pi x}} \sin x$

(ii) $J_{-1/2}(x) = \sqrt{\dfrac{2}{\pi x}} \cos x$

(iii) $J_{3/2}(x) = \sqrt{\dfrac{2}{\pi x}} \left(\dfrac{\sin x}{x} - \cos x\right)$

(iv) $J_{-3/2}(x) = \sqrt{\dfrac{2}{\pi x}} \left(-\dfrac{\cos x}{x} - \sin x\right)$

(v) $\displaystyle\int_0^x t^{n+1} J_n(t)dt = x^{n+1} J_{n+1}(x), \quad n > -1.$

9.3. In a study of planetary motion Bessel encountered the integral

$$y(x) = \frac{1}{\pi} \int_0^\pi \cos(n\theta - x \sin\theta)d\theta.$$

Show that $y(x) = J_n(x)$.

9.4. Show that for any $a \geq 0$, the function

$$y(x) = x^a \int_0^\pi \cos(x \cos\theta) \sin^{2a}\theta d\theta$$

is a solution to the Bessel DE (2.15).

9.5. Show that the transformation $x = \alpha t^\beta$, $y = t^\gamma w$, where α, β and γ are constants, converts Bessel's DE (2.15) to

$$t^2 \frac{d^2 w}{dt^2} + (2\gamma+1)t\frac{dw}{dt} + (\alpha^2\beta^2 t^{2\beta} + \gamma^2 - a^2\beta^2)w = 0. \tag{9.19}$$

For $\beta = 1$, $\gamma = 0$, $a = n$ (nonnegative integer) (9.19) reduces to

$$t^2 \frac{d^2 w}{dt^2} + t\frac{dw}{dt} + (\alpha^2 t^2 - n^2)w = 0. \tag{9.20}$$

(i) Show that the solution of the DE

$$\frac{d^2 w}{dt^2} + t^m w = 0, \quad m > 0$$

can be expressed as

$$w(t) = t^{1/2}\left(A J_{1/(m+2)}\left(\frac{2}{m+2}t^{(m+2)/2}\right) + B J_{-1/(m+2)}\left(\frac{2}{m+2}t^{(m+2)/2}\right)\right).$$

(ii) Express the solution of DE

$$\frac{d^2 w}{dt^2} + w = 0$$

in terms of Bessel functions and deduce the relations given in Problem 9.2 Parts (i) and (ii)

9.6. Show that the general solution of the DE

$$y' = x^2 + y^2$$

is given by

$$y(x) = x\,\frac{J_{-3/4}\left(\frac{1}{2}x^2\right) + cJ_{3/4}\left(\frac{1}{2}x^2\right)}{cJ_{-1/4}\left(\frac{1}{2}x^2\right) - J_{1/4}\left(\frac{1}{2}x^2\right)},$$

where c is an arbitrary constant.

9.7. Consider a long, flat triangular piece of metal whose ends are joined by an inextensible piece of string of length slightly less than that of the piece of metal (see Figure 9.1). The line of the string is taken as the x-axis, with the left end as the origin. The deflection $y(x)$ of the piece of metal from horizontal at x satisfies the DE

$$EIy'' + Ty = 0,$$

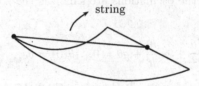

string

Figure 9.1

where T is the tension in the string, E is Young's modulus, and I is the moment of inertia of a cross section of the piece of metal. Since the metal is triangular, $I = \alpha x$ for some constant $\alpha > 0$. The above equation thus can be written as

$$xy'' + k^2 y = 0,$$

where $k^2 = T/E\alpha$. Find the general solution of the above equation.

9.8. Consider a vertical column of length ℓ, such as a telephone pole (see Figure 9.2). In certain cases it leads to the DE

$$\frac{d^2y}{dx^2} + \frac{P}{EI}e^{kx/\ell}(y - a) = 0,$$

where E is the modulus of elasticity, I the moment of inertia at the base of the column about an axis perpendicular to the plane of bending, and P and k are constants.

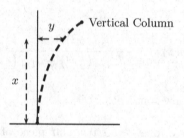

Figure 9.2

Find the general solutions of the above equation.

9.9. A simple pendulum initially has the length ℓ_0 and makes an angle θ_0 with the vertical line. It is then released from this position. If the length of the pendulum increases with time t according to $\ell = \ell_0 + \epsilon t$, where ϵ is a small constant, then the angle θ that the pendulum makes with the vertical line, assuming that the oscillations are small, is the solution of the initial value problem

$$(\ell_0 + \epsilon t)\frac{d^2\theta}{dt^2} + 2\epsilon\frac{d\theta}{dt} + g\theta = 0, \quad \theta(0) = \theta_0, \quad \frac{d\theta}{dt}(0) = 0,$$

where g is the acceleration due to gravity. Find the solution of the above initial value problem.

9.10. The DE

$$x^2\frac{d^2 E}{dx^2} - \mu x\frac{dE}{dx} - k^2 x^\nu E = 0$$

occurs in the study of the flow of current through electrical transmission lines; here μ, ν are positive constants and E represents the potential difference (with respect to one end of the line) at a point a distance x from that end of the line. Find its general solution.

9.11. Consider the wedge-shaped cooling fin as shown in Figure 9.3. The DE which describes the heat flow through and from this wedge is

$$x^2\frac{d^2 y}{dx^2} + x\frac{dy}{dx} - \mu x y = 0,$$

where x is distance from tip to fin; T, temperature of fin at x; T_0, constant temperature of surrounding air; $y = T - T_0$; h, heat-transfer coefficient from outside surface of fin to the surrounding air; k, thermal conductivity of fin material; ℓ, length of fin; w, thickness of fin at its base; θ, one-half the wedge angle; and $\mu = 2h\sec(\ell\theta)/kw$.

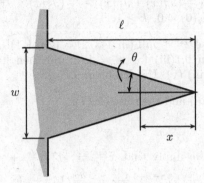

Figure 9.3

Find the general solutions of the above equation.

9.12. Consider a horizontal beam of length 2ℓ supported at both ends, and laterally fixed at its left end. The beam carries a uniform load w per unit length and is subjected to a constant end-pull P from the right (see Figure 9.4). Suppose that the moment of inertia of a cross section of the beam at a distance s from its left end is $I = 2(s+1)$. If origin is the middle of the bar, then the vertical deflection $y(x)$ at x is governed by the nonhomogeneous DE

$$2E(x+1+\ell)\frac{d^2y}{dx^2} - Py = \frac{1}{2}w(x+\ell)^2 - w(x+\ell),$$

where E is Young's modulus.

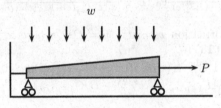

Figure 9.4

Find the general solutions of the above equation.

Answers or Hints

9.1. (i) See the proof of Theorem 9.2 (ii) Use (9.18) and (i) (iii) Use (9.18) and (i) (iv) Multiply the relation (9.18) by x^{-n-1} (v) Multiply the relation in (i) by x^{n-1} (vi) Use (9.18) and (i) (vii) Proceed as in (vi)

(viii) Use (vii) to obtain $\frac{d}{dx}(J_0^2(x) + J_1^2(x)) + \frac{d}{dx}(J_1^2(x) + J_2^2(x)) + \cdots = 0$. Finally use $J_0(0) = 1$, $J_k(0) = 0$, $k \geq 1$.

9.2. For (i) and (ii) use $2^{2m+1} m! \Gamma(m + 3/2) = (2m + 1)! \sqrt{\pi}$ (iii) Verify directly, or use Problem 9.1(iii) for $n = 1/2$ and then use (i) and (ii) (iv) Similar to that of (iii) (v) Use Problem 9.1(v).

9.3. Set $t = e^{i\theta}$ in (9.16) and then use (9.17).

9.4. Verify directly.

9.5. $x = \alpha t^\beta$, $y = t^\gamma w$ imply that $x\frac{dy}{dx} = \frac{1}{\beta} t^\gamma \left(t\frac{dw}{dt} + \gamma w \right)$, $x^2 \frac{d^2 y}{dx^2} = \frac{1}{\beta^2} t^{\beta+1} \left((\gamma - \beta + 1) t^{\gamma-\beta} \frac{dw}{dt} + t^{\gamma-\beta+1} \frac{d^2 w}{dt^2} + \gamma \frac{dw}{dt} t^{\gamma-\beta} + \gamma(\gamma - \beta) w t^{\gamma-\beta-1} \right)$
(i) Compare $t^2 \frac{d^2 w}{dt^2} + t^{m+2} w = 0$ with (9.19) (ii) Compare $t^2 \frac{d^2 w}{dt^2} + t^2 w = 0$ with (9.19).

9.6. The transformation $y = -u'/u$ converts the DE $y' = x^2 + y^2$ to $u'' + x^2 u = 0$. Now use Problem 9.5(i), Problem 9.1(i) and (9.18).

9.7. $x^{1/2} \left(AJ_1 \left(2kx^{1/2} \right) + BJ_{-1} \left(2kx^{1/2} \right) \right)$.

9.8. Use $z = y - a$, $2\mu = k/\ell$, $\nu = \sqrt{P/EI}$ to convert the DE as $\frac{d^2 z}{dx^2} + \nu^2 e^{2\mu x} z = 0$. Now use $t = \nu e^{\mu x}$ to obtain $t^2 \frac{d^2 z}{dt^2} + t\frac{dz}{dt} + \frac{t^2}{\mu^2} z = 0$. Finally, compare this with (9.19).

9.9. $\theta(t) = \frac{1}{\sqrt{\ell_0 + \epsilon t}} \left[AJ_1 \left(\frac{2\sqrt{g}}{\epsilon} \sqrt{\ell_0 + \epsilon t} \right) + BJ_{-1} \left(\frac{2\sqrt{g}}{\epsilon} \sqrt{\ell_0 + \epsilon t} \right) \right]$ where $A = \frac{\sqrt{\ell_0} J'_{-1}(c) - (\epsilon/2\sqrt{g}) J_{-1}(c)}{J_1(c) J'_{-1}(c) - J_{-1}(c) J'_1(c)} \theta_0$, $B = \frac{(\epsilon/2\sqrt{g}) J_1(c) - \sqrt{\ell_0} J'_1(c)}{J_1(c) J'_{-1}(c) - J_{-1}(c) J'_1(c)} \theta_0$ and $c = \frac{2\sqrt{g\ell_0}}{\epsilon}$.

9.10. The transformation $x = \alpha t^\beta$, $y = t^\gamma w$, where α, β and γ are constants, converts (9.15) to
$$t^2 \frac{d^2 w}{dt^2} + (2\gamma + 1) t\frac{dw}{dt} + (-\alpha^2 \beta^2 t^{2\beta} + \gamma^2 - n^2 \beta^2) w = 0. \tag{9.21}$$
$E(x) = x^{(1+\mu)/2} \left[AI_{(1+\mu)/\nu} \left(\frac{2k}{\nu} x^{\nu/2} \right) + BK_{(1+\mu)/\nu} \left(\frac{2k}{\nu} x^{\nu/2} \right) \right]$.

9.11. Compare with (9.21) to get $y(x) = AI_0 \left(2\sqrt{\mu} x^{1/2} \right) + BK_0 \left(2\sqrt{\mu} x^{1/2} \right)$.

9.12. $y(x) = x^{1/2} \left[AI_1 \left(\sqrt{\frac{2P}{E}} x^{1/2} \right) + BK_1 \left(\sqrt{\frac{2P}{E}} x^{1/2} \right) \right] - \frac{w}{2P} (x + \ell)^2 + \frac{w}{P^2} (P - 2E)(x + \ell) - \frac{2E}{P^2} w$.

Lecture 10
Hypergeometric Functions

In this lecture we shall study the hypergeometric DE (3.17) and its solutions, known as hypergeometric functions. This DE finds applications in several problems of mathematical physics, quantum mechanics, and fluid dynamics.

A series of the form

$$1 + \frac{ab}{c\,1!}x + \frac{a(a+1)b(b+1)}{c(c+1)\,2!}x^2 + \frac{a(a+1)(a+2)b(b+1)(b+2)}{c(c+1)(c+2)\,3!}x^3 + \cdots$$

$$= \left(\sum_{m=0}^{\infty} \frac{\Gamma(a+m)\,\Gamma(b+m)}{\Gamma(c+m)\,m!}x^m \right) \frac{\Gamma(c)}{\Gamma(a)\,\Gamma(b)} \tag{10.1}$$

is called a *hypergeometric series*.

The ratio of the coefficients of x^{m+1} and x^m in the series (10.1) is

$$\frac{(a+m)(b+m)}{(c+m)(m+1)}, \tag{10.2}$$

which tends to 1 uniformly as $m \to \infty$, regardless of the values of a, b and c. Hence, by the ratio test, the series (10.1) has unit radius of convergence in every case. Also, since (10.2) can be written as

$$1 - \frac{1+c-a-b}{m} + O\left(\frac{1}{m^2} \right),$$

the series (10.1) converges absolutely at $x = \pm 1$ by the Gauss test if $c > a + b$.

The hypergeometric series is commonly designated by $F(a,b,c,x)$ and in this form it is called a *hypergeometric function*.

In the hypergeometric DE (3.17), a, b, and c are parameters. It is clear that $x = 0$ and 1 are regular singular points of (3.17), whereas all other points are ordinary points. Also, in the neighborhood of zero we have $p_0 = c$, $q_0 = 0$ and the indicial equation $r(r-1) + cr = 0$ has the roots $r_1 = 0$ and $r_2 = 1 - c$.

R.P. Agarwal, D. O'Regan, *Ordinary and Partial Differential Equations*,
Universitext, DOI 10.1007/978-0-387-79146-3_10,
© Springer Science+Business Media, LLC 2009

On substituting directly $y(x) = x^r \sum_{m=0}^{\infty} c_m x^m$ in the equation (3.17), we obtain the recurrence relation

$$(r+1+m)(r+c+m)c_{m+1} = (r+a+m)(r+b+m)c_m, \quad m = 0, 1, \cdots. \quad (10.3)$$

For the exponent $r_1 = 0$, the recurrence relation (10.3) leads to the solution $c_0 F(a, b, c, x)$. Taking $c_0 = 1$, we find the first solution of the DE (3.17) as

$$y(x) = F(a, b, c, x). \quad (10.4)$$

The second solution with the exponent $r_2 = 1 - c$ when c is neither zero nor a negative integer can be obtained as follows: In the DE (3.17) using the substitution $y = x^{1-c}w$, we obtain

$$x(1 - x)w'' + [c_1 - (a_1 + b_1 + 1)x]w' - a_1 b_1 w = 0, \quad (10.5)$$

where $c_1 = 2 - c$, $a_1 = a - c + 1$, $b_1 = b - c + 1$. This DE has a series solution $w(x) = F(a_1, b_1, c_1, x)$, and hence the second solution of the DE (3.17) is

$$y(x) = x^{1-c}F(a - c + 1, b - c + 1, 2 - c, x). \quad (10.6)$$

The general solution of the DE (3.17) in the neighborhood of $x = 0$ is a linear combination of the two solutions (10.4) and (10.6).

Solutions of the DE (3.17) at the singular point $x = 1$ can be obtained directly or may be deduced from the preceding solutions by a change of independent variable $t = 1 - x$. Indeed, with this substitution DE (3.17) reduces to

$$t(1 - t)\frac{d^2 y}{dt^2} + [c_1 - (a + b + 1)t]\frac{dt}{dt} - aby = 0, \quad (10.7)$$

where $c_1 = a + b - c + 1$.

Thus, we have the solutions

$$y(x) = F(a, b, a + b - c + 1, 1 - x) \quad (10.8)$$

and

$$y(x) = (1 - x)^{c-a-b}F(c - b, c - a, c - a - b + 1, 1 - x) \quad (10.9)$$

provided $c - a - b$ is not a positive integer.

The hypergeometric equation (3.17) has an interesting property that the derivative of a solution satisfies an associated hypergeometric DE. Indeed, differentiating (3.17), we obtain

$$x(1 - x)y''' + [c + 1 - (a + 1 + b + 1 + 1)x]y'' - (a + 1)(b + 1)y' = 0,$$

which is a hypergeometric DE in y' in which a, b, c have been replaced by $a+1$, $b+1$, $c+1$, respectively. Thus, it follows that

$$\frac{d}{dx}F(a,b,c,x) = AF(a+1,b+1,c+1,x), \qquad (10.10)$$

where A is a constant.

Comparing the constant terms on both sides of (10.10), it follows that $A = ab/c$. Hence, the following relation holds

$$\frac{d}{dx}F(a,b,c,x) = \frac{1}{c}ab\, F(a+1,b+1,c+1,x). \qquad (10.11)$$

In general for the mth derivative it is easy to find

$$\frac{d^m}{dx^m}F(a,b,c,x) = \frac{\Gamma(a+m)\,\Gamma(b+m)}{\Gamma(c+m)}\frac{\Gamma(c)}{\Gamma(a)\,\Gamma(b)}F(a+m,b+m,c+m,x). \qquad (10.12)$$

Like Bessel functions (cf. Problem 9.3), hypergeometric functions can also be represented in terms of suitable integrals. For this, we shall prove the following theorem.

Theorem 10.1 (Integral Representation).

$$F(a,b,c,x) = \frac{1}{B(b,c-b)}\int_0^1 (1-t)^{c-b-1}t^{b-1}(1-xt)^{-a}dt, \quad c > b > 0. \qquad (10.13)$$

Proof. From the relation (3.7) and the definition of Beta function (3.6), we have

$$\begin{aligned}
F(a,b,c,x) &= \sum_{m=0}^{\infty} \frac{\Gamma(a+m)}{\Gamma(a)\,m!}\left(\frac{\Gamma(b+m)\,\Gamma(c-b)}{\Gamma(c+m)} \Big/ \frac{\Gamma(b)\,\Gamma(c-b)}{\Gamma(c)}\right)x^m \\
&= \sum_{m=0}^{\infty} \frac{\Gamma(a+m)}{\Gamma(a)\,m!}\frac{B(b+m,c-b)}{B(b,c-b)}x^m \\
&= \frac{1}{B(b,c-b)}\sum_{m=0}^{\infty} \frac{\Gamma(a+m)}{\Gamma(a)\,m!}x^m \int_0^1 (1-t)^{c-b-1}t^{b+m-1}dt.
\end{aligned}$$

Thus, on interchanging the order in which the operations of summation and integration are performed, we obtain

$$F(a,b,c,x) = \frac{1}{B(b,c-b)}\int_0^1 (1-t)^{c-b-1}t^{b-1}\left\{\sum_{m=0}^{\infty}\frac{\Gamma(a+m)}{\Gamma(a)\,m!}(xt)^m\right\}dt. \qquad (10.14)$$

However, since

$$\sum_{m=0}^{\infty} \frac{\Gamma(a+m)}{\Gamma(a)\,m!}(xt)^m = (1-xt)^{-a}$$

the relation (10.14) immediately leads to (10.13).   ∎

As a consequence of (10.13), we find

$$
\begin{aligned}
F(a,b,c,1) &= \frac{1}{B(b,c-b)}\int_0^1 (1-t)^{c-a-b-1}t^{b-1}dt \\
&= \frac{B(c-a-b,b)}{B(b,c-b)} = \frac{\Gamma(c-a-b)\,\Gamma(c)}{\Gamma(c-a)\,\Gamma(c-b)}
\end{aligned}
$$

provided $c - a > b > 0$.

Next we shall prove the following relation.

Theorem 10.2 (Recurrence Relation).

$$F(a+1,b,c,x) - F(a,b,c,x) = \frac{1}{c}bx\,F(a+1,b+1,c+1,x). \qquad (10.15)$$

Proof. From (10.1), we have

$$
\begin{aligned}
F(a+1,&b,c,x) - F(a,b,c,x) \\
&= \sum_{m=0}^{\infty} \frac{\Gamma(b+m)\,\Gamma(c)}{\Gamma(c+m)\,m!\,\Gamma(b)}\left\{\frac{\Gamma(a+m+1)}{\Gamma(a+1)} - \frac{\Gamma(a+m)}{\Gamma(a)}\right\}x^m \\
&= \sum_{m=0}^{\infty} \frac{\Gamma(a+m)\,\Gamma(b+m)\,\Gamma(c)}{\Gamma(c+m)\,m!\,\Gamma(b)\,\Gamma(a)}\frac{m}{a}x^m \\
&= x\sum_{m=1}^{\infty} \frac{\Gamma(a+m)\,\Gamma(b+m)\,\Gamma(c)}{\Gamma(c+m)\,(m-1)!\,\Gamma(b)\,\Gamma(a+1)}x^{m-1} \\
&= \frac{1}{c}bx\sum_{m=0}^{\infty} \frac{\Gamma(a+m+1)\,\Gamma(b+m+1)\,\Gamma(c+1)}{\Gamma(c+m+1)\,m!\,\Gamma(b+1)\,\Gamma(a+1)}x^m \\
&= \frac{1}{c}bx\,F(a+1,b+1,c+1,x),
\end{aligned}
$$

which is the right side of (10.15).   ∎

Legendre polynomials can be expressed in terms of hypergeometric functions. We provide this relation in the following result.

Theorem 10.3. The following relation holds:

$$P_n(x) = F(-n,n+1,1,(1-x)/2). \qquad (10.16)$$

Proof. From the Rodrigues formula for Legendre polynomials (7.9), we have

$$P_n(x) = \frac{1}{2^n\,n!}\frac{d^n}{dx^n}(x^2-1)^n = \frac{1}{n!}\frac{d^n}{dx^n}\left[(x-1)^n\left\{\frac{1}{2}(x+1)\right\}^n\right]$$

$$= \frac{1}{n!}\frac{d^n}{dx^n}\left[(x-1)^n\left\{1-\frac{1}{2}(1-x)\right\}^n\right]$$

$$= \frac{(-1)^n}{n!}\frac{d^n}{dx^n}\left[(1-x)^n\left\{1-\frac{n}{2}(1-x)+\frac{n(n-1)}{2^2\,2!}(1-x)^2-\cdots\right\}\right]$$

$$= \frac{(-1)^n}{n!}\frac{d^n}{dx^n}\left[(1-x)^n-\frac{n}{2}(1-x)^{n+1}+\frac{n(n-1)}{2^2\,2!}(1-x)^{n+2}-\cdots\right]$$

$$= \frac{(-1)^n}{n!}\left[(-1)^n n!-\frac{n}{2}(-1)^n\frac{(n+1)!}{1!}(1-x)\right.$$
$$\left.+\frac{n(n-1)}{2^2\,2!}(-1)^n\frac{(n+2)!}{2!}(1-x)^2-\cdots\right]$$

$$= 1+\frac{(-n)(n+1)}{1\cdot 1}\left(\frac{1-x}{2}\right)+\frac{(-n)(-n+1)(n+1)(n+2)}{1\cdot 2\cdot 1\cdot 2}\left(\frac{1-x}{2}\right)^2+\cdots$$

$$= F(-n,n+1,1,(1-x)/2). \qquad \blacksquare$$

In the DE (3.17) if we use the substitution $x = t/b$, then the hypergeometric function $F(a,b,c,t/b)$ is a solution of the DE

$$t\left(1-\frac{t}{b}\right)\frac{d^2y}{dt^2}+\left[c-\left(1+\frac{a+1}{b}\right)t\right]\frac{dy}{dt}-ay = 0.$$

Thus, letting $b \to \infty$, we see that the function $\lim_{b\to\infty} F(a,b,c,t/b)$ is a solution of the DE

$$t\frac{d^2y}{dt^2}+(c-t)\frac{dy}{dt}-ay = 0. \tag{10.17}$$

Next from (10.1), we have

$$F\left(a,b,c,\frac{t}{b}\right) = \left(\sum_{m=0}^{\infty}\frac{\Gamma(a+m)}{\Gamma(c+m)\,m!}\left\{\frac{\Gamma(b+m)}{b^m\,\Gamma(b)}\right\}t^m\right)\frac{\Gamma(c)}{\Gamma(a)}$$

and since

$$\lim_{b\to\infty}\frac{\Gamma(b+m)}{b^m\,\Gamma(b)} = 1,$$

it follows that

$$\lim_{b\to\infty}F\left(a,b,c,\frac{t}{b}\right) = \left(\sum_{m=0}^{\infty}\frac{\Gamma(a+m)}{\Gamma(c+m)\,m!}t^m\right)\frac{\Gamma(c)}{\Gamma(a)} = F(a,c,t), \quad \text{say.}$$

$$\tag{10.18}$$

The function $F(a, c, t)$ defined in (10.18) is called a *confluent hypergeometric function*, and the DE (10.17) is known as *confluent hypergeometric equation*.

Finally, we note that the second solution of the DE (10.17) is $t^{1-c}F(a - c + 1, 2 - c, t)$.

Problems

10.1. Express the solutions of the following DEs in terms of $F(a, b, c, x)$

(i) $x(1 - x)y'' - 3y' + 2y = 0$

(ii) $2x(1 - x)y'' + (1 - x)y' + 3y = 0$

(iii) $x(1 - x)y'' + (2 - 3x)y' - y = 0$

(iv) $(2x + 2x^2)y'' + (1 + 5x)y' + y = 0.$

10.2. Show that in the DE (3.17) the change of the dependent variable $y = (1 - x)^{c-a-b}u$ leads to a hypergeometric equation having the solutions

$$y(x) = (1 - x)^{c-a-b}F(c - a, c - b, c, x)$$

and

$$y(x) = x^{1-c}(1 - x)^{c-a-b}F(1 - a, 1 - b, 2 - c, x).$$

Further, show that these are identical with (10.4) and (10.6), respectively.

10.3. Setting $x = 1/t$ in the DE (3.17), find the exponents at infinity, and derive the solutions

$$y(x) = x^{-a}F(a, a - c + 1, a - b + 1, 1/x)$$

and

$$y(x) = x^{-b}F(b, b - c + 1, b - a + 1, 1/x).$$

10.4. Show that the DE

$$(x^2 + a_1 x + a_2)y'' + (a_3 x + a_4)y' + a_5 y = 0$$

can be reduced to the hypergeometric form by a suitable change of variable provided the roots of $x^2 + a_1 x + a_2 = 0$ are distinct.

10.5.   Use the result of Problem 10.4 to show that the Legendre equation (3.19) with $a = n$ can be reduced to the hypergeometric form

$$t(1 - t)y'' + (1 - 2t)y' + n(n + 1)y = 0$$

by the substitution $x - 1 = -2t$. Hence, establish the relation (10.16).

10.6. Use the transformation $t = (1-x)/2$ to convert the Chebyshev equation (3.15) into a hypergeometric equation and show that its general solution near $x = 1$ is given by

$$y(x) = c_0 F\left(a, -a, \frac{1}{2}, \frac{1-x}{2}\right) + c_1 \left(\frac{1-x}{2}\right)^{1/2} F\left(a + \frac{1}{2}, -a + \frac{1}{2}, \frac{3}{2}, \frac{1-x}{2}\right).$$

10.7. Show that *Jacobi's DE*

$$x(1-x)y'' + [a - (1+b)x]y' + n(b+n)y = 0$$

is a hypergeometric equation, and hence its general solution near $x = 0$ can be written as

$$y(x) = c_1 F(n+b, -n, a, x) + c_2 x^{1-a} F(n+b-a+1, 1-n-a, 2-a, x).$$

10.8. Prove the following recurrence relations

(i) $F(a, b, c, x) - F(a, b, c-1, x) = -\dfrac{abx}{c(c-1)} F(a+1, b+1, c+1, x)$

(ii) $F(a, b+1, c+1, x) - F(a, b, c, x) = \dfrac{a(c-b)x}{c(c+1)} F(a+1, b+1, c+2, x)$

(iii) $F(a-1, b+1, c, x) - F(a, b, c, x) = \dfrac{(a-b-1)x}{c} F(a, b+1, c+1, x).$

10.9. In the analysis of a deformation due to a uniform load on a certain circular plate, the following equation occurs:

$$x^2(1 - (\epsilon x)^k)\phi'' + x(1 - (\epsilon x)^k - 3k(\epsilon x)^k)\phi' - (1 - (\epsilon x)^k + 3kv(\epsilon x)^k)\phi = 0, \tag{10.19}$$

where ϵ, k, and v are constants; x is proportional to the distance from the center of the plate, and ϕ is the angle between the normal to the deformed surface of the plate and the normal to that surface at the center of the plate. Show that the successive substitutions $z = \epsilon x$, $\phi = z\psi$, $z^k = \sigma$ transform (10.19) to the hypergeometric equation

$$\sigma(1-\sigma)\frac{d^2\psi}{d\sigma^2} + \left[\frac{k+2}{k} - \left(1 + \frac{2+3k}{k}\right)\sigma\right]\frac{d\psi}{d\sigma} - \frac{3(v+1)}{k}\psi = 0$$

with $c = (k+2)/k$, $a+b = (2+3k)/k$, $ab = 3(v+1)/k$. Hence, if α and β are the roots of $k\lambda^2 - (2+3k)\lambda + 3(v+1) = 0$, the solution of (10.19) can be written as $\phi(x) = \epsilon x F(\alpha, \beta, (k+2)/k, \epsilon^k x^k)$.

Answers or Hints

10.1. (i) $F(-2, 1, -3, x)$ (ii) $F(1, -3/2, 1/2, x)$ (iii) $F(1, 1, 2, x)$

(iv) $F(1/2, 1, 1/2, -x)$.

10.2. Substitute $y = (1-x)^{c-a-b}u$ in the DE (3.17) to get

$$x(1-x)u'' + [c - (c - a + c - b + 1)x]u' - (c-a)(c-b)u = 0.$$

The identities follow from the uniqueness of the initial value problems for (3.17).

10.3. Substitute $x = 1/t$ in the DE (3.17) to get

$$\frac{d^2y}{dt^2} + [(a+b-1) - (c-2)t]\frac{1}{t(t-1)}\frac{dy}{dt} - \frac{ab}{t^2(t-1)}y = 0.$$

For this DE $t = 0$ is a regular singular point and the exponents are a and b.

10.4. If $x^2 + a_1 x + a_2 = (x-a)(x-b)$, then use the substitution $t = (x-a)/(b-a)$.

10.5. Problem 10.4 suggests the substitution $t = \frac{x-1}{-2}$.

10.6. For this also Problem 10.4 suggests the substitution $t = \frac{x-1}{-2}$.

10.7. Compare the coefficients to get $a = n + b, b = -n, c = a$.

10.8. Use (10.1) directly.

10.9. Verify directly.

Lecture 11
Piecewise Continuous and Periodic Functions

Mathematical problems describing real world situations often have solutions which are not even continuous. Thus, to analyze such problems we need to work in a set which is bigger than the set of continuous functions. In this lecture we shall introduce the sets of piecewise continuous and piecewise smooth functions, which are quite adequate to deal with a wide variety of applied problems. Here we shall also define periodic functions, and introduce even and odd extensions. These terms will be used in later lectures repeatedly.

Recall that for a given function $f(x)$ the right-hand and the left-hand limits at the point x_0 are defined as follows:

$$f(x_0+) = \lim_{\substack{x \to x_0 \\ x > x_0}} f(x) = \lim_{x \to x_0+} f(x),$$

and

$$f(x_0-) = \lim_{\substack{x \to x_0 \\ x < x_0}} f(x) = \lim_{x \to x_0-} f(x).$$

We say $f(x)$ has the limit at $x = x_0$ provided $f(x_0+) = f(x_0-)$. The function $f(x)$ is said to be continuous at a point x_0 provided $\lim_{x \to x_0} f(x) = f(x_0)$, and it is continuous in an interval $\alpha < x < \beta$ if it is continuous at each point x for $\alpha < x < \beta$ (see Figure 11.1).

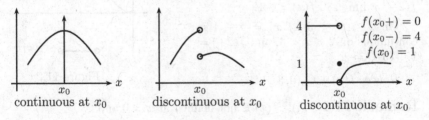

Figure 11.1

As an illustration, consider the function $f(x) = \cos 1/x$, $0 < x < 1$. For this function $f(0+)$ does not exist. In fact, as $x \to 0+$, $f(x)$ oscillates between -1 and 1. To see this let $x_n = 1/(2n\pi)$, $n = 1, 2, \cdots$. Then,

R.P. Agarwal, D. O'Regan, *Ordinary and Partial Differential Equations*,
Universitext, DOI 10.1007/978-0-387-79146-3_11,
© Springer Science+Business Media, LLC 2009

$x_n \to 0$ as $n \to \infty$, but $f(x_n) = \cos 1/x_n = \cos(2n\pi) = 1$. Also, let $y_n = 1/[(2n+1)\pi]$, $n = 1, 2, \cdots$. Then, $y_n \to 0$ as $n \to \infty$ but $f(y_n) = \cos 1/y_n = \cos(2n+1)\pi = -1$. As another example, for the function $f(x) = 1/x$, $0 < x < 1$, $f(0+) = \lim_{x\to 0+} 1/x = +\infty$.

Definition 11.1. A function $f(x)$ is called *piecewise continuous (sectionally continuous)* on an interval $\alpha < x < \beta$ if there are finitely many points $\alpha = x_0 < x_1 < \cdots < x_n = \beta$ such that

(i). $f(x)$ is continuous on each subinterval $x_0 < x < x_1$, $x_1 < x < x_2$, $\cdots$, $x_{n-1} < x < x_n$, and

(ii). on each subinterval (x_{k-1}, x_k) both $f(x_{k-1}+)$ and $f(x_k-)$ exist, i.e., are finite.

Note that the function $f(x)$ need not be defined at the points x_k.

We shall denote the class of piecewise continuous functions on $\alpha < x < \beta$ by $C_p(\alpha, \beta)$.

Example 11.1. Consider the function

$$f(x) = \begin{cases} x^2, & 0 < x \le 1 \\ -1, & 1 < x < 2 \\ x - 1, & 2 < x \le 3. \end{cases}$$

We shall show that $f(x)$ is piecewise continuous on $0 < x < 3$. Note that for this function $x_0 = 0$, $x_1 = 1$, $x_2 = 2$, $x_3 = 3$ (see Figure 11.2).

On $0 < x < 1$: $f(x) = x^2$ is continuous on $0 < x < 1$, and $f(0+) = \lim_{x\to 0+} x^2 = 0$, $f(1-) = \lim_{x\to 1-} x^2 = 1$.

On $1 < x < 2$: $f(x) = -1$ is continuous on $1 < x < 2$, and $f(1+) = \lim_{x\to 1+} f(x) = \lim_{x\to 1+}(-1) = -1$, $f(2-) = \lim_{x\to 2-} f(x) = -1$.

On $2 < x < 3$: $f(x) = x - 1$ is continuous on $2 < x < 3$, and $f(2+) = 1$, $f(3-) = 2$.

Hence, $f(x) \in C_p(0, 3)$.

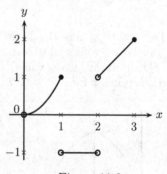

Figure 11.2

Remark 11.1. (i). $f(x)$ is not defined at $x = 0$ and 2.

(ii). One can integrate a piecewise continuous function interval-wise, e.g., for the above function:

$$\int_0^3 f(x)dx = \int_0^1 x^2 dx + \int_1^2 -1dx + \int_2^3 (x-1)dx = \frac{5}{6}.$$

Example 11.2. The function $f(x) = \cos 1/x$, $0 < x < 1$ is continuous on $0 < x < 1$. However, since $\lim_{x \to 0+} f(x)$ does not exist, $f(x) \notin C_p(0,1)$. Similarly, $f(x) = 1/x \notin C_p(0,1)$.

Example 11.3. For the function $f(x) = x^2 \sin 1/x$, $0 < x < 1$ we have $x_0 = 0$, $x_1 = 1$. Clearly, on $0 < x < 1$ this function is continuous, and $f(1-) = \lim_{x \to 1-} x^2 \sin 1/x = 1^2 \sin 1/1 = \sin 1$. We shall show that $f(0+) = \lim_{x \to 0+} x^2 \sin 1/x = 0$. For this note that for $x > 0$, $-1 \leq \sin 1/x \leq 1$, and hence

$$-x^2 \leq x^2 \sin \frac{1}{x} \leq x^2,$$

so by the sandwich theorem $\lim_{x \to 0+} x^2 \sin 1/x = 0$. Thus, $f(x) \in C_p(0,1)$.

Definition 11.2. A function $f(x)$, $\alpha < x < \beta$ is said to be *piecewise smooth* (*sectionally smooth*) if both $f(x)$ and $f'(x)$ are piecewise continuous on $\alpha < x < \beta$. The class of piecewise smooth functions on $\alpha < x < \beta$ is denoted by $C_p^1(\alpha, \beta)$.

Note that $f'(x)$ is piecewise continuous means that $f'(x)$ is continuous except at $s_0, \cdots, s_m$ (these points include $x_0, \cdots, x_n$ where $f(x)$ is not continuous) and on each subinterval (s_{k-1}, s_k) both $f'(s_{k-1}+)$ and $f'(s_k-)$ exist. Here $f'(s_j+) = \lim_{x \to s_j+} f'(x)$ and $f'(s_j-) = \lim_{x \to s_j-} f'(x)$.

Example 11.4. Consider the function

$$f(x) = \begin{cases} x+1, & -1 < x < 0 \\ \sin x, & 0 < x < \pi/2. \end{cases}$$

Clearly, $f(x) \in C_p(-1, \pi/2)$, $x_0 = -1$, $x_1 = 0$, $x_2 = \pi/2$ (see Figure 11.3(a)). Since

$$f'(x) = \begin{cases} 1, & -1 < x < 0 \\ \cos x, & 0 < x < \pi/2 \end{cases}$$

we have $s_0 = -1$, $s_1 = 0$, $s_2 = \pi/2$ (see Figure 11.3(b)).

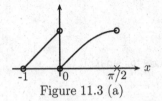
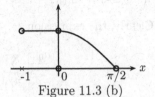

Figure 11.3 (a)   Figure 11.3 (b)

On $-1 < x < 0$: $f'(x) = 1$ is continuous, and $f'(-1+) = 1$, $f'(0-) = 1$.

On $0 < x < \pi/2$: $f'(x) = \cos x$ is continuous, and $f'(0+) = 1$, $f'(\pi/2-) = 0$.

Thus, $f'(x) \in C_p(-1, \pi/2)$. In conclusion, $f(x) \in C_p^1(-1, -\pi/2)$.

Example 11.5. Consider the function

$$f(x) = \begin{cases} 0.01, & -0.1 < x < 0 \\ x^2 \cos 1/x, & 0 < x < 0.1 \end{cases}$$

(see Figure 11.4).

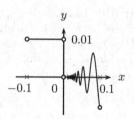

Clearly, $f(x) \in C_p(-0.1, 0.1)$, and

$$f'(x) = \begin{cases} 0, & -0.1 < x < 0 \\ 2x \cos 1/x + \sin 1/x, & 0 < x < 0.1. \end{cases}$$

Figure 11.4

If $f'(x) \in C_p(-0.1, 0.1)$, $f'(0+)$ must exist. But, $f'(0+)$ does not exist. In fact, $\lim_{x \to 0+} 2x \cos 1/x = 0$ by the sandwich theorem, but $\lim_{x \to 0+} \sin 1/x$ does not exist, because $\sin 1/x$ oscillates between -1 and 1 as $x \to 0$. For this, let $x_n = 1/[(2n + \frac{1}{2})\pi]$, $y_n = 1/[(2n + \frac{3}{2})\pi]$ so that $x_n \to 0+$, $y_n \to 0+$ as $n \to \infty$. Clearly, $\sin 1/x_n = 1$, $\sin 1/y_n = -1$. Thus, $f'(x) \notin C_p(-0.1, 0.1)$, and hence $f(x) \notin C_p^1(-0.1, 0.1)$.

Rule of Thumb. If $f(x)$ is defined in pieces, where each piece is differentiable twice in a slightly larger interval, (e.g., $(-\infty, \infty)$), such as polynomials in x, e^x, $\sin x$, etc., then $f(x)$ is piecewise smooth. For example, the function

$$f(x) = \begin{cases} 4x + 2, & 0 < x < 1 \\ e^x, & 1 < x < 2 \end{cases}$$

is piecewise smooth.

Definition 11.3. A function $f(x)$ is said to be *periodic with period* ω if $f(x + \omega) = f(x)$ for all x. For example, $f(x) = \cos x$ is periodic with period 2π. In fact, $f(x + 2\pi) = \cos(x + 2\pi) = \cos x = f(x)$ for all x. For $n \geq 1$ the function $f(x) = \sin nx$ is periodic with period 2π, $f(x + 2\pi) = \sin n(x + 2\pi) = \sin(nx + 2n\pi) = \sin nx = f(x)$.

Clearly, the expression

$$\frac{a_0}{2} + \sum_{n=1}^{\infty} (a_n \cos nx + b_n \sin nx)$$

is periodic with period 2π.

Any function $f(x)$ defined on $-\pi < x < \pi$ can be extended to a periodic function $F(x)$ with period 2π. The function $F(x)$ is called the *periodic extension* of $f(x)$.

Example 11.6. For the function

$$f(x) = \begin{cases} 0, & -\pi < x < 0 \\ x, & 0 \le x < \pi \end{cases}$$

the extension $F(x)$ is given in Figure 11.5.

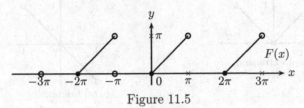

Figure 11.5

Remark 11.2. If a function $f(x)$, $-\pi \le x \le \pi$ can be extended to a periodic function with period 2π, we must have $f(-\pi) = f(\pi)$

Example 11.7. For the function $f(x) = x^2$, $-\pi \le x \le \pi$ the extension $F(x)$ is given in Figure 11.6.

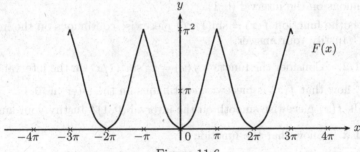

Figure 11.6

Finally, we recall that if a function $g(x)$ is *odd*, i.e., $g(-x) = -g(x)$ then $\int_{-\alpha}^{\alpha} g(x)dx = 0$, and if $g(x)$ is *even*, i.e., $g(-x) = g(x)$ then $\int_{-\alpha}^{\alpha} g(x)dx = 2\int_0^{\alpha} g(x)dx$. Further, if $g(x)$ is defined only on $(0, \alpha)$ we can extend it as follows:

$$g_o(x) = \begin{cases} g(x), & 0 < x < \alpha \\ -g(-x), & -\alpha < x < 0. \end{cases}$$

Clearly, $g_o(x)$ is an odd function, and hence it is called an *odd extension* of $g(x)$. We can also extend $g(x)$ as

$$g_e(x) = \begin{cases} g(x), & 0 < x < \alpha \\ g(-x), & -\alpha < x < 0. \end{cases}$$

Since, $g_e(x)$ is even it is called an *even extension* of $g(x)$.

Graphically, the even extension is made by reflecting the graph in the vertical axis. The odd extension is made by-reflecting first in the vertical axis and then in the horizontal axis (see Figure 11.7 for the function x, $0 < x < 1$).

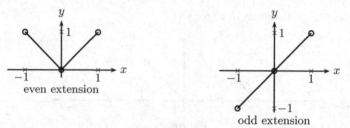

Figure 11.7

Problems

11.1. (i) Show that the function $f(x) = \sqrt{x}\cos(1/x^2)$ is piecewise continuous on the interval $(0, 1)$.

(ii) Is the function $f(x) = \sin(1/x^2)$ piecewise continuous on the interval $(0, 1)$? Justify your answer.

11.2. Consider the function $f(x) = x^2 \sin^2(1/x)$ on the interval $(0, 1)$.

(i) Show that $f(x)$ is piecewise continuous on the interval $(0, 1)$.

(ii) Is $f(x)$ piecewise smooth on the interval $(0, 1)$? Justify your answer.

11.3. Show that the function

$$f(x) = \begin{cases} \sqrt{1 - x^2} & \text{for } 0 < x < 1 \\ x & \text{for } 1 < x < 2 \end{cases}$$

is piecewise continuous but not piecewise smooth on $(0, 2)$.

11.4. Using l'Hospital's rule find $f(0+)$ and $f'(0+)$ for the function

$$f(x) = \begin{cases} (1 - e^x)/x, & x \neq 0 \\ 0, & x = 0. \end{cases}$$

Deduce that $f(x)$ is piecewise smooth on the interval $(0, 1)$.

11.5. Consider the function

$$f(x) = \begin{cases} x(\cos(\ln x) + \sin(\ln x)), & 0 < x < 1 \\ 0, & x = 0. \end{cases}$$

Compute $f'(x)$ for $0 < x < 1$ and deduce that $f(x)$ is not piecewise smooth on the interval $(0,1)$.

11.6.   Prove that if $f(x)$ and $f'(x)$ are both even functions, then $f(x)$ is a constant.

Answers or Hints

11.1. (i) See Example 11.3 (ii) Take $x_n = [(2n + \frac{1}{2})\pi]^{-1/2}$, $y_n = [(2n - \frac{1}{2})\pi]^{-1/2}$.

11.2. (i) $f(0+) = 0$ (ii) No. For $f'(x) = 2x \sin^2(1/x) - \sin(2/x)$ take $x_n = 2[(2n + \frac{1}{2})\pi]^{-1}$, $y_n = 2[(2n - \frac{1}{2})\pi]^{-1}$.

11.3. $f'(1)$ is undefined.

11.4. $f(0+) = -1$, $f'(0+) = -1/2$, $f(x) \in C_p^1(0,1)$.

11.5. $f'(x) = 2\cos(\ln x)$ oscillates between -2 and 2 as $x \to 0+$. Take $x_n = e^{-2n\pi}$, $y_n = e^{-(2n+1)\pi}$.

11.6. Differentiate $f(x) = f(-x)$ and use $f'(x) = f'(-x)$.

Lecture 12
Orthogonal Functions
and Polynomials

In this lecture first we shall introduce orthogonality of functions, which is a generalization of orthogonality of vectors in the sense that the sum of products in the scalar multiplication (dot product) of vectors will be replaced by the integral of products, and then show that the Legendre, Chebyshev, and Hermite polynomials are orthogonal. The orthogonality of Laguerre polynomials and Bessel functions will be shown in the next lecture. Orthogonality of functions plays a fundamental role in constructing Fourier series, which we shall discuss in detail later.

We begin with the following definition.

Definition 12.1. The set of functions $\{\phi_n(x) : n = 0, 1, \cdots\}$ each of which is piecewise continuous in an infinite or a finite interval $[\alpha, \beta]$, is said to be *orthogonal* in $[\alpha, \beta]$ with respect to the nonnegative function $r(x)$ if

$$\prec \phi_m, \phi_n \succ = \int_\alpha^\beta r(x)\phi_m(x)\phi_n(x)dx = 0 \quad \text{for all} \quad m \neq n$$

and

$$\int_\alpha^\beta r(x)\phi_n^2(x)dx \neq 0 \quad \text{for all} \quad n.$$

The function $r(x)$ is called the *weight function*.

We shall always assume that $r(x)$ has only a finite number of zeros in $[\alpha, \beta]$ and the integrals $\int_\alpha^\beta r(x)\phi_n(x)dx$, $n = 0, 1, \cdots$ exist.

The orthogonal set $\{\phi_n(x) : n = 0, 1, \cdots\}$ in $[\alpha, \beta]$ with respect to the weight function $r(x)$ is said to be *orthonormal* if

$$\int_\alpha^\beta r(x)\phi_n^2(x)dx = 1 \quad \text{for all} \quad n.$$

Thus, orthonormal functions have the same properties as orthogonal functions, but, in addition, they have been normalized, i.e., each function $\phi_n(x)$ of the orthogonal set has been divided by the norm of that function, which is defined as

$$\|\phi_n\| = \left(\int_\alpha^\beta r(x)\phi_n^2(x)dx \right)^{1/2}.$$

R.P. Agarwal, D. O'Regan, *Ordinary and Partial Differential Equations*,
Universitext, DOI 10.1007/978-0-387-79146-3_12,
© Springer Science+Business Media, LLC 2009

Example 12.1. The set

$$\{1, \ \cos nx, \ n = 1, 2, \cdots\}$$

is orthogonal on $0 < x < \pi$ with $r(x) = 1$. Indeed, for $m \neq n$, we have

$$
\begin{aligned}
\int_0^\pi \cos mx \cos nx \, dx &= \frac{1}{2} \int_0^\pi [\cos(m-n)x + \cos(m+n)x] dx \\
&= \frac{1}{2} \left[\frac{\sin(m-n)x}{m-n} + \frac{\sin(m+n)x}{m+n} \right] \Big|_0^\pi \\
&= \frac{1}{2} \left[\frac{\sin(m-n)\pi}{m-n} - 0 + \frac{\sin(m+n)\pi}{m+n} - 0 \right] = 0.
\end{aligned}
$$

Now since

$$
\begin{aligned}
\|\phi_0\| &= \left(\int_0^\pi 1^2 dx \right)^{1/2} = \sqrt{\pi} \\
\|\phi_n\| &= \left(\int_0^\pi \cos^2 nx \, dx \right)^{1/2} = \left(\int_0^\pi \frac{1 + \cos 2nx}{2} dx \right)^{1/2} = \sqrt{\frac{\pi}{2}},
\end{aligned}
$$

it follows that the set

$$\left\{ \frac{1}{\sqrt{\pi}}, \ \sqrt{\frac{2}{\pi}} \cos nx, \ n = 1, 2, \cdots \right\}$$

is orthonormal on $0 < x < \pi$ with $r(x) = 1$.

Example 12.2. The set

$$\left\{ \sqrt{\frac{2}{\pi}} \sin nx, \ n = 1, 2, \cdots \right\}$$

is orthonormal on $0 < x < \pi$ with $r(x) = 1$.

Example 12.3. The set

$$\left\{ \frac{1}{\sqrt{2\pi}}, \ \frac{1}{\sqrt{\pi}} \cos nx, \ \frac{1}{\sqrt{\pi}} \sin nx, \ n = 1, 2, \cdots \right\}$$

is orthonormal on $-\pi < x < \pi$ with $r(x) = 1$. For this we need to check

$$\int_{-\pi}^\pi \left(\frac{1}{\sqrt{2\pi}} \right)^2 dx = 1, \quad \int_{-\pi}^\pi \left(\frac{1}{\sqrt{\pi}} \cos nx \right)^2 dx = 1, \quad \int_{-\pi}^\pi \left(\frac{1}{\sqrt{\pi}} \sin nx \right)^2 dx = 1,$$

$$\int_{-\pi}^\pi \frac{1}{\sqrt{2\pi}} \times \frac{1}{\sqrt{\pi}} \cos nx \, dx = 0, \quad \int_{-\pi}^\pi \frac{1}{\sqrt{2\pi}} \times \frac{1}{\sqrt{\pi}} \sin nx \, dx = 0,$$

$$\int_{-\pi}^{\pi} \frac{1}{\sqrt{\pi}} \cos nx \frac{1}{\sqrt{\pi}} \cos mx dx = 0, \quad \int_{-\pi}^{\pi} \frac{1}{\sqrt{\pi}} \sin nx \frac{1}{\sqrt{\pi}} \sin mx dx = 0, \quad m \neq n$$

and

$$\int_{-\pi}^{\pi} \frac{1}{\sqrt{\pi}} \cos nx \ \frac{1}{\sqrt{\pi}} \sin mx dx = 0, \quad m \neq n.$$

Orthogonality of Legendre polynomials. We shall use Rodrigues' formula (7.9) to show the orthogonality of Legendre polynomials in $[-1, 1]$ with $r(x) = 1$. Indeed, from (7.9), we have

$$2^n n! \int_{-1}^{1} P_m(x) P_n(x) dx = \int_{-1}^{1} P_m(x) \frac{d^n}{dx^n} (x^2 - 1)^n dx.$$

Now an integration by parts gives

$$\int_{-1}^{1} P_m(x) \frac{d^n}{dx^n} (x^2 - 1)^n dx$$

$$= P_m(x) \frac{d^{n-1}}{dx^{n-1}} (x^2 - 1)^n \Big|_{-1}^{1} - \int_{-1}^{1} \frac{d}{dx} P_m(x) \frac{d^{n-1}}{dx^{n-1}} (x^2 - 1)^n dx.$$

However, since $d^{n-1}(x^2 - 1)^n / dx^{n-1}$ contains a factor $(x^2 - 1)$, it follows that

$$2^n n! \int_{-1}^{1} P_m(x) P_n(x) dx = - \int_{-1}^{1} \frac{d}{dx} P_m(x) \frac{d^{n-1}}{dx^{n-1}} (x^2 - 1)^n dx.$$

We can integrate the right side once again, and continue until we have performed n such integrations. At this stage, we find

$$2^n n! \int_{-1}^{1} P_m(x) P_n(x) dx = (-1)^n \int_{-1}^{1} \left(\frac{d^n}{dx^n} P_m(x) \right) (x^2 - 1)^n dx. \quad (12.1)$$

There is no loss of generality if we assume that $m \leq n$. If $m < n$, then $d^n P_m(x)/dx^n = 0$ and it follows that

$$\int_{-1}^{1} P_m(x) P_n(x) dx = 0.$$

Further, if $m = n$, then once again from (7.9), we have

$$\frac{d^n}{dx^n} P_n(x) = \frac{1}{2^n \ n!} \frac{d^{2n}}{dx^{2n}} (x^2 - 1)^n = \frac{(2n)!}{2^n \ n!}$$

and (12.1) gives

$$\int_{-1}^{1} P_n^2(x) dx = \frac{(-1)^n}{2^{2n}} \frac{(2n)!}{(n!)^2} \int_{-1}^{1} (x^2 - 1)^n dx. \quad (12.2)$$

Let $y = x^2$ in the integral $\int_{-1}^{1}(x^2 - 1)^n dx$, to obtain

$$\int_{-1}^{1}(x^2-1)^n dx = 2\int_{0}^{1}(x^2-1)^n dx = (-1)^n\int_{0}^{1}(1-y)^n y^{-1/2} dy$$

$$= (-1)^n B\left(n+1, \frac{1}{2}\right) = (-1)^n \frac{n!\,\Gamma\left(\frac{1}{2}\right)}{\Gamma\left(n+\frac{3}{2}\right)}.$$

Thus, (12.2) is the same as

$$\int_{-1}^{1} P_n^2(x) dx = \frac{(2n)!}{2^{2n}\,(n!)^2} \frac{n!\,\Gamma\left(\frac{1}{2}\right)}{\left(n+\frac{1}{2}\right)\left(n-\frac{1}{2}\right)\cdots\frac{1}{2}\Gamma\left(\frac{1}{2}\right)} = \frac{2}{2n+1}.$$

Orthogonality of Chebyshev polynomials. We shall show that Chebyshev polynomials of the first kind $T_n(x) = \cos(n\cos^{-1}x)$, $n = 0, 1, \cdots$ are orthogonal in $[-1, 1]$ with $r(x) = (1 - x^2)^{-1/2}$. For this, it suffices to note that

$$
\begin{aligned}
I &= \int_{-1}^{1}(1-x^2)^{-1/2}T_m(x)T_n(x)dx \\
&= \int_{-1}^{1}(1-x^2)^{-1/2}\cos(m\cos^{-1}x)\cos(n\cos^{-1}x)dx \\
&= \int_{0}^{\pi}\cos m\theta \, \cos n\theta\, d\theta \quad (x = \cos\theta) \\
&= \begin{cases} 0, & m \neq n \\ \pi/2, & m = n \neq 0 \\ \pi, & m = n = 0. \end{cases}
\end{aligned}
$$

Similarly, it follows that Chebyshev polynomials of the second kind $U_n(x)$, $n = 0, 1, \cdots$ are orthogonal in $[-1, 1]$ with $r(x) = (1 - x^2)^{1/2}$. In fact, we have

$$\int_{-1}^{1}(1-x^2)^{1/2}U_m(x)U_n(x)dx = \begin{cases} 0, & m \neq n \\ \pi/2, & m = n. \end{cases}$$

Orthogonality of Hermite polynomials. We shall show that the set of Hermite polynomials $\{H_n(x), \ n = 0, 1, \cdots\}$ is orthogonal in $(-\infty, \infty)$ with $r(x) = e^{-x^2}$. Since $H_n(x)$ is a solution of the DE (8.9), we have

$$\left(e^{-x^2}H_n'(x)\right)' + 2ne^{-x^2}H_n(x) = 0. \tag{12.3}$$

Multiplying (12.3) by $H_m(x)$ and integrating from $-\infty$ to ∞, we find

$$\int_{-\infty}^{\infty}\left(e^{-x^2}H_n'(x)\right)' H_m(x)dx = -2n\int_{-\infty}^{\infty}e^{-x^2}H_n(x)H_m(x)dx,$$

which is the same as

$$e^{-x^2} H_n'(x) H_m(x) \Big|_{-\infty}^{\infty} - \int_{-\infty}^{\infty} e^{-x^2} H_n'(x) H_m'(x) dx$$

$$= -2n \int_{-\infty}^{\infty} e^{-x^2} H_n(x) H_m(x) dx$$

and hence

$$\int_{-\infty}^{\infty} e^{-x^2} H_n'(x) H_m'(x) dx = 2n \int_{-\infty}^{\infty} e^{-x^2} H_n(x) H_m(x) dx. \qquad (12.4)$$

Interchanging m and n in (12.4) and subtracting the resulting equation from (12.4), we obtain

$$(2n - 2m) \int_{-\infty}^{\infty} e^{-x^2} H_n(x) H_m(x) dx = 0.$$

Thus, if $m \neq n$, we get

$$\int_{-\infty}^{\infty} e^{-x^2} H_n(x) H_m(x) dx = 0.$$

Next we shall show that

$$I_n = \int_{-\infty}^{\infty} e^{-x^2} H_n^2(x) dx = 2^n \, n! \, \sqrt{\pi}. \qquad (12.5)$$

In view of Problem 8.2(iii), we have

$$
\begin{aligned}
I_n &= \int_{-\infty}^{\infty} e^{-x^2} H_n(x) H_n(x) dx = \int_{-\infty}^{\infty} e^{-x^2} H_n(x) (-1)^n e^{x^2} \frac{d^n}{dx^n} e^{-x^2} dx \\
&= (-1)^n \int_{-\infty}^{\infty} H_n(x) \frac{d^n}{dx^n} e^{-x^2} dx \\
&= (-1)^n \left[H_n(x) \frac{d^{n-1}}{dx^{n-1}} e^{-x^2} \Big|_{-\infty}^{\infty} - \int_{-\infty}^{\infty} H_n'(x) \frac{d^{n-1}}{dx^{n-1}} e^{-x^2} dx \right] \\
&= (-1)^{n+1} \int_{-\infty}^{\infty} 2n H_{n-1}(x) \frac{d^{n-1}}{dx^{n-1}} e^{-x^2} dx \\
&= 2n I_{n-1} = 2n(2n-2) I_{n-2} = \cdots = 2n(2n-2) \cdots 2 \, I_0 \\
&= 2^n \, n! \, I_0.
\end{aligned}
$$

$$(12.6)$$

Now since $I_0 = \int_{-\infty}^{\infty} e^{-x^2} dx = \sqrt{\pi}$, (12.5) follows immediately from (12.6).

Lecture 13
Orthogonal Functions
and Polynomials (Cont'd.)

As mentioned in the previous lecture here first we shall establish the orthogonality of Laguerre polynomials and Bessel functions and then prove a fundamental property about the zeros of orthogonal polynomials.

Orthogonality of Laguerre polynomials. The Laguerre DE (8.10) can be written as

$$\left(x^{a+1}e^{-x}y'\right)' + nx^a e^{-x}y = 0. \tag{13.1}$$

Since $L_n^{(a)}(x)$ is a solution of (13.1), it follows that

$$\left(x^{a+1}e^{-x}L_n^{(a)'}(x)\right)' + nx^a e^{-x}L_n^{(a)}(x) = 0.$$

The rest of the proof of

$$\int_0^\infty e^{-x}x^a L_m^{(a)}(x)L_n^{(a)}(x)dx = 0 \quad \text{for} \quad m \neq n$$

is similar to that for Hermite polynomials.

Next, from Problem 8.3(i), we have

$$
\begin{aligned}
\int_0^\infty e^{-x}x^a L_n^{(a)}(x)L_n^{(a)}(x)dx &= \frac{1}{n!}\int_0^\infty L_n^{(a)}(x)\frac{d^n}{dx^n}\left(e^{-x}x^{n+a}\right)dx \\
&= -\frac{1}{n!}\int_0^\infty L_n^{(a)'}(x)\frac{d^{n-1}}{dx^{n-1}}\left(e^{-x}x^{n+a}\right)dx \\
&\quad \cdots \\
&= \frac{(-1)^n}{n!}\int_0^\infty \left(L_n^{(a)}(x)\right)^{(n)}e^{-x}x^{n+a}dx.
\end{aligned}
$$

However, since in view of (8.11), $\left(L_n^{(a)}(x)\right)^{(n)} = (-1)^n$, we find

$$\int_0^\infty e^{-x}x^a\left(L_n^{(a)}(x)\right)^2 dx = \frac{1}{n!}\int_0^\infty e^{-x}x^{n+a}dx = \frac{\Gamma(n+a+1)}{n!}.$$

R.P. Agarwal, D. O'Regan, *Ordinary and Partial Differential Equations*,
Universitext, DOI 10.1007/978-0-387-79146-3_13,
© Springer Science+Business Media, LLC 2009

Orthogonality of Bessel functions. In Lecture 3, as an application of Theorem 3.1, we have seen that for all a every solution of Bessel's DE (2.15) has an infinite number of zeros in $J = (0, \infty)$. This in turn implies that for every fixed a the solution $J_a(x)$ of the Bessel equation has an infinite number of zeros in $J = (0, \infty)$. Let these zeros be $\{b_n, \ n = 0, 1, \cdots\}$, where $b_0 < b_1 < \cdots$. We shall show that $\{J_a(b_n x), \ n = 0, 1, \cdots\}$ is an orthogonal set in $[0, 1]$ with respect to the weight function $r(x) = x$. In fact, we shall prove that

$$\int_0^1 x J_a(b_p x) J_a(b_q x) dx = \begin{cases} 0 & \text{if } b_p \neq b_q \\ \dfrac{1}{2} J_{a+1}^2(b_p) & \text{if } b_p = b_q. \end{cases} \tag{13.2}$$

For this, first we use the substitution $x = pt$ ($p > 0$ is a constant) in the Bessel DE (2.15). Since

$$\frac{dy}{dx} = \frac{dy}{dt}\frac{dt}{dx} = \frac{1}{p}\frac{dy}{dt} \quad \text{and} \quad \frac{d^2y}{dx^2} = \frac{1}{p^2}\frac{d^2y}{dt^2},$$

we have

$$p^2 t^2 \frac{1}{p^2}\frac{d^2y}{dt^2} + pt\frac{1}{p}\frac{dy}{dt} + (p^2t^2 - a^2)y = 0,$$

or

$$t^2\frac{d^2y}{dt^2} + t\frac{dy}{dt} + (p^2t^2 - a^2)y = 0.$$

Hence, if $y = J_a(x)$ is a solution of

$$y'' + \frac{1}{x}y' + \left(1 - \frac{a^2}{x^2}\right)y = 0$$

then it follows that the functions $u = J_a(\mu x)$ and $v = J_a(\nu x)$ (μ and ν are distinct positive constants) satisfy the equations

$$u'' + \frac{1}{x}u' + \left(\mu^2 - \frac{a^2}{x^2}\right)u = 0 \tag{13.3}$$

and

$$v'' + \frac{1}{x}v' + \left(\nu^2 - \frac{a^2}{x^2}\right)v = 0. \tag{13.4}$$

Multiplying (13.3) by v and (13.4) by u and subtracting, we obtain

$$\frac{d}{dx}(u'v - v'u) + \frac{1}{x}(u'v - v'u) = (\nu^2 - \mu^2)uv,$$

which is the same as

$$\frac{d}{dx}(x(u'v - v'u)) = x(\nu^2 - \mu^2)uv$$

and hence

$$(\nu^2 - \mu^2) \int_0^1 xuvdx = (x(u'v - v'u))\Big|_0^1 = u'(1)J_a(\nu) - v'(1)J_a(\mu). \quad (13.5)$$

It follows that the integral on the left side is zero if μ and ν are distinct positive zeros b_p and b_q of $J_a(x)$; i.e., then we have

$$\int_0^1 xJ_a(b_px)J_a(b_qx)dx = 0. \quad (13.6)$$

Now we multiply (13.3) by $2x^2u'$, to obtain

$$\frac{d}{dx}(x^2u'^2 + \mu^2x^2u^2 - a^2u^2) = 2\mu^2xu^2$$

and hence

$$2\mu^2 \int_0^1 xu^2dx = (x^2u'^2 + \mu^2x^2u^2 - a^2u^2)\Big|_0^1.$$

When $x = 0$ the expression in the bracket vanishes ($a^2u^2 = 0$, since $u = J_a(\mu x)$ and $J_a(0) = 0$ if $a > 0$), and since $u'(1) = \mu J_a'(\mu)$, we have

$$\int_0^1 xu^2dx = \frac{1}{2}(J_a'(\mu))^2 + \frac{1}{2}\left(1 - \frac{a^2}{\mu^2}\right)(J_a(\mu))^2,$$

which from the recurrence relation (9.18) is the same as

$$\int_0^1 xJ_a^2(b_px)dx = \frac{1}{2}J_{a+1}^2(b_p). \quad (13.7)$$

The following table contains the values of the p–th positive zero $b_{n,p}$ of the Bessel function $J_n(x)$:

p	1	2	3	4	5
$b_{0,p}$	2.4048	5.5201	8.6537	11.7915	14.9309
$b_{1,p}$	3.8317	7.0156	10.1735	13.3237	16.4706
$b_{2,p}$	5.1356	8.4172	11.6198	14.7960	17.9598
$b_{3,p}$	6.3802	9.7610	13.0152	16.2235	19.4094
$b_{4,p}$	7.5883	11.0647	14.3725	17.6160	20.8269
$b_{5,p}$	8.7714	12.3386	15.7002	18.9801	22.2178

Zeros of orthogonal polynomials. Now we shall consider a fixed set of orthogonal polynomials $\{p_n(x), n = 0, 1, \cdots\}$ in the interval $[\alpha, \beta]$ with respect to the weight function $r(x)$. We shall represent the polynomial $p_n(x)$ as

$$p_n(x) = \sum_{i=0}^{n} b_{ni}x^i, \quad n = 0, 1, \cdots \quad (13.8)$$

where $b_{nn} \neq 0$. Although there are numerous properties of orthogonal polynomials, we shall prove only that the zeros of each polynomial $p_n(x)$, $n \geq 1$ are real, simple, and lie in the open interval (α, β). For this, we need the following theorem.

Theorem 13.1. Any polynomial

$$Q_n(x) = \sum_{i=0}^{n} a_i^0 x^i$$

has a unique representation of the form

$$Q_n(x) = \sum_{i=0}^{n} c_i p_i(x). \tag{13.9}$$

Proof. We define the following sequence of polynomials

$$Q_{n-1}(x) = Q_n(x) - \frac{a_n^0}{b_{nn}} p_n(x) = \sum_{i=0}^{n-1} a_i^1 x^i$$

$$Q_{n-2}(x) = Q_{n-1}(x) - \frac{a_{n-1}^1}{b_{n-1,n-1}} p_{n-1}(x) = \sum_{i=0}^{n-2} a_i^2 x^i$$

$$\cdots$$

$$Q_0(x) = Q_1(x) - \frac{a_1^{n-1}}{b_{11}} p_1(x) = a_0^n = \frac{a_0^n}{b_{00}} p_0(x).$$

Now summing these relations, we obtain

$$\sum_{i=0}^{n-1} Q_i(x) = \sum_{i=1}^{n} Q_i(x) - \sum_{i=1}^{n} \frac{a_i^{n-i}}{b_{ii}} p_i(x),$$

which is the same as

$$Q_n(x) = \sum_{i=1}^{n} \frac{a_i^{n-i}}{b_{ii}} p_i(x) + Q_0(x) = \sum_{i=0}^{n} \frac{a_i^{n-i}}{b_{ii}} p_i(x).$$

Thus, in (13.9) the constants c_i, $0 \leq i \leq n$ are determined successively by the relations

$$c_{n-i} = \frac{a_{n-i}^i}{b_{n-i,n-i}}, \quad i = 0, 1, \cdots, n.$$

To show the uniqueness, let us assume that

$$Q_n(x) = \sum_{i=0}^{n} c_i p_i(x) = \sum_{i=0}^{n} d_i p_i(x)$$

and hence

$$\sum_{i=0}^{n}(c_i - d_i)p_i(x) = 0.$$

However, from the orthogonality of the polynomials $p_n(x)$, we find

$$0 = \int_{\alpha}^{\beta} r(x) \left(\sum_{i=0}^{n}(c_i - d_i)p_i(x) \right) p_j(x)dx = (c_j - d_j)\int_{\alpha}^{\beta} r(x)p_j^2(x)dx,$$

which implies that $c_j = d_j$, $0 \leq j \leq n$. ∎

Corollary 13.2. Let $Q_k(x)$ be any polynomial of degree $k < n$. Then,

$$\int_{\alpha}^{\beta} r(x)p_n(x)Q_k(x)dx = 0.$$

Proof. From Theorem 13.1 there exist constants c_i, $i = 0, 1, \cdots, k$ such that $Q_k(x) = \sum_{i=0}^{k} c_i p_i(x)$. Now the result immediately follows from the orthogonality of the polynomials $p_n(x)$. ∎

Theorem 13.3. The zeros of each orthogonal polynomial $p_n(x)$, $n \geq 1$ are real, simple, and lie in the open interval (α, β).

Proof. Let $p_n(x)$ be of fixed sign in (α, β), then

$$0 \neq \int_{\alpha}^{\beta} r(x)p_n(x)dx = \frac{1}{p_0(x)} \int_{\alpha}^{\beta} r(x)p_n(x)p_0(x)dx.$$

However, this contradicts the definition of the orthogonality. Thus, $p_n(x_1) = 0$ for some $x_1 \in (\alpha, \beta)$. Next let $x_1 \in (\alpha, \beta)$ be such that $p_n^{(i)}(x_1) = 0$, $0 \leq i \leq k - 1$ ($2 \leq k \leq n$), then $Q_{n-k}(x) = p_n(x)/(x - x_1)^k$ will be a polynomial of degree $n - k$. Hence, from Corollary 13.2, we have $\int_{\alpha}^{\beta} r(x)p_n(x)Q_{n-k}(x)dx = 0$. But, for k even

$$\int_{\alpha}^{\beta} r(x)p_n(x)Q_{n-k}(x)dx = \int_{\alpha}^{\beta} r(x)\frac{p_n^2(x)}{(x - x_1)^k}dx \neq 0.$$

This contradiction shows that the roots of $p_n(x)$ in (α, β) cannot have even multiplicity. Finally, let $x_1, x_2, \cdots, x_r \in (\alpha, \beta)$, $1 \leq r < n$ be the only zeros of $p_n(x)$; then

$$p_n(x) = (x - x_1)(x - x_2)\cdots(x - x_r)Q_{n-r}(x),$$

where $Q_{n-r}(x)$ is a polynomial of degree $n - r$ having fixed sign in (α, β). Thus,

$$(x - x_1)^2(x - x_2)^2 \cdots (x - x_r)^2 Q_{n-r}(x)$$

is of fixed sign in (α, β), and hence

$$\int_\alpha^\beta r(x)(x-x_1)^2(x-x_2)^2\cdots(x-x_r)^2 Q_{n-r}(x)dx$$

$$= \int_\alpha^\beta r(x)p_n(x)(x-x_1)(x-x_2)\cdots(x-x_r)dx \neq 0,$$

which contradicts Corollary 13.2 and so $r = n$.   ∎

As a consequence of Theorem 13.3, we have

(i)   The Legendre polynomial $P_n(x)$, $n \geq 1$ has n real simple zeros in $(-1,1)$.

(ii)   The Chebyshev polynomial of the first kind $T_n(x)$, $n \geq 1$ has n real simple zeros in $(-1,1)$.

(iii) The Chebyshev polynomial of the second kind $U_n(x)$, $n \geq 1$ has n real simple zeros in $(-1,1)$.

(iv)   The Laguerre polynomial $L_n^{(a)}(x)$, $n \geq 1$ has n real simple zeros in $(0, \infty)$.

(v)   The Hermite polynomial $H_n(x)$, $n \geq 1$ has n real simple zeros in $(-\infty, \infty)$.

We conclude this lecture by stating the following theorem.

Theorem 13.4.   If $x_1 < x_2 < \cdots < x_n$ are the zeros of $p_n(x)$, and $y_1 < y_2 < \cdots < y_{n+1}$ are those of $p_{n+1}(x)$, then

$$\alpha < y_1 < x_1 < y_2 < x_2 < \cdots < y_n < x_n < y_{n+1} < \beta.$$

Problems

13.1.   (i) Show that the functions $\phi_1(x) = 1$ and $\phi_2(x) = 2x - 1$ are orthogonal on the interval $0 < x < 1$ with $r(x) = 1$. Further, determine constants A and B so that the function $\phi_3(x) = 1 + Ax + Bx^2$ is orthogonal to both $\phi_1(x)$ and $\phi_2(x)$.

(ii)   From the orthogonal functions in part (i) find three orthonormal functions on the interval $0 < x < 1$ with $r(x) = 1$.

13.2.   Consider the functions $\phi_1(x) = \sqrt{2}$, $\phi_2(x) = x$ and $\phi_3(x) = x^2 + Ax + B$. It is given that $\phi_3(x)$ is orthogonal to both $\phi_1(x)$ and $\phi_2(x)$ on the interval $0 < x < 1$ with $r(x) = 1$. Find A and B.

13.3.   Find constants a_i, $i = 1, 2, 3, 4, 5$ so that the set of three functions $\{a_1, a_2x, a_3x^2 + a_4x + a_5\}$ forms an orthogonal set on $[-1, 1]$ with weight function $w(x) = 1$.

13.4. Verify that the set of functions given in Example 12.2 is orthonormal on the interval $0 < x < \pi$ with $r(x) = 1$.

13.5. Show that the relation (13.6) holds if μ and ν are one of the following:

(i)   zeros of $J'_n(x)$

(ii)   zeros of $xJ'_n(x) + hJ_n(x)$, where h is any constant

(iii) zeros of $J_{n+1}(x)$

(iv) zeros of $J_{n-1}(x)$.

13.6. For the Legendre polynomials show that

(i) $\displaystyle \int_{-1}^{1} x^n P_n(x)dx = \frac{2^{n+1}(n!)^2}{(2n+1)!}$

(ii) $\displaystyle \int_{-1}^{1} x^2 P_{n+1}(x)P_{n-1}(x)dx = \frac{2n(n+1)}{(2n-1)(2n+1)(2n+3)}$

(iii) $\displaystyle \int_{-1}^{1} x^2 P_n^2(x)dx = \frac{1}{8(2n-1)} + \frac{3}{4(2n+1)} + \frac{1}{8(2n+3)}$

(iv) $\displaystyle (m+n+1)\int_{0}^{1} x^m P_n(x)dx = m\int_{0}^{1} x^{m-1}P_{n-1}(x)dx$

$$= (m-n+2)\int_{0}^{1} x^m P_{n-2}(x)dx$$

(v) $\displaystyle \int_{-1}^{1} x^m P_n(x)dx = \begin{cases} 0 & \text{if } m < n \\[2mm] \dfrac{m! \, \Gamma\left(\frac{1}{2}m - \frac{1}{2}n + \frac{1}{2}\right)}{2^n \, (m-n)! \, \Gamma\left(\frac{1}{2}m + \frac{1}{2}n + \frac{3}{2}\right)} & \\ \qquad\qquad \text{if } m-n \, (\geq 0) \text{ is even} \\[2mm] 0 & \text{if } m-n \, (> 0) \text{ is odd} \end{cases}$

(vi) $\displaystyle \int_{-1}^{1} (1-x^2)P'_m(x)P'_n(x)dx = 0, \quad m \neq n$

(vii) $\displaystyle \int_{-1}^{1} (x^2-1)P_{n+1}(x)P'_n(x)dx = \frac{2n(n+1)}{(2n+1)(2n+3)}$.

13.7. Let $x_1, \cdots, x_n$ be the zeros of the Legendre polynomial $P_n(x)$, $n \geq 1$. Show that

(i) $\displaystyle \Pi(x) = (x-x_1)\cdots(x-x_n) = \frac{2^n(n!)^2}{(2n)!}P_n(x)$

(ii) $\displaystyle \int_{-1}^{1} \Pi^2(x)dx = \frac{2^{2n+1}(n!)^4}{(2n+1)((2n)!)^2}$.

13.8. Find results similar to those in Problem 13.7 for the polynomials

$T_n(x)$, $U_n(x)$, $L_n^{(a)}(x)$ and $H_n(x)$, $n \geq 1$.

13.9.   For the associated Legendre functions $P_n^m(x)$ of the first kind defined in Problem 7.12 show that

$$\int_{-1}^1 P_n^m(x) P_k^m(x) dx = \begin{cases} 0, & k \neq n \\ \dfrac{2}{2n+1} \dfrac{(n+m)!}{(n-m)!}, & k = n. \end{cases}$$

13.10.   A set $\{w_n(x)\}$ of complex-valued functions of a real variable x is orthogonal in the *Hermitian sense* in an interval (α, β) if

$$\int_\alpha^\beta w_m(x) \overline{w}_n(x) dx = 0, \quad m \neq n.$$

Show that the set

$$\left\{ \frac{1}{\sqrt{\beta - \alpha}} \exp\left(i \frac{2\pi n x}{\beta - \alpha} \right), \quad n = 0, \pm 1, \pm 2, \cdots \right\}$$

is orthogonal in the Hermitian sense in (α, β).

Answers or Hints

13.1.   (i) $A = -6$, $B = 6$ (ii) 1, $\sqrt{3}(2x - 1)$, $\sqrt{5}(6x^2 - 6x + 1)$.

13.2. $A = -1$, $B = 1/6$.

13.3. $a_1 \neq 0$, $a_2 \neq 0$ but arbitrary, $a_3 = -3a_5$, $a_4 = 0$.

13.4.   Use $\cos(m - n)x - \cos(m + n)x = 2 \sin mx \sin nx$.

13.5.   Relation (13.5) is the same as

$$(\nu^2 - \mu^2) \int_0^1 x J_n(\mu x) J_n(\nu x) dx = \mu J_n'(\mu) J_n(\nu) - \nu J_n'(\nu) J_n(\mu). \quad (13.10)$$
(i)   Use (13.10) (ii)   Use (13.10) (iii)   Use (13.10) and (9.18) (iv)   Use (13.10) and Problem 9.1(i)

13.6.   (i)   From (7.8), $P_n(x) = \frac{(2n)!}{2^n (n!)^2} x^n + Q_{n-1}(x)$, where $Q_{n-1}(x)$ is a polynomial of degree at most $n-1$ (ii)   Use $x^2 P_{n-1}(x) = \frac{(2n-2)!}{2^{n-1}((n-1)!)^2} x^{n+1}$ $+ Q_n(x)$ and (i) (iii)   Use (7.13) (iv)   Use Problem 7.8(i) and then Problem 7.8(iii) (v)   Use Corollary 13.2 and (7.9) (vi)   Integrate by parts and use (3.19) with $a = n$ (vii)   Use Problem 7.8(iv) and (7.13).

13.7.   From (7.8), $P_n(x) = \frac{(2n)!}{2^n (n!)^2} (x - x_1) \cdots (x - x_n)$.

13.8. $\Pi(x) = \frac{1}{2^{n-1}}T_n(x)$, $\quad \int_{-1}^{1}(1-x^2)^{-1/2}\Pi^2(x)dx = \frac{\pi}{2^{2n-1}}$

$\Pi(x) = \frac{1}{2^n}U_n(x)$, $\quad \int_{-1}^{1}(1-x^2)^{1/2}\Pi^2(x)dx = \frac{\pi}{2^{2n+1}}$

$\Pi(x) = (-1)^n\, n!L_n^{(a)}(x)$, $\quad \int_0^{\infty} e^{-x}x^a\Pi^2(x)dx = n!\,\Gamma(n+a+1)$

$\Pi(x) = \frac{1}{2^n}H_n(x)$, $\quad \int_{-\infty}^{\infty} e^{-x^2}\Pi^2(x)dx = \frac{n!\sqrt{\pi}}{2^n}$.

13.9. See the orthogonality of Legendre polynomials.

13.10. Verify directly.

Lecture 14
Boundary Value Problems

So far, we have concentrated only on initial value problems, in which for a given DE the supplementary conditions on the unknown function and its derivatives are prescribed at a fixed value x_0 of the independent variable x. However, there are a variety of other possible conditions that are important in applications. In many practical problems the additional requirements are given in the form of boundary conditions: the unknown function and some of its derivatives are fixed at more than one value of the independent variable x. The DE together with the boundary conditions is referred to as a *boundary value problem*. In this lecture we shall provide a necessary and sufficient condition so that a given boundary value problem has a unique solution.

We shall consider the second-order linear DE

$$p_0(x)y'' + p_1(x)y' + p_2(x)y = r(x), \quad x \in J = [\alpha, \beta] \tag{14.1}$$

where the functions $p_0(x)$, $p_1(x)$, $p_2(x)$ and $r(x)$ are continuous in J. Together with the DE (14.1) we shall consider the boundary conditions of the form

$$\begin{aligned}
\ell_1[y] &= a_0 y(\alpha) + a_1 y'(\alpha) + b_0 y(\beta) + b_1 y'(\beta) = A \\
\ell_2[y] &= c_0 y(\alpha) + c_1 y'(\alpha) + d_0 y(\beta) + d_1 y'(\beta) = B,
\end{aligned} \tag{14.2}$$

where a_i, b_i, c_i, d_i, $i = 0, 1$ and A, B are given constants. Throughout, we shall assume that these are essentially two conditions, i.e., there does not exist a constant c such that $(a_0\ a_1\ b_0\ b_1) = c(c_0\ c_1\ d_0\ d_1)$. The boundary value problem (14.1), (14.2) is called a *nonhomogeneous two-point linear boundary value problem*, whereas the homogeneous DE

$$p_0(x)y'' + p_1(x)y' + p_2(x)y = 0, \quad x \in J \tag{14.3}$$

together with the homogeneous boundary conditions

$$\ell_1[y] = 0, \quad \ell_2[y] = 0 \tag{14.4}$$

will be called a *homogeneous two–point linear boundary value problem*.

Boundary conditions (14.2) are quite general and in particular include the

R.P. Agarwal, D. O'Regan, *Ordinary and Partial Differential Equations*,
Universitext, DOI 10.1007/978-0-387-79146-3_14,
© Springer Science+Business Media, LLC 2009

(i)   first boundary conditions,

$$y(\alpha) = A, \quad y(\beta) = B; \tag{14.5}$$

(ii)   second boundary conditions,

$$y(\alpha) = A, \quad y'(\beta) = B, \tag{14.6}$$

or

$$y'(\alpha) = A, \quad y(\beta) = B; \tag{14.7}$$

(iii) separated boundary conditions (third boundary conditions), also known as *Sturm-Liouville conditions*,

$$a_0 y(\alpha) + a_1 y'(\alpha) = A$$
$$d_0 y(\beta) + d_1 y'(\beta) = B, \tag{14.8}$$

where $a_0^2 + a_1^2 \neq 0$ and $d_0^2 + d_1^2 \neq 0$; and

(iv)   periodic boundary conditions,

$$y(\alpha) = y(\beta), \quad y'(\alpha) = y'(\beta). \tag{14.9}$$

The boundary value problem (14.1), (14.2) is called *regular* if both α and β are finite, and the function $p_0(x) \neq 0$ for all $x \in J$. If $\alpha = -\infty$ and/or $\beta = \infty$ and/or $p_0(x) = 0$ for at least one point x in J, then the problem (14.1), (14.2) is said to be *singular*. We shall consider only regular boundary value problems.

By a *solution* of the boundary value problem (14.1), (14.2) we mean a solution of the DE (14.1) satisfying the boundary conditions (14.2).

The existence and uniqueness theory for the boundary value problems is more difficult than that of initial value problems. In fact, in the case of boundary value problems a slight change in the boundary conditions can lead to significant changes in the behavior of the solutions. For example, the initial value problem $y'' + y = 0$, $y(0) = c_1$, $y'(0) = c_2$ has a unique solution $y(x) = c_1 \cos x + c_2 \sin x$ for any set of values c_1, c_2. However, the boundary value problem $y'' + y = 0$, $y(0) = 0$, $y(\pi) = \epsilon(\neq 0)$ has no solution; the problem $y'' + y = 0$, $y(0) = 0$, $y(\beta) = \epsilon$, $0 < \beta < \pi$ has a unique solution $y(x) = \epsilon \sin x / \sin \beta$; while the problem $y'' + y = 0$, $y(0) = 0$, $y(\pi) = 0$ has an infinite number of solutions $y(x) = c \sin x$, where c is an arbitrary constant.

Obviously, for the homogeneous problem (14.3), (14.4) the trivial solution always exists. However, from the above example it follows that besides the trivial solution homogeneous boundary value problems may have non-trivial solutions also. Out first result provides a necessary and sufficient condition so that the problem (14.3), (14.4) has only the trivial solution.

Theorem 14.1. Let $y_1(x)$ and $y_2(x)$ be any two linearly independent solutions of the DE (14.3). Then, the homogeneous boundary value problem (14.3), (14.4) has only the trivial solution if and only if

$$\Delta = \begin{vmatrix} \ell_1[y_1] & \ell_1[y_2] \\ \ell_2[y_1] & \ell_2[y_2] \end{vmatrix} \neq 0. \tag{14.10}$$

Proof. Any solution of the DE (14.3) can be written as

$$y(x) = c_1 y_1(x) + c_2 y_2(x).$$

This is a solution of the problem (14.3), (14.4) if and only if

$$\begin{aligned} \ell_1[c_1 y_1 + c_2 y_2] &= c_1 \ell_1[y_1] + c_2 \ell_1[y_2] = 0 \\ \ell_2[c_1 y_1 + c_2 y_2] &= c_1 \ell_2[y_1] + c_2 \ell_2[y_2] = 0. \end{aligned} \tag{14.11}$$

However, the system (14.11) has only the trivial solution if and only if $\Delta \neq 0$. ∎

Clearly, Theorem 14.1 is independent of the choice of the solutions $y_1(x)$ and $y_2(x)$. Thus, for convenience we can always take $y_1(x)$ and $y_2(x)$ to be the solutions of (14.3) satisfying the initial conditions

$$y_1(\alpha) = 1, \quad y_1'(\alpha) = 0 \tag{14.12}$$

and

$$y_2(\alpha) = 0, \quad y_2'(\alpha) = 1. \tag{14.13}$$

Corollary 14.2. The homogeneous boundary value problem (14.3), (14.4) has an infinite number of nontrivial solutions if and only if $\Delta = 0$.

The following examples illustrate how easily Theorem 14.1 and Corollary 14.2 are applicable in practice.

Example 14.1. Consider the boundary value problem

$$xy'' - y' - 4x^3 y = 0 \tag{14.14}$$

$$\begin{aligned} \ell_1[y] &= y(1) = 0 \\ \ell_2[y] &= y(2) = 0. \end{aligned} \tag{14.15}$$

For the DE (14.14), $y_1(x) = \cosh(x^2 - 1)$ and $y_2(x) = (1/2)\sinh(x^2 - 1)$ are two linearly independent solutions. Further, for the boundary conditions (14.15), we have

$$\Delta = \begin{vmatrix} 1 & 0 \\ \cosh 3 & (1/2)\sinh 3 \end{vmatrix} \neq 0.$$

Thus, in view of Theorem 14.1, the problem (14.14), (14.15) has only the trivial solution.

Example 14.2. Consider once again the DE (14.14) together with the boundary conditions

$$\ell_1[y] = y'(1) = 0$$
$$\ell_2[y] = y'(2) = 0.$$

(14.16)

Since $y_1'(x) = 2x \sinh(x^2 - 1)$ and $y_2'(x) = x \cosh(x^2 - 1)$, for the boundary conditions (14.16), we find

$$\Delta = \begin{vmatrix} 0 & 1 \\ 4\sinh 3 & 2\cosh 3 \end{vmatrix} \neq 0.$$

Thus, again in view of Theorem 14.1, the problem (14.14), (14.16) has only the trivial solution.

Example 14.3. Consider the boundary value problem

$$y'' + 2y' + 5y = 0$$

(14.17)

$$\ell_1[y] = y(0) = 0$$
$$\ell_2[y] = y(\pi/2) = 0.$$

(14.18)

For the DE (14.17), $y_1(x) = e^{-x}\cos 2x$ and $y_2(x) = e^{-x}\sin 2x$ are two linearly independent solutions. Further, since for the boundary conditions (14.18),

$$\Delta = \begin{vmatrix} 1 & 0 \\ -e^{-\pi/2} & 0 \end{vmatrix} = 0$$

the problem (14.17), (14.18) besides the trivial solution also has nontrivial solutions. Indeed, from Corollary 14.2 if follows that it has an infinite number of solutions $y(x) = ce^{-x}\sin 2x$, where c is an arbitrary constant.

The following result provides a necessary and sufficient condition for the existence of a unique solution of the boundary value problem (14.1), (14.2).

Theorem 14.3. The nonhomogeneous boundary value problem (14.1), (14.2) has a unique solution if and only if the homogeneous boundary value problem (14.3), (14.4) has only the trivial solution.

Proof. Let $y_1(x)$ and $y_2(x)$ be any two linearly independent solutions of the DE (14.3) and $z(x)$ be a particular solution of (14.1). Then, the general solution of (14.1) can be written as

$$y(x) = c_1 y_1(x) + c_2 y_2(x) + z(x).$$

(14.19)

This is a solution of the problem (14.1), (14.2) if and only if

$$\ell_1[c_1y_1 + c_2y_2 + z] = c_1\ell_1[y_1] + c_2\ell_1[y_2] + \ell_1[z] = A$$
$$\ell_2[c_1y_1 + c_2y_2 + z] = c_1\ell_2[y_1] + c_2\ell_2[y_2] + \ell_2[z] = B. \tag{14.20}$$

The nonhomogeneous system (14.20) has a unique solution if and only if $\Delta \neq 0$, i.e., if and only if the homogeneous system (14.11) has only the trivial solution. From Theorem 14.1, $\Delta \neq 0$ is equivalent to the homogeneous boundary value problem (14.3), (14.4) having only the trivial solution. ∎

Example 14.4. Consider the boundary value problem

$$xy'' - y' - 4x^3y = 1 + 4x^4 \tag{14.21}$$

$$\ell_1[y] = y(1) = 0$$
$$\ell_2[y] = y(2) = 1. \tag{14.22}$$

Since the corresponding homogeneous problem (14.14), (14.15) has only the trivial solution, Theorem 14.3 implies that the problem (14.21), (14.22) has a unique solution. Further, to find this solution once again we choose the linearly independent solutions of (14.14) to be $y_1(x) = \cosh(x^2 - 1)$ and $y_2(x) = (1/2)\sinh(x^2 - 1)$, and note that $z(x) = -x$ is a particular solution of (14.21). Thus, the system (14.20) for the boundary conditions (14.22) reduces to

$$c_1 - 1 = 0$$
$$\cosh 3 \, c_1 + (1/2)\sinh 3 \, c_2 - 2 = 1.$$

This system can be easily solved to obtain $c_1 = 1$ and $c_2 = 2(3 - \cosh 3)/\sinh 3$. Now substituting these quantities in (14.19), we find the solution of (14.21), (14.22) as

$$y(x) = \cosh(x^2 - 1) + \frac{(3 - \cosh 3)}{\sinh 3}\sinh(x^2 - 1) - x.$$

Lecture 15
Boundary Value Problems
(Cont'd.)

In this lecture we shall formulate some boundary value problems with engineering applications, and show that often solutions of these problems can be written in terms of Bessel functions.

Example 15.1. Consider a string of length a with constant linear density ρ which is stretched along the x-axis and fixed at $x = 0$ and $x = a$. Suppose the string is then rotated about that axis at a constant speed ω. This is similar to two persons holding a jump rope and then twirling it in a synchronous manner. We shall find the differential equation which defines the shape (deflection from the initial position) $y(x)$ of the string. For this, we consider the portion of the string on the interval $[x, x + \Delta x]$, where Δx is small. In what follows, for simplicity, we assume that the magnitude T of the tension $\mathbf{T}$ acting tangential to the string is constant along the string. Now from Figure 15.1 it is clear that the net vertical force F acting on the string on the interval $[x, x + \Delta x]$ is

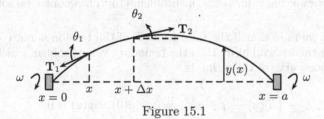

Figure 15.1

$$F = T \sin \theta_2 - T \sin \theta_1.$$

If the angles θ_1 and θ_2 (measured in radians) are small, then we have

$$\sin \theta_2 \simeq \tan \theta_2 \simeq y'(x + \Delta x) \quad \text{and} \quad \sin \theta_1 \simeq \tan \theta_1 \simeq y'(x)$$

and hence

$$F \simeq T[y'(x + \Delta x) - y'(x)]. \tag{15.1}$$

The net force F can also be given by Newton's second law as $F = ma$. Clearly, the mass of the string on the interval $[x, x + \Delta x]$ is $m = \rho \, \Delta x$, and the centripetal acceleration of a point rotating with angular speed ω in a

R.P. Agarwal, D. O'Regan, *Ordinary and Partial Differential Equations*,
Universitext, DOI 10.1007/978-0-387-79146-3_15,
© Springer Science+Business Media, LLC 2009

circle of radius r is $a = r\omega^2$. Since Δx is small, we can assume that $r = y$. Thus, another formulation of the net force is

$$F \simeq -(\rho\,\Delta x)y\omega^2, \tag{15.2}$$

where the minus sign indicates the fact that the acceleration points in the direction opposite to the positive y direction. From (15.1) and (15.2), we get

$$T[y'(x + \Delta x) - y'(x)] \simeq -(\rho\,\Delta x)y\omega^2,$$

or

$$T\frac{y'(x + \Delta x) - y'(x)}{\Delta x} \simeq -\rho\omega^2 y,$$

which as $\Delta x \to 0$ leads to the differential equation

$$T\frac{d^2 y}{dt^2} = -\rho\omega^2 y. \tag{15.3}$$

Since the string is fixed at the ends, the solution $y(x)$ of (15.3) must also satisfy the boundary conditions $y(0) = 0$, $y(a) = 0$. Thus, the shape of the string $y(x)$ can be determined by solving the boundary value problem

$$\frac{d^2 y}{dx^2} + \frac{\rho\omega^2}{T}y = 0, \quad y(0) = y(a) = 0. \tag{15.4}$$

Clearly, $y(x) \equiv 0$ is a solution of (15.4). However, in Lecture 19 we shall see that for some special values of ω the problem (15.4) has nontrivial solutions also.

Finally, we note that if the magnitude T of the tension is not constant throughout the interval $[0, a]$, then the boundary value problem which gives the deflection curve of the string is

$$\frac{d}{dx}\left(T(x)\frac{dy}{dx}\right) + \rho\omega^2 y = 0, \quad y(0) = y(a) = 0. \tag{15.5}$$

Example 15.2. Consider the problem of a vertical column of uniform material and cross section, bent by its own weight. Let a long thin rod be set up in a vertical plane so that the lower end is constrained to remain vertical (Figure 15.2). Suppose the rod is of length a and weight W, and has the coefficient of flexural rigidity B (> 0). Then, if $p = dy/dx$, the equation describing this system can be written as

$$\frac{d^2 p}{dx^2} + \frac{W}{B}\frac{(a - x)}{a}p = 0 \tag{15.6}$$

$$p(0) = 0 = p'(a). \tag{15.7}$$

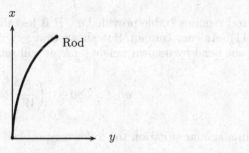

Figure 15.2

One possibility that is always present is that the rod does not bend at all, which is just another way of saying that the problem has only the trivial solution; i.e., $p(x) \equiv 0$. One would expect that if the rod is short enough (just how short it would need to be depends on the constants W, and B, of course) the rod *cannot* bend at all, which is to say that the trivial solution is the only solution of the problem, and the problem is accordingly said to be *stable*. However, for all sufficiently large a, the rod can bend and the problem has a nontrivial solution. Clearly, then uniqueness no longer holds for the boundary value problem.

Equation (15.6) can be transformed into Bessel's equation by the substitution

$$\xi = \frac{2}{3}\left(\frac{W}{aB}\right)^{1/2}(a-x)^{3/2}, \quad p = \eta(a-x)^{1/2}. \tag{15.8}$$

In fact, it leads to the equation

$$\frac{d^2\eta}{d\xi^2} + \frac{1}{\xi}\frac{d\eta}{d\xi} + \left(1 - \frac{1}{9\xi^2}\right)\eta = 0, \tag{15.9}$$

whose solution can be written as

$$\eta(\xi) = AJ_{1/3}(\xi) + BJ_{-1/3}(\xi),$$

and hence the solution of (15.6) is

$$p(\xi) = (a-x)^{1/2}[AJ_{1/3}(\xi) + BJ_{-1/3}(\xi)]. \tag{15.10}$$

Now it a simple matter to see that $p'(a) = 0$ only if $A = 0$, and $p(0) = 0$ provided $J_{-1/3}(\xi) = 0$ at $\xi = (2a/3)(W/B)^{1/2}$. Since

$$J_{-1/3}(\xi) = 1 - \frac{1}{3.2}\frac{a^2W}{B} + \frac{1}{3\cdot6\cdot2\cdot5}\left(\frac{a^2W}{B}\right)^2$$

$$+ \cdots + (-1)^n\frac{1}{3\cdot6\cdots(3n)\cdot2\cdot5\cdots(3n-1)}\left(\frac{a^2W}{B}\right)^n + \cdots \tag{15.11}$$

the rod remains stable provided a^2/B is less than the first zero, say, ξ_1 of (15.11). An easy computation shows that $\xi_1 = 7.84\cdots$, and hence the rod will not bend by its own weight—i.e., it will remain stable—provided

$$a < (2.80\cdots)\left(\frac{B}{W}\right)^{1/2}. \tag{15.12}$$

In a similar situation the following problem occurs:

$$\frac{d^2\phi}{ds^2} + \frac{R^2}{AC}(a-s)^2\phi = 0 \tag{15.13}$$

$$\phi(0) = 0 = \phi'(a). \tag{15.14}$$

For this problem equations (15.8)–(15.12) take the following form:

$$\xi = \frac{1}{2}\frac{R}{\sqrt{AC}}(a-x)^2, \quad \phi = \eta(a-s)^{1/2}, \tag{15.15}$$

$$\frac{d^2\eta}{d\xi^2} + \frac{1}{\xi}\frac{d\eta}{d\xi} + \left(1 - \frac{1}{16\xi^2}\right)\eta = 0, \tag{15.16}$$

$$\phi(\xi) = (a-s)^{1/2}[AJ_{1/4}(\xi) + BJ_{-1/4}(\xi)], \tag{15.17}$$

$$J_{-1/4}(\xi) = 1 - \frac{1}{2.6}\frac{R^2a^4}{AC} + \frac{1}{2\cdot4\cdot6\cdot14}\left(\frac{R^2a^4}{AC}\right)^2$$
$$+\cdots+(-1)^n\frac{1}{2\cdot4\cdots(2n)\cdot6\cdot14\cdots(8n-2)}\left(\frac{R^2a^4}{AC}\right)^n+\cdots \tag{15.18}$$

$$a < \gamma\frac{(AC)^{1/4}}{\sqrt{R}}, \tag{15.19}$$

where γ is a number very close to 2.

Example 15.3. Consider a wedge-shaped canal of uniform depth ℓ that empties into the open sea (see Figure 15.3). Assume that the water level at the mouth of the canal varies harmonically, i.e., the depth at the mouth of the canal is given by $H\cos\omega t$, where H and ω are positive constants. This assumption simulates the motion of the tides. Now the function $h(x,t)$, which gives the depth at a distance x from the inland end of the canal at time t, has the form $h(x,t) = y(x)\cos\omega t$, where $y(x)$ satisfies the DE

$$x^2y'' + xy' + k^2x^2y = 0, \tag{15.20}$$

where $k > 0$ is a constant.

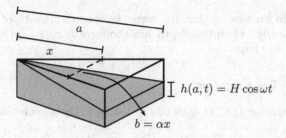

Figure 15.3

Comparing (15.20) with (9.20), we find that its solution can be written as

$$y(x) = AJ_0(kx) + BJ^0(kx).$$ (15.21)

Since at $x = 0$ the depth of the water must be finite at all times, $\lim_{x \to 0} y(x)$ has to be finite. However, since $\lim_{x \to 0} |J^0(kx)| = \infty$, we need to assume that $B = 0$. Hence, the depth $h(x, t)$ can be written as

$$h(x, t) = AJ_0(kx) \cos \omega t.$$ (15.22)

Next using the condition that the depth at the mouth of the canal is $H \cos \omega t$, we have

$$H \cos \omega t = h(a, t) = AJ_0(ka) \cos \omega t$$

and hence $A = H/J_0(ka)$ provided $J_0(ka) \neq 0$. Thus, the depth h appears as

$$h(x, t) = H \frac{J_0(kx)}{J_0(ka)} \cos \omega t.$$ (15.23)

Clearly, from (9.8) we have $J_0(0) = 1$ and hence $J_0(x) \neq 0$ at least for sufficiently small $x > 0$. Thus, the solution (15.23) is meaningful as long as ka is sufficiently small.

Example 15.4. Now in Example 15.3 we assume that the depth of the canal is not uniform, but varies according as $\ell(x) = \beta x$. Figure 15.4 shows the lengthwise cross section of the canal.

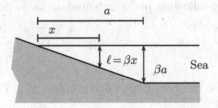

Figure 15.4

Again we assume that the water level at the mouth of the canal varies harmonically. Then the depth has the form $h(x,t) = y(x)\cos\omega t$, where y satisfies the DE

$$x^2 y'' + 2xy' + k^2 xy = 0; \qquad (15.24)$$

here $k > 0$ is a constant.

Comparing (15.24) with (9.19), we find that its solution can be written as

$$y(x) = Ax^{-1/2} J_1\left(2kx^{1/2}\right) + Bx^{-1/2} J_{-1}\left(2kx^{1/2}\right). \qquad (15.25)$$

Since at $x = 0$ the depth of the water must be finite at all times, we need to assume that $B = 0$. At $t = 0$ we have $h(x,0) = \beta x$. In particular, $\beta a = h(a,0) = y(a) = a^{-1/2} A J_1(2ka^{1/2})$ and hence $A = \beta a^{3/2}/J_1(2ka^{1/2})$. Thus, the solution appears as

$$h(x,t) = \beta a^{3/2} x^{-1/2} \frac{J_1\left(2kx^{1/2}\right)}{J_1\left(2ka^{1/2}\right)} \cos\omega t,$$

which is meaningful as long as $2ka^{1/2}$ is small.

Problems

15.1. Solve the following boundary value problems:

(i) $\begin{array}{l} y'' - y = 0 \\ y(0) = 0, \ y(1) = 1 \end{array}$
(ii) $\begin{array}{l} y'' + 4y' + 7y = 0 \\ y(0) = 0, \ y'(1) = 1 \end{array}$

(iii) $\begin{array}{l} y'' - 6y' + 25y = 0 \\ y'(0) = 1, \ y(\pi/4) = 0 \end{array}$
(iv) $\begin{array}{l} x^2 y'' + 7xy' + 3y = 0 \\ y(1) = 1, \ y(2) = 2 \end{array}$

(v) $\begin{array}{l} y'' + y = 0 \\ y(0) + y'(0) = 10 \\ y(1) + 3y'(1) = 4 \end{array}$
(vi) $\begin{array}{l} y'' + y = x^2 \\ y(0) = 0, \ y(\pi/2) = 1 \end{array}$

(vii) $\begin{array}{l} y'' + 2y' + y = x \\ y(0) = 0, \ y(2) = 3 \end{array}$
(viii) $\begin{array}{l} y'' + y' + y = x \\ y(0) + 2y'(0) = 1 \\ y(1) - y'(1) = 8. \end{array}$

15.2. Solve the following periodic boundary value problems:

(i) $\begin{array}{l} y'' + 2y' + 10y = 0 \\ y(0) = y(\pi/6) \\ y'(0) = y'(\pi/6) \end{array}$
(ii) $\begin{array}{l} y'' + \pi^2 y = 0 \\ y(-1) = y(1) \\ y'(-1) = y'(1). \end{array}$

15.3. Show that the boundary value problem $y'' = r(x)$, (14.8) has a unique solution if and only if

$$\Delta = a_0 d_0 (\beta - \alpha) + a_0 d_1 - a_1 d_0 \neq 0.$$

15.4. Determine the values of the constants β, A, and B so that the boundary value problem $y'' + 2py' + qy = 0$, $y(0) = A$, $y(\beta) = B$ with $p^2 - q < 0$ has only one solution.

15.5. Show that the boundary value problem $y'' + p(x)y = q(x)$, (14.5) where $p(x) \leq 0$ in $[\alpha, \beta]$, has a unique solution.

15.6. Let $z(x)$ be the solution of the initial value problem (14.1), $z(\alpha) = A$, $z'(\alpha) = 0$, and $y_2(x)$ be the solution of the initial value problem (14.3), (14.13). Show that the boundary value problem (14.1), (14.5) has a unique solution $y(x)$ if and only if $y_2(\beta) \neq 0$ and it can be written as

$$y(x) = z(x) + \frac{(B - z(\beta))}{y_2(\beta)} y_2(x).$$

15.7. Let $y_1(x)$ and $y_2(x)$ be the solutions of the initial value problems (14.3), $y_1(\alpha) = a_1$, $y_1'(\alpha) = -a_0$ and (14.3), $y_2(\beta) = -d_1$, $y_2'(\beta) = d_0$, respectively. Show that the boundary value problem (14.3), (14.8) has a unique solution if and only if $W(y_1, y_2)(\alpha) \neq 0$.

15.8. Let $y_1(x)$ and $y_2(x)$ be the solutions of the boundary value problems (14.3), (14.2) and (14.1), (14.4), respectively. Show that $y(x) = y_1(x) + y_2(x)$ is a solution of the problem (14.1), (14.2).

15.9. For the homogeneous DE

$$\mathcal{L}_2[y] = (x^2 + 1)y'' - 2xy' + 2y = 0 \qquad (15.26)$$

x and $(x^2 - 1)$ are two linearly independent solutions. Use this information to show that the boundary value problem

$$\mathcal{L}_2[y] = 6(x^2 + 1)^2, \quad y(0) = 1, \quad y(1) = 2 \qquad (15.27)$$

has a unique solution, and find it.

15.10. A telephone cable stretched tightly with constant tension T between supports at $x = 0$ and $x = 1$ hangs at rest under its own weight. For small displacements y the equation of equilibrium and the boundary conditions are

$$y'' = -mg/T, \quad 0 < x < 1, \quad y(0) = 0 = y(1), \qquad (15.28)$$

where m is the mass per unit length of the cable, and g is the gravitational constant. Show that the solution of (15.28) can be written as $y(x) = mgx(1 - x)/(2T)$, i.e., the telephone cable hangs in a parabolic arc.

15.11. In the construction of large buildings a long beam is often needed to span a given distance. To decide the size of the beam, the architect needs to calculate the amount of bending the beam will undergo due to its own weight. If E represents the modulus of elasticity of the beam material, I is the moment of inertia of a cross section about its center axis, $2a$ is the length of the beam, and W is the weight per unit length, then the differential equation used to find the sag curve for a beam supported at both ends is

$$EIy'' = aWx - \frac{Wx^2}{2},$$

where y denotes the vertical sag distance per horizontal x unit. If the beam is resting on two supports at its ends (*simple beam*), then the natural boundary conditions are $y(0) = 0$, $y(2a) = 0$. Show that the solution of this boundary value problem is

$$y(x) = \frac{W}{EI}\left[\frac{a}{6}x^3 - \frac{1}{24}x^4 - \frac{a^3}{3}x\right].$$

Verify that $y'(a) = 0$ and give its interpretation.

15.12. The equation of equilibrium of a tightly stretched and initially straight elastic string embedded in an elastic foundation of modulus $k > 0$ is given by

$$y'' - (k/T)y = 0,$$

where y is the deflection of the string. Here the weight of the string is neglected, the deflections are assumed to be small, and the tension T is considered as a constant. The end $x = 0$ of the string is fixed, i.e., $y(0) = 0$, and at the end $x = a$ there is a displacement given by $y(a) = \beta > 0$. Show that $y(x) = \beta \sinh(\sqrt{k/T})x / \sinh(\sqrt{k/T})a$ is the solution of this boundary value problem, and $\max_{0 \le x \le a} y(x) = y(a) = \beta$.

15.13. A gas diffuses into a liquid in a narrow pipe. Let $y(x)$ denote the concentration of the gas at the distance x in the pipe. The gas is absorbed by the liquid at a rate proportional to $y'(x)$, and the gas reacts chemically with the liquid and as a result disappears at a rate proportional to $y(x)$. This leads to the balance equation

$$y'' - (k/D)y = 0,$$

where k is the reaction rate and D is the diffusion coefficient. If the initial concentration is α, i.e., $y(0) = \alpha$ and at $x = a$ the gas is completely absorbed by the liquid, i.e., $y(a) = 0$, show that $y(x) = \alpha \sinh[\sqrt{k/D}(a - x)] / \sinh[\sqrt{k/D}a]$.

15.14. A long river flows through a populated region with uniform velocity u. Sewage continuously enters at a constant rate at the beginning of the river $x = 0$. The sewage is convected down the river by the flow and it is simultaneously decomposed by bacteria and other biological activities.

Assume that the river is sufficiently narrow so that the concentration y of sewage is uniform over the cross section and that the polluting has been going on for a long time, so that y is a function only of the distance x downstream from the sewage plant. If the rate of decomposition at x is proportional to the concentration $y(x)$ and k is the proportionality constant, then y satisfies the DE

$$y'' - \beta y' - \alpha^2 y = 0,$$

where $\beta = u/D$ and $\alpha^2 = k/(AD)$, A is the cross-sectional area of the river, and $D > 0$ is a constant. If the concentrations at $x = 0$ and $x = a$ are known to be $y(0) = y_0$, $y(a) = y_1$ ($< y_0$), then show that the concentration in the stream for $0 \le x \le a$ is

$$y(x) = e^{\beta x/2} \left[y_0 \cosh \theta x + \frac{y_1 e^{-\beta a/2} - y_0 \cosh \theta a}{\sinh \theta a} \sinh \theta x \right],$$

where $\theta = \sqrt{\beta^2 + 4\alpha^2}/2$.

15.15. Suppose a hollow spherical shell has an inner radius $r = \alpha$ and outer radius $r = \beta$, and the temperature at the inner and outer surfaces are u_α and u_β, respectively. The temperature u at a distance r from the center ($\alpha \le r \le \beta$) is determined by the boundary value problem

$$r \frac{d^2 u}{dr^2} + 2 \frac{du}{dr} = 0, \quad u(\alpha) = u_\alpha, \quad u(\beta) = u_\beta.$$

Show that

$$u(r) = \frac{u_\beta \alpha^{-1} - u_\alpha \beta^{-1}}{\alpha^{-1} - \beta^{-1}} + \left(\frac{u_\alpha - u_\beta}{\alpha^{-1} - \beta^{-1}} \right) r^{-1}.$$

15.16. A steam pipe has temperature u_α at its inner surface $r = \alpha$ and temperature u_β at its outer surface $r = \beta$. The temperature u at a distance r from the center ($\alpha \le r \le \beta$) is determined by the boundary value problem

$$r \frac{d^2 u}{dr^2} + \frac{du}{dr} = 0, \quad u(\alpha) = u_\alpha, \quad u(\beta) = u_\beta.$$

Show that

$$u(r) = \frac{u_\alpha \ln(r/\beta) - u_\beta \ln(r/\alpha)}{\ln(\alpha/\beta)}.$$

15.17. For the telephone cable considered in Problem 15.10, the large displacements y are governed by the equation and boundary conditions

$$y'' = -\frac{mg}{T} \sqrt{1 + (y')^2}, \quad 0 < x < 1, \quad y(0) = 0 = y(1). \qquad (15.29)$$

Show that the solution of (15.29) can be written as

$$y(x) = \frac{T}{mg}\left\{\cosh\frac{mg}{2T} - \cosh\left[\frac{mg}{T}\left(x - \frac{1}{2}\right)\right]\right\};$$

i.e., the telephone cable hangs in a catenary.

Answers or Hints

15.1. (i) $\frac{\sinh x}{\sinh 1}$ (ii) $\frac{e^{2(1-x)}\sin\sqrt{3}x}{(\sqrt{3}\cos\sqrt{3}-2\sin\sqrt{3})}$ (iii) $\frac{1}{4}e^{3x}\sin 4x$ (iv) $\frac{1}{(2^{\sqrt{6}}-2^{-\sqrt{6}})}\times$
$[(16-2^{-\sqrt{6}})x^{-3+\sqrt{6}}+(2^{\sqrt{6}}-16)x^{-3-\sqrt{6}}]$ (v) $\frac{1}{2\sin 1+\cos 1}[\{5(\sin 1+3\cos 1)-2\}\cos x+\{5(3\sin 1-\cos 1)+2\}\sin x]$ (vi) $2\cos x+\left(3-\frac{\pi^2}{4}\right)\sin x+x^2-2$
(vii) $e^{-x}\left[2+\left(\frac{3}{2}e^2-1\right)x\right]+x-2$ (viii) $\frac{18}{3\cos\frac{\sqrt{3}}{2}+\sqrt{3}\sin\frac{\sqrt{3}}{2}}e^{(1-x)/2}\cos\frac{\sqrt{3}}{2}x$
$+x-1$.

15.2. (i) Trivial solution (ii) $c_1\cos\pi x + c_2\sin\pi x$, where c_1 and c_2 are arbitrary constants.

15.3. For the DE $y'' = 0$ two linearly independent solutions are 1, x. Now apply Theorem 14.3.

15.4. $\beta \neq \frac{n\pi}{\sqrt{q-p^2}}$, $e^{-px}\left[A\cos\sqrt{q-p^2}x + \frac{Be^{p\beta}-A\cos\sqrt{q-p^2}\beta}{\sin\sqrt{q-p^2}\beta}\sin\sqrt{q-p^2}x\right]$.

15.5. Let $y(x)$ be a nonnegative solution of $y'' + p(x)y = 0$, $y(\alpha) = 0 = y(\beta)$. Then, at the point $x_1 \in (\alpha, \beta)$ where $y(x)$ attains its maximum $y''(x_1) + p(x_1)y(x_1) < 0$.

15.6. The function $y(x) = z_1(x) + cy_1(x)$ is a solution of the DE (14.1).

15.7. Use Theorem 14.3.

15.8. Verify directly.

15.9. Use variation of parameters to find the particular solution $z(x) = x^4 + 3x^2$. The solution of (15.27) is $x^4 + 2x^2 - 2x + 1$.

15.11. $|y|_{\max} = |y(a)| = \frac{5}{24}a^4$.

Lecture 16
Green's Functions

The function $H(x,t)$ defined in (2.12) is a solution of the homogeneous DE (2.1) and it helps in finding an explicit representation of a particular solution of the nonhomogeneous DE (2.8). In this lecture, we shall find an analog of this function called a *Green's function* $G(x,t)$ for the homogeneous boundary value problem (14.3), (14.4) and show that the solution of the nonhomogeneous boundary value problem (14.1), (14.4) can be explicitly expressed in terms of $G(x,t)$. The solution of the problem (14.1), (14.2) then can be obtained easily as an application of Problem 15.8.

In what follows throughout we shall assume that the problem (14.3), (14.4) has only the trivial solution. Green's function $G(x,t)$ for the boundary value problem (14.3), (14.4) is defined in the square $[\alpha, \beta] \times [\alpha, \beta]$ and possesses the following fundamental properties:

(i) $G(x,t)$ is continuous in $[\alpha, \beta] \times [\alpha, \beta]$,

(ii) $\partial G(x,t)/\partial x$ is continuous in each of the triangles $\alpha \leq x \leq t \leq \beta$ and $\alpha \leq t \leq x \leq \beta$; moreover,

$$\frac{\partial G}{\partial x}(t^+, t) - \frac{\partial G}{\partial x}(t^-, t) = \frac{1}{p_0(t)},$$

where

$$\frac{\partial G}{\partial x}(t^+, t) = \lim_{\substack{x \to t \\ x > t}} \frac{\partial G(x,t)}{\partial x} \quad \text{and} \quad \frac{\partial G}{\partial x}(t^-, t) = \lim_{\substack{x \to t \\ x < t}} \frac{\partial G(x,t)}{\partial x},$$

(iii) for every $t \in [\alpha, \beta]$, $z(x) = G(x,t)$ is a solution of the DE (14.3) in each of the intervals $[\alpha, t)$ and $(t, \beta]$,

(iv) for every $t \in [\alpha, \beta]$, $z(x) = G(x,t)$ satisfies the boundary conditions (14.4).

These properties completely characterize Green's function $G(x,t)$. To show this, let $y_1(x)$ and $y_2(x)$ be two linearly independent solutions of the DE (14.3). From the property (iii) there exist four functions, say, $\lambda_1(t)$, $\lambda_2(t)$, $\mu_1(t)$, and $\mu_2(t)$ such that

$$G(x,t) = \begin{cases} y_1(x)\lambda_1(t) + y_2(x)\lambda_2(t), & \alpha \leq x \leq t \\ y_1(x)\mu_1(t) + y_2(x)\mu_2(t), & t \leq x \leq \beta. \end{cases} \tag{16.1}$$

R.P. Agarwal, D. O'Regan, *Ordinary and Partial Differential Equations*,
Universitext, DOI 10.1007/978-0-387-79146-3_16,
© Springer Science+Business Media, LLC 2009

Now using properties (i) and (ii), we obtain the following two equations:

$$y_1(t)\lambda_1(t) + y_2(t)\lambda_2(t) = y_1(t)\mu_1(t) + y_2(t)\mu_2(t) \qquad (16.2)$$

$$y_1'(t)\mu_1(t) + y_2'(t)\mu_2(t) - y_1'(t)\lambda_1(t) - y_2'(t)\lambda_2(t) = \frac{1}{p_0(t)}. \qquad (16.3)$$

Let $\nu_1(t) = \mu_1(t) - \lambda_1(t)$ and $\nu_2(t) = \mu_2(t) - \lambda_2(t)$, so that (16.2) and (16.3) can be written as

$$y_1(t)\nu_1(t) + y_2(t)\nu_2(t) = 0 \qquad (16.4)$$

$$y_1'(t)\nu_1(t) + y_2'(t)\nu_2(t) = \frac{1}{p_0(t)}. \qquad (16.5)$$

Since $y_1(x)$ and $y_2(x)$ are linearly independent the Wronskian $W(y_1, y_2)(t)$ $\neq 0$ for all $t \in [\alpha, \beta]$. Thus, the relations (16.4), (16.5) uniquely determine $\nu_1(t)$ and $\nu_2(t)$.

Now using the relations $\mu_1(t) = \lambda_1(t) + \nu_1(t)$ and $\mu_2(t) = \lambda_2(t) + \nu_2(t)$, Green's function can be written as

$$G(x,t) = \begin{cases} y_1(x)\lambda_1(t) + y_2(x)\lambda_2(t), & \alpha \leq x \leq t \\ y_1(x)\lambda_1(t) + y_2(x)\lambda_2(t) + y_1(x)\nu_1(t) + y_2(x)\nu_2(t), & t \leq x \leq \beta. \end{cases}$$
$$(16.6)$$

Finally, using the property (iv), we find

$$\begin{aligned} \ell_1[y_1]\lambda_1(t) + \ell_1[y_2]\lambda_2(t) &= -b_0(y_1(\beta)\nu_1(t) + y_2(\beta)\nu_2(t)) \\ &\quad -b_1(y_1'(\beta)\nu_1(t) + y_2'(\beta)\nu_2(t)) \\ \ell_2[y_1]\lambda_1(t) + \ell_2[y_2]\lambda_2(t) &= -d_0(y_1(\beta)\nu_1(t) + y_2(\beta)\nu_2(t)) \\ &\quad -d_1(y_1'(\beta)\nu_1(t) + y_2'(\beta)\nu_2(t)). \end{aligned} \qquad (16.7)$$

Since the problem (14.3), (14.4) has only the trivial solution, from Theorem 14.1 it follows that the system (16.7) uniquely determines $\lambda_1(t)$ and $\lambda_2(t)$.

From the above construction it is clear that no other function exists which has properties (i)–(iv); i.e., Green's function $G(x,t)$ of the boundary value problem (14.3), (14.4) is unique.

As mentioned earlier, we shall now show that the unique solution $y(x)$ of the problem (14.1), (14.4) can be represented in terms of $G(x,t)$ as follows:

$$y(x) = \int_\alpha^\beta G(x,t)r(t)dt = \int_\alpha^x G(x,t)r(t)dt + \int_x^\beta G(x,t)r(t)dt. \qquad (16.8)$$

Since $G(x,t)$ is differentiable with respect to x in each of the intervals, we find

$$y'(x) = G(x,x)r(x) + \int_\alpha^x \frac{\partial G(x,t)}{\partial x} r(t)dt - G(x,x)r(x) + \int_x^\beta \frac{\partial G(x,t)}{\partial x} r(t)dt$$

$$= \int_\alpha^x \frac{\partial G(x,t)}{\partial x} r(t)dt + \int_x^\beta \frac{\partial G(x,t)}{\partial x} r(t)dt$$

$$= \int_\alpha^\beta \frac{\partial G(x,t)}{\partial x} r(t)dt.$$

(16.9)

Next since $\partial G(x,t)/\partial x$ is a continuous function of (x,t) in the triangles $\alpha \leq t \leq x \leq \beta$ and $\alpha \leq x \leq t \leq \beta$, for any point (s,s) on the diagonal of the square, i.e., $t = x$ it is necessary that

$$\frac{\partial G}{\partial x}(s,s^-) = \frac{\partial G}{\partial x}(s^+,s) \quad \text{and} \quad \frac{\partial G}{\partial x}(s,s^+) = \frac{\partial G}{\partial x}(s^-,s). \quad (16.10)$$

Now differentiating the relation (16.9), we obtain

$$y''(x) = \frac{\partial G(x,x^-)}{\partial x}r(x) + \int_\alpha^x \frac{\partial^2 G(x,t)}{\partial x^2} r(t)dt$$

$$- \frac{\partial G(x,x^+)}{\partial x}r(x) + \int_x^\beta \frac{\partial^2 G(x,t)}{\partial x^2} r(t)dt,$$

which in view of (16.10) is the same as

$$y''(x) = \left[\frac{\partial G(x^+,x)}{\partial x} - \frac{\partial G(x^-,x)}{\partial x}\right] r(x) + \int_\alpha^\beta \frac{\partial^2 G(x,t)}{\partial x^2} r(t)dt.$$

Using property (ii) this relation gives

$$y''(x) = \frac{r(x)}{p_0(x)} + \int_\alpha^\beta \frac{\partial^2 G(x,t)}{\partial x^2} r(t)dt. \quad (16.11)$$

Thus, from (16.8), (16.9), and (16.11), and the property (iii), we get

$$p_0(x)y''(x) + p_1(x)y'(x) + p_2(x)y(x)$$

$$= r(x) + \int_\alpha^\beta \left[p_0(x)\frac{\partial^2 G(x,t)}{\partial x^2} + p_1(x)\frac{\partial G(x,t)}{\partial x} + p_2(x)G(x,t)\right] r(t)dt$$

$$= r(x),$$

i.e., $y(x)$ as given in (16.8) is a solution of the DE (14.1).

Finally, since

$$y(\alpha) = \int_\alpha^\beta G(\alpha,t)r(t)dt, \qquad y(\beta) = \int_\alpha^\beta G(\beta,t)r(t)dt$$

$$y'(\alpha) = \int_\alpha^\beta \frac{\partial G(\alpha,t)}{\partial x} r(t)dt, \qquad y'(\beta) = \int_\alpha^\beta \frac{\partial G(\beta,t)}{\partial x} r(t)dt,$$

it is easy to see that

$$\ell_1[y] = \int_\alpha^\beta \ell_1[G(x,t)]r(t)dt = 0 \quad \text{and} \quad \ell_2[y] = \int_\alpha^\beta \ell_2[G(x,t)]r(t)dt = 0$$

and hence $y(x)$ as given in (16.8) satisfies the boundary conditions (14.4) as well.

We summarize these results in the following theorem.

Theorem 16.1. Let the homogeneous problem (14.3), (14.4) have only the trivial solution. Then, the following hold:

(i)   there exists a unique Green's function $G(x,t)$ for the problem (14.3), (14.4),

(ii)   the unique solution $y(x)$ of the nonhomogeneous problem (14.1), (14.4) can be represented by (16.8).

Example 16.1. We shall construct Green's function of the problem

$$y'' = 0 \tag{16.12}$$

$$\begin{aligned} a_0 y(\alpha) + a_1 y'(\alpha) &= 0 \\ d_0 y(\beta) + d_1 y'(\beta) &= 0. \end{aligned} \tag{16.13}$$

For the DE (16.12) two linearly independent solutions are $y_1(x) = 1$ and $y_2(x) = x$. Hence, in view of Theorem 14.1 the problem (16.12), (16.13) has only the trivial solution if and only if

$$\Delta = \begin{vmatrix} a_0 & a_0\alpha + a_1 \\ d_0 & d_0\beta + d_1 \end{vmatrix} = a_0 d_0(\beta - \alpha) + a_0 d_1 - a_1 d_0 \neq 0$$

(see Problem 15.3). Further, equalities (16.4) and (16.5) reduce to

$$\nu_1(t) + t\nu_2(t) = 0 \quad \text{and} \quad \nu_2(t) = 1.$$

Thus, $\nu_1(t) = -t$ and $\nu_2(t) = 1$.

Next for (16.12), (16.13) the system (16.7) reduces to

$$\begin{aligned} a_0\lambda_1(t) + (a_0\alpha + a_1)\lambda_2(t) &= 0 \\ d_0\lambda_1(t) + (d_0\beta + d_1)\lambda_2(t) &= -d_0(-t + \beta) - d_1, \end{aligned}$$

which easily determines $\lambda_1(t)$ and $\lambda_2(t)$ as

$$\lambda_1(t) = \frac{1}{\Delta}(a_0\alpha + a_1)(d_0\beta - d_0 t + d_1) \quad \text{and} \quad \lambda_2(t) = \frac{1}{\Delta}a_0(d_0 t - d_0\beta - d_1).$$

Substituting these functions in (16.6), we get the required Green's function

$$G(x,t) = \frac{1}{\Delta} \begin{cases} (d_0\beta - d_0 t + d_1)(a_0\alpha - a_0 x + a_1), & \alpha \le x \le t \\ (d_0\beta - d_0 x + d_1)(a_0\alpha - a_0 t + a_1), & t \le x \le \beta, \end{cases} \quad (16.14)$$

which is symmetric, i.e., $G(x,t) = G(t,x)$.

Example 16.2. Consider the periodic boundary value problem

$$y'' + k^2 y = 0, \quad k > 0 \tag{16.15}$$

$$\begin{aligned} y(0) &= y(\omega) \\ y'(0) &= y'(\omega), \quad \omega > 0. \end{aligned} \tag{16.16}$$

For the DE (16.15) two linearly independent solutions are $y_1(x) = \cos kx$ and $y_2(x) = \sin kx$. Hence, in view of Theorem 14.1 the problem (16.15), (16.16) has only the trivial solution if and only if

$$\Delta = 4k \sin^2 \frac{k\omega}{2} \ne 0, \quad \text{i.e.,} \quad \omega \in (0, 2\pi/k).$$

Further, equalities (16.4) and (16.5) reduce to

$$\cos kt\, \nu_1(t) + \sin kt\, \nu_2(t) = 0$$
$$-k \sin kt\, \nu_1(t) + k \cos kt\, \nu_2(t) = 1.$$

These relations easily give

$$\nu_1(t) = -\frac{1}{k} \sin kt \quad \text{and} \quad \nu_2(t) = \frac{1}{k} \cos kt.$$

Next for (16.15), (16.16) the system (16.7) reduces to

$$(1 - \cos k\omega)\lambda_1(t) - \sin k\omega\, \lambda_2(t) = \frac{1}{k} \sin k(\omega - t)$$

$$\sin k\omega\, \lambda_1(t) + (1 - \cos k\omega)\lambda_2(t) = \frac{1}{k} \cos k(\omega - t),$$

which determines $\lambda_1(t)$ and $\lambda_2(t)$ as

$$\lambda_1(t) = \frac{1}{2k \sin \frac{k}{2}\omega} \cos k\left(t - \frac{\omega}{2}\right) \quad \text{and} \quad \lambda_2(t) = \frac{1}{2k \sin \frac{k}{2}\omega} \sin k\left(t - \frac{\omega}{2}\right).$$

Substituting these functions in (16.6), we get Green's function of the boundary value problem (16.15), (16.16) as

$$G(x,t) = \frac{1}{2k \sin \frac{k}{2}\omega} \begin{cases} \cos k\left(x - t + \frac{\omega}{2}\right), & 0 \le x \le t \\ \cos k\left(t - x + \frac{\omega}{2}\right), & t \le x \le \omega \end{cases} \tag{16.17}$$

which as expected is symmetric.

Problems

16.1. Show that

$$G(x,t) = \begin{cases} -\cos t \sin x, & 0 \le x \le t \\ -\sin t \cos x, & t \le x \le \pi/2 \end{cases}$$

is Green's function of the problem $y'' + y = 0,\ y(0) = y(\pi/2) = 0$. Hence, solve the boundary value problem

$$y'' + y = 1 + x, \quad y(0) = y(\pi/2) = 1.$$

16.2. Show that

$$G(x,t) = \frac{1}{\sinh 1} \begin{cases} \sinh(t-1)\sinh x, & 0 \le x \le t \\ \sinh t \sinh(x-1), & t \le x \le 1 \end{cases}$$

is Green's function of the problem $y'' - y = 0,\ y(0) = y(1) = 0$. Hence, solve the boundary value problem

$$y'' - y = 2\sin x, \quad y(0) = 0, \quad y(1) = 2.$$

16.3. Construct Green's function for each of the boundary value problems given in Problem 15.1, parts (vi) and (vii), and then find their solutions.

16.4. Verify that Green's function of the problem (15.26), $y(0) = 0,\ y(1) = 0$ is

$$G(x,t) = \begin{cases} \dfrac{t(x^2 - 1)}{(t^2 + 1)^2}, & 0 \le t \le x \\ \dfrac{x(t^2 - 1)}{(t^2 + 1)^2}, & x \le t \le 1. \end{cases}$$

Hence, solve the boundary value problem (15.27).

16.5. Show that the solution of the boundary value problem

$$y'' - \frac{1}{x}y' = r(x), \quad y(0) = 0, \quad y(1) = 0$$

can be written as

$$y(x) = \int_0^1 G(x,t)r(t)dt,$$

where

$$G(x,t) = \begin{cases} -\dfrac{(1-t^2)x^2}{2t}, & x \le t \\ -\dfrac{t(1-x^2)}{2}, & x \ge t. \end{cases}$$

16.6. Show that the solution of the boundary value problem

$$y'' - y = r(x), \quad y(-\infty) = 0, \quad y(\infty) = 0$$

can be written as

$$y(x) = \frac{1}{2} \int_{-\infty}^{\infty} e^{-|x-t|} r(t) dt.$$

16.7. Consider the nonlinear DE

$$y'' = f(x, y, y') \tag{16.18}$$

together with the boundary conditions (14.5). Show that $y(x)$ is a solution of this problem if and only if

$$y(x) = \frac{(\beta - x)}{(\beta - \alpha)} A + \frac{(x - \alpha)}{(\beta - \alpha)} B + \int_{\alpha}^{\beta} G(x,t) f(t, y(t), y'(t)) dt,$$

where $G(x,t)$ is Green's function of the problem $y'' = 0$, $y(\alpha) = y(\beta) = 0$ and is given by

$$G(x,t) = \frac{1}{(\beta - \alpha)} \begin{cases} (\beta - t)(\alpha - x), & \alpha \le x \le t \\ (\beta - x)(\alpha - t), & t \le x \le \beta. \end{cases}$$

Also establish that

(i) $G(x,t) \le 0$ in $[\alpha, \beta] \times [\alpha, \beta]$

(ii) $|G(x,t)| \le \dfrac{1}{4}(\beta - \alpha)$

(iii) $\displaystyle\int_{\alpha}^{\beta} |G(x,t)| dt = \frac{1}{2}(\beta - x)(x - \alpha) \le \frac{1}{8}(\beta - \alpha)^2$

(iv) $\displaystyle\int_{\alpha}^{\beta} |G(x,t)| \sin \frac{\pi(t - \alpha)}{(\beta - \alpha)} dt = \frac{(\beta - \alpha)^2}{\pi^2} \sin \frac{\pi(x - \alpha)}{(\beta - \alpha)}$

(v) $\displaystyle\int_{\alpha}^{\beta} \left| \frac{\partial G(x,t)}{\partial x} \right| dt = \frac{(x - \alpha)^2 + (\beta - x)^2}{2(\beta - \alpha)} \le \frac{1}{2}(\beta - \alpha).$

16.8. Consider the boundary value problem (16.18), (14.6). Show that $y(x)$ is a solution of this problem if and only if

$$y(x) = A + (x - \alpha)B + \int_{\alpha}^{\beta} G(x,t) f(t, y(t), y'(t)) dt,$$

where $G(x,t)$ is Green's function of the problem $y'' = 0$, $y(\alpha) = y'(\beta) = 0$
and is given by

$$G(x,t) = \begin{cases} (\alpha - x), & \alpha \le x \le t \\ (\alpha - t), & t \le x \le \beta. \end{cases}$$

Also establish that

(i) $G(x,t) \le 0$ in $[\alpha, \beta] \times [\alpha, \beta]$

(ii) $|G(x,t)| \le (\beta - \alpha)$

(iii) $\displaystyle\int_\alpha^\beta |G(x,t)|dt = \frac{1}{2}(x - \alpha)(2\beta - \alpha - x) \le \frac{1}{2}(\beta - \alpha)^2$

(iv) $\displaystyle\int_\alpha^\beta \left| \frac{\partial G(x,t)}{\partial x} \right| dt = (\beta - x) \le (\beta - \alpha).$

16.9.   Consider the nonlinear DE

$$y'' - ky = f(x, y, y'), \quad k > 0$$

together with the boundary conditions (14.5). Show that $y(x)$ is a solution
of this problem if and only if

$$y(x) = \frac{\sinh \sqrt{k}(\beta - x)}{\sinh \sqrt{k}(\beta - \alpha)} A + \frac{\sinh \sqrt{k}(x - \alpha)}{\sinh \sqrt{k}(\beta - \alpha)} B + \int_\alpha^\beta G(x,t)f(t, y(t), y'(t))dt,$$

where $G(x,t)$ is Green's function of the problem $y'' - ky = 0$, $y(\alpha) = y(\beta) = 0$ and is given by

$$G(x,t) = \frac{-1}{\sqrt{k}\sinh\sqrt{k}(\beta-\alpha)} \begin{cases} \sinh\sqrt{k}(x - \alpha)\sinh\sqrt{k}(\beta - t), & \alpha \le x \le t \\ \sinh\sqrt{k}(t - \alpha)\sinh\sqrt{k}(\beta - x), & t \le x \le \beta. \end{cases}$$

Also establish that

(i) $G(x,t) \le 0$ in $[\alpha, \beta] \times [\alpha, \beta]$

(ii) $\displaystyle\int_\alpha^\beta |G(x,t)|dt = \frac{1}{k}\left(1 - \frac{\cosh\sqrt{k}\left(\frac{\beta+\alpha}{2} - x\right)}{\cosh\sqrt{k}\left(\frac{\beta-\alpha}{2}\right)}\right) \le \frac{1}{k}\left(1 - \frac{1}{\cosh\sqrt{k}\left(\frac{\beta-\alpha}{2}\right)}\right).$

16.10.   Show that

(i)   if we multiply the DE (14.3) by the integrating factor

$$\mu(x) = \frac{1}{p_0(x)}\exp\left(\int^x \frac{p_1(t)}{p_0(t)}dt\right)$$

then it can be written in the *self-adjoint* form

$$L[y] = (p(x)y')' + q(x)y = 0, \tag{16.19}$$

(ii)   if $y_1(x)$ and $y_2(x)$ are two linearly independent solutions of (16.19) in $[\alpha, \beta]$, then $p(x)W(y_1, y_2)(x) = C$, where $C \neq 0$ is a constant,

(iii)   if $y_1(x)$ and $y_2(x)$ are solutions of (16.19) satisfying the same initial conditions as in Problem 15.7, then Green's function of the problem (16.19), (16.13) can be written as

$$G(x,t) = \frac{1}{C} \begin{cases} y_2(t)y_1(x), & \alpha \leq x \leq t \\ y_1(t)y_2(x), & t \leq x \leq \beta \end{cases} \tag{16.20}$$

which is also symmetric.

Answers or Hints

16.1. $1 + x - \frac{\pi}{2}\sin x$.

16.2. $\frac{(2+\sin 1)}{\sinh 1}\sinh x - \sin x$.

16.3. The associated Green's functions are—

for Problem 15.1(vi) $G(x,t) = \begin{cases} -\cos t \sin x, & 0 \leq x \leq t \\ -\sin t \cos x, & t \leq x \leq \pi/2 \end{cases}$

for Problem 15.1(vii) $G(x,t) = \begin{cases} -\frac{x}{2}(2-t)e^{-(x-t)}, & 0 \leq x \leq t \\ -\frac{t}{2}(2-x)e^{-(x-t)}, & t \leq x \leq 2. \end{cases}$

16.4. Verify directly. $x^4 + 2x^2 - 2x + 1$.

16.5. Verify directly.

16.6. Verify directly.

16.7. Verify directly. For Part (ii) note that $|G(x,t)| \leq (\beta-x)(x-\alpha)/(\beta-\alpha)$.

16.8. Verify directly.

16.9. Verify directly.

16.10.(i)   Verify directly (ii) $y_2(py_1')' - y_1(py_2')' = (y_2py_1' - y_1py_2')' = 0$ (iii)   From Problem 15.7 the homogeneous problem (16.19), (16.13) has only the trivial solution; from the same problem it also follows that $y_1(x)$ and $y_2(x)$ are linearly independent solutions of (16.19). Thus, in view of (2.13) and (ii) the general solution of nonhomogeneous self-adjoint equation

$$L[y] = (p(x)y')' + q(x)y = r(x), \tag{16.21}$$

can be written as

$$y(x) = c_1 y_1(x) + c_2 y_2(x) + \frac{1}{C} \int_\alpha^x [y_1(t)y_2(x) - y_2(t)y_1(x)]r(t)dt.$$

This solution also satisfies the boundary conditions (16.13) if and only if $c_1 = \frac{1}{C} \int_\alpha^\beta y_2(t)r(t)dt$ and $c_2 = 0$.

Lecture 17
Regular Perturbations

In Theorem 4.1 we have obtained series solution of the second-order initial value problem (2.1), (4.1) whose radius of convergence is at least as large as that for both the functions $p_1(x)$ and $p_2(x)$. However, in many problems the functions $p_1(x)$ and $p_2(x)$ are not necessarily analytic; moreover, we often need to find at least an approximate solution which is meaningful for all x in a given interval. In this lecture we shall discuss the *regular perturbation technique* which relates the unknown solution of (2.1), (4.1) with the known solutions of infinite related initial value problems.

The essential ideas of regular perturbation technique can be exhibited as follows: Suppose that the auxiliary DE

$$y'' + \overline{p}_1(x)y' + \overline{p}_2(x)y = 0 \tag{17.1}$$

together with the initial conditions (4.1) can be solved explicitly to obtain its solution $y_0(x)$. We write the DE (2.1) in the form

$$y'' + (\overline{p}_1(x) + p_1(x) - \overline{p}_1(x))y' + (\overline{p}_2(x) + p_2(x) - \overline{p}_2(x))y = 0,$$

which is the same as

$$y'' + \overline{p}_1(x)y' + \overline{p}_2(x)y = q_1(x)y' + q_2(x)y, \tag{17.2}$$

where $q_1(x) = \overline{p}_1(x) - p_1(x)$ and $q_2(x) = \overline{p}_2(x) - p_2(x)$. We introduce a parameter ϵ and consider the new DE

$$y'' + \overline{p}_1(x)y' + \overline{p}_2(x)y = \epsilon(q_1(x)y' + q_2(x)y). \tag{17.3}$$

Obviously, if $\epsilon = 1$, then this new DE (17.3) is the same as (17.2). We look for the solution of (17.3), (4.1) having the form

$$y(x) = \sum_{n=0}^{\infty} \epsilon^n y_n(x) = y_0(x) + \epsilon y_1(x) + \epsilon^2 y_2(x) + \cdots. \tag{17.4}$$

For this, it is necessary to have

$$\sum_{n=0}^{\infty} \epsilon^n (y_n''(x) + \overline{p}_1(x)y_n'(x) + \overline{p}_2(x)y_n(x)) = \epsilon \sum_{n=0}^{\infty} \epsilon^n (q_1(x)y_n'(x) + q_2(x)y_n(x))$$

R.P. Agarwal, D. O'Regan, *Ordinary and Partial Differential Equations*,
Universitext, DOI 10.1007/978-0-387-79146-3_17,
© Springer Science+Business Media, LLC 2009

and

$$\sum_{n=0}^{\infty} \epsilon^n y_n(x_0) = c_0, \quad \sum_{n=0}^{\infty} \epsilon^n y_n'(x_0) = c_1.$$

Thus, on equating the coefficients of ϵ^n, $n = 0, 1, \cdots$ we find the infinite system of initial value problems

$$y_0''(x) + \bar{p}_1(x)y_0'(x) + \bar{p}_2(x)y_0(x) = 0, \quad y_0(x_0) = c_0, \quad y_0'(x_0) = c_1 \quad (17.5)$$

$$y_n''(x) + \bar{p}_1(x)y_n'(x) + \bar{p}_2(x)y_n(x) = q_1(x)y_{n-1}'(x) + q_2(x)y_{n-1}(x)$$
$$y_n(x_0) = y_n'(x_0) = 0, \quad n = 1, 2, \cdots. \quad (17.6)_n$$

This infinite system can be solved recursively. Indeed, from our initial assumption the solution $y_0(x)$ of the initial value problem (17.5) can be obtained explicitly, and thus the term $q_1(x)y_0'(x) + q_2(x)y_0(x)$ in $(17.6)_1$ is known; consequently the solution $y_1(x)$ of the nonhomogeneous initial value problem $(17.6)_1$ can be obtained by the method of variation of parameters. Continuing in this way the functions $y_2(x), y_3(x), \cdots$ can similarly be obtained. Finally, the solution of the original problem is obtained by summing the series (17.4) for $\epsilon = 1$.

The above formal perturbative procedure is not only applicable for the initial value problem (2.1), (4.1) but can also be applied to a variety of linear as well as nonlinear problems. The implementation of this powerful technique consists the following three basic steps:

(i) Conversion of the given problem into a perturbation problem by introducing the small parameter ϵ.

(ii) Assumption of the solution in the form of a perturbation series and computation of the coefficients of that series.

(iii) Finally, obtaining the solution of the original problem by summing the perturbation series for the appropriate value of ϵ.

It is clear that the parameter ϵ in the original problem can be introduced in an infinite number of ways; however, the perturbed problem is meaningful only if the zero-th order solution, i.e., $y_0(x)$ is obtainable explicitly. Further, in a large number of applied problems this parameter occurs naturally, representing such diverse physical quantities as Planck's constant, a coupling coefficient, the intensity of a shock, the reciprocal of the speed of light, or the amplitude of a forcing term.

The perturbation method naturally leads to the question, under what conditions does the perturbation series converge and actually represent a solution of the original problem? Unfortunately, often perturbation series are divergent; however, this is not necessarily bad because a good approximation to the solution when ϵ is very small can be obtained by summing only the first few terms of the series.

We shall illustrate this fruitful technique in the following examples.

Example 17.1. The initial value problem

$$y'' - |x|y = 0, \quad y(0) = 1, \quad y'(0) = 0 \qquad (17.7)$$

has a unique solution in $\mathbb{R}$. However, in any interval containing zero the series solution method cannot be employed because the function $|x|$ is not analytic. We convert (17.7) into a perturbation problem

$$y'' = \epsilon |x| y, \quad y(0) = 1, \quad y'(0) = 0 \qquad (17.8)$$

and assume that the solution of (17.8) can be written as perturbation series (17.4). This leads to an infinite system of initial value problems

$$y_0''(x) = 0, \quad y_0(0) = 1, \quad y_0'(0) = 0 \qquad (17.9)$$
$$y_n''(x) = |x| y_{n-1}(x), \quad y_n(0) = y_n'(0) = 0, \quad n = 1, 2, \cdots \qquad (17.10)$$

which can be solved recursively, to obtain

$$y_0(x) = 1, \quad y_n(x) = \frac{1.4.7. \cdots (3n - 2)}{(3n)!} \begin{cases} x^{3n-1}|x|, & \text{if } n \text{ odd} \\ x^{3n}, & \text{if } n \text{ even.} \end{cases}$$

Thus, the solution $y(x, \epsilon)$ of the perturbation problem (17.8) appears as

$$y(x, \epsilon) = 1 + \sum_{n=1}^{\infty} \epsilon^n \frac{1.4.7. \cdots (3n - 2)}{(3n)!} \begin{cases} x^{3n-1}|x|, & \text{if } n \text{ odd} \\ x^{3n}, & \text{if } n \text{ even.} \end{cases} \qquad (17.11)$$

Hence, the solution $y(x) = y(x, 1)$ of the initial value problem (17.7) can be written as

$$y(x) = 1 + \sum_{n=1}^{\infty} \frac{1.4.7 \cdots (3n - 2)}{(3n)!} \begin{cases} x^{3n-1}|x|, & \text{if } n \text{ odd} \\ x^{3n}, & \text{if } n \text{ even.} \end{cases} \qquad (17.12)$$

From (17.12) it is clear that for the problem $y'' - xy = 0$, $y(0) = 1$, $y'(0) = 0$ the perturbation method as well as its series solution leads to the same Airy function.

Example 17.2. Consider the initial value problem

$$y'' + y = 2x - 1, \quad y(1) = 1, \quad y'(1) = 3. \qquad (17.13)$$

We convert (17.13) into a perturbation problem

$$y'' = \epsilon(-y + 2x - 1), \quad y(1) = 1, \quad y'(1) = 3 \qquad (17.14)$$

and assume that the solution of (17.14) can be written as perturbation series (17.4). This leads to the system

$$y_0''(x) = 0, \quad y_0(1) = 1, \quad y_0'(1) = 3$$
$$y_1''(x) = -y_0(x) + 2x - 1, \quad y_1(1) = y_1'(1) = 0$$
$$y_n''(x) = -y_{n-1}(x), \quad y_n(1) = y_n'(1) = 0, \quad n = 2, 3, \cdots.$$

Thus, the solution $y(x)$ of the problem (17.13) appears as

$$y(x) = 3x - 2 + \sum_{n=1}^{\infty} (-1)^n \frac{(x-1)^{2n+1}}{(2n+1)!} = 2x - 1 + \sin(x-1).$$

Example 17.3. In the van der Pol's equation (4.18) we consider μ as the perturbing parameter, and seek its solution in the form $y(x) = \sum_{n=0}^{\infty} \mu^n y_n(x)$. For this, we must have

$$\sum_{n=0}^{\infty} \mu^n (y_n''(x) + y_n(x)) = \mu \left(1 - \left(\sum_{n=0}^{\infty} \mu^n y_n(x) \right)^2 \right) \sum_{n=0}^{\infty} \mu^n y_n'(x),$$

which leads to the system

$$y_0''(x) + y_0(x) = 0 \tag{17.15}$$
$$y_1''(x) + y_1(x) = (1 - y_0^2(x)) y_0'(x) \tag{17.16}$$
$$y_2''(x) + y_2(x) = (1 - y_0^2(x)) y_1'(x) - 2y_0(x) y_1(x) y_0'(x) \tag{17.17}$$
$$\cdots.$$

The general solution of (17.15) is readily available, and we prefer to write it as $y_0(x) = a\cos(x + b)$, where a and b are arbitrary constants. Substituting $y_0(x)$ in (17.16), one obtains

$$\begin{aligned} y_1''(x) + y_1(x) &= -(1 - a^2 \cos^2(x + b)) a \sin(x + b) \\ &= \frac{a^3 - 4a}{4} \sin(x+b) + \frac{1}{4} a^3 \sin 3(x+b), \end{aligned}$$

which easily determines $y_1(x)$ as

$$y_1(x) = -\frac{a^3 - 4a}{8} x \cos(x+b) - \frac{1}{32} a^3 \sin 3(x+b).$$

With $y_0(x)$ and $y_1(x)$ known, the right side of (17.17) is known. Thus, $y_2(x)$ can be determined from (17.17) in a similar fashion. Certainly, for a small μ the solution $y(x)$ of (4.18) is better approximated by the function

$$a\cos(x+b) - \mu \left(\frac{a^3 - 4a}{8} x \cos(x+b) + \frac{1}{32} a^3 \sin 3(x+b) \right)$$

compared to just $a\cos(x+b)$.

Example 17.4. *Duffing's equation*

$$my'' + ay + by^3 = 0 \tag{17.18}$$

models the free velocity vibrations of a mass m on a nonlinear spring, where the term ay represents the force exerted by a linear spring, whereas the term by^3 represents the nonlinearity of an actual spring. For simplicity, in (17.18) let $m = a = 1$ so that (17.18) reduces to

$$y'' + y + by^3 = 0. \tag{17.19}$$

We shall consider (17.19) together with the initial conditions

$$y(0) = y_0, \quad y'(0) = 0. \tag{17.20}$$

In (17.19) let b be the perturbing parameter. We seek the solution of (17.19), (17.20) in the form $y(x) = \sum_{n=0}^{\infty} b^n y_n(x)$. This leads to the system

$$y_0''(x) + y_0(x) = 0, \quad y_0(0) = y_0, \quad y_0'(0) = 0 \tag{17.21}$$
$$y_1''(x) + y_1(x) = -y_0^3(x), \quad y_1(0) = y_1'(0) = 0 \tag{17.22}$$
$$\cdots.$$

From (17.21) and (17.22), it is easy to obtain the functions

$$y_0(x) = y_0 \cos x$$
$$y_1(x) = -\frac{3}{8}y_0^3 x \sin x + \frac{1}{32}y_0^3(\cos 3x - \cos x).$$

Thus, the solution $y(x)$ of the problem (17.19), (17.20) can be written as

$$y(x) = y_0 \cos x + by_0^3 \left(-\frac{3}{8}x \sin x + \frac{1}{32}(\cos 3x - \cos x) \right) + O(b^2).$$

Example 17.5. The boundary value problem

$$y'' = -2yy', \quad y(0) = 1, \quad y(1) = 1/2 \tag{17.23}$$

has a unique solution $y(x) = 1/(1 + x)$. We convert (17.23) into a perturbation problem

$$y'' = \epsilon(-2yy'), \quad y(0) = 1, \quad y(1) = 1/2 \tag{17.24}$$

and assume that the solution of (17.24) can be written as perturbation series (17.4). This leads to the system

$$y_0'' = 0, \quad y_0(0) = 1, \quad y_0(1) = 1/2 \tag{17.25}$$
$$y_1'' = -2y_0y_0', \quad y_1(0) = y_1(1) = 0 \tag{17.26}$$
$$y_2'' = -2(y_0y_1' + y_1y_0'), \quad y_2(0) = y_2(1) = 0 \tag{17.27}$$
$$\cdots.$$

From (17.25)–(17.27), we find the functions

$$y_0(x) = \frac{1}{2}(2 - x), \quad y_1(x) = \frac{1}{12}(-5x + 6x^2 - x^3)$$
$$y_2(x) = \frac{1}{180}(-17x + 75x^2 - 85x^3 + 30x^4 - 3x^5).$$

Thus, an approximation to the solution of the boundary value problem (17.23) can be taken as

$$\bar{y}(x) \;=\; y_0(x) + y_1(x) + y_2(x)$$
$$\;=\; \frac{1}{180}(180 - 182x + 165x^2 - 100x^3 + 30x^4 - 3x^5).$$

In Table 17.1 we compare this approximate solution $\bar{y}(x)$ with the exact solution $y(x)$.

Table 17.1

x	Exact solution	Approximate solution	Difference
0.0	1.000000	1.000000	0.000000
0.1	0.909091	0.907517	0.001574
0.2	0.833333	0.830261	0.003072
0.3	0.769231	0.765476	0.003755
0.4	0.714286	0.710763	0.003523
0.5	0.666667	0.664062	0.002605
0.6	0.625000	0.623637	0.001363
0.7	0.588235	0.588049	0.000186
0.8	0.555556	0.556139	-0.000583
0.9	0.526316	0.527008	-0.000692
1.0	0.500000	0.500000	0.000000

Problems

17.1. The initial value problem

$$(1 + \epsilon\theta)\frac{d\theta}{d\tau} + \theta = 0, \quad \theta(0) = 1$$

occurs in cooling of a lumped system. Show that

$$\theta(\tau) = e^{-\tau} + \epsilon\left(e^{-\tau} - e^{-2\tau}\right) + \epsilon^2\left(e^{-\tau} - 2e^{-2\tau} + \frac{3}{2}e^{-3\tau}\right) + O(\epsilon^3).$$

Compare this approximation with the exact solution $\ln\theta + \epsilon(\theta - 1) = -\tau$.

17.2. The initial value problem

$$\frac{d\theta}{d\tau} + \theta + \epsilon\theta^4 = 0, \quad \theta(0) = 1$$

occurs in cooling of a lumped system. Show that

$$\theta(\tau) = e^{-\tau} + \epsilon\frac{1}{3}\left(e^{-4\tau} - e^{-\tau}\right) + \epsilon^2\frac{2}{9}\left(e^{-\tau} - 2e^{-4\tau} + e^{-7\tau}\right) + O(\epsilon^3).$$

Compare this approximation with the exact solution

$$\frac{1}{3}\ln\frac{1 + \epsilon\theta^3}{(1 + \epsilon)\theta^3} = \tau.$$

17.3. For the initial value problem

$$y'' + (1 - \epsilon x)y = 0, \quad y(0) = 1, \quad y'(0) = 0$$

show that

$$y(x) = \cos x + \epsilon\left(\frac{1}{4}x^2\sin x + \frac{1}{4}x\cos x - \frac{1}{4}\sin x\right)$$

$$+\epsilon^2\left(-\frac{1}{32}x^4\cos x + \frac{5}{48}x^3\sin x + \frac{7}{16}x^2\cos x - \frac{7}{16}x\sin x\right) + O(\epsilon^3).$$

17.4. Consider the case of dropping a stone from the height h. Let $r = r(t)$ denote the distance of the stone from the surface at time t. Then, the equation of motion is

$$\frac{d^2r}{dt^2} = -\frac{\gamma M}{(R + r)^2}, \quad r(0) = h, \quad r'(0) = 0, \tag{17.28}$$

where R and M are the radius and the mass of the earth. Let $\epsilon = 1/R$ in (17.28), to obtain

$$\frac{d^2r}{dt^2} = -\frac{\gamma M\epsilon^2}{(1 + \epsilon r)^2}, \quad r(0) = h, \quad r'(0) = 0. \tag{17.29}$$

In (17.29) use the expansion $r(t) = \sum_{i=0}^4 \epsilon^i r_i(t)$ to show that

$$r(t) = h - \frac{\gamma M}{R^2}\left(1 - \frac{2h}{R}\right)\frac{t^2}{2} + O\left(\frac{1}{R^4}\right).$$

17.5. Consider the *satellite equation*

$$\frac{d^2y}{dt^2} + y = ky^2$$

together with the initial conditions $y(0) = A, \ y'(0) = 0$. Show that

$$y(t) = A\cos t + kA^2\left(\frac{1}{2} - \frac{1}{3}\cos t - \frac{1}{6}\cos 2t\right) + k^2A^3\left(-\frac{1}{3} + \frac{29}{144}\cos t\right.$$

$$\left. + \frac{5}{12}t\sin t + \frac{1}{9}\cos 2t + \frac{1}{48}\cos 3t\right) + O(k^3).$$

17.6. For the harmonically *forced Duffing's equation*

$$my'' + ay + by^3 = A \cos \Omega x$$

show that the periodic solution $y(x)$ of period $T = 2\pi/\Omega$ can be written as

$$y(x) = \frac{F}{\omega^2 - \Omega^2} \cos \Omega x + \epsilon \left\{ \frac{3F^3}{4(\omega^2 - \Omega^2)^4} \cos \Omega x \right.$$

$$\left. + \frac{F^3}{4(\omega^2 - \Omega^2)^3(\omega^2 - 3\Omega^2)} \cos 3\Omega x \right\} + O(\epsilon^2),$$

where $\omega^2 = a/m \neq n^2\Omega^2$, $\epsilon = b/m$ and $F = A/m$.

17.7. The boundary value problem

$$\frac{d^2\theta}{dX^2} - \epsilon\theta^4 = 0, \quad \theta'(0) = 0, \quad \theta(1) = 1$$

occurs in heat transfer. Show that

$$\theta(X) = 1 + \epsilon \frac{1}{2}(X^2 - 1) + \epsilon^2 \frac{1}{6}(X^4 - 6X^2 + 5) + O(\epsilon^3).$$

17.8. The boundary value problem

$$(1 + \epsilon\theta)\frac{d^2\theta}{dX^2} + \epsilon\left(\frac{d\theta}{dX}\right)^2 - N^2\theta = 0, \quad \theta'(0) = 0, \quad \theta(1) = 1$$

occurs in heat transfer. Show that

$$\theta(X) = \operatorname{sech} N \cosh NX + \epsilon\frac{1}{3}\operatorname{sech}^2 N(\cosh 2N \operatorname{sech} N \cosh NX$$

$$- \cosh 2NX) + \epsilon^2\frac{1}{6}\operatorname{sech}^3 N \left[\left(\frac{4}{3}\operatorname{sech}^2 N \cosh^2 2N - \frac{1}{2}N\tanh N \right. \right.$$

$$\left. - \frac{9}{8}\operatorname{sech} N \cosh 3N \right) \cosh NX - \frac{4}{3}\operatorname{sech} N \cosh 2N \cosh 2NX$$

$$\left. + \frac{9}{8}\cosh 3NX + \frac{1}{2}NX \sinh NX \right] + O(\epsilon^3).$$

17.9. The boundary value problem

$$\frac{d^2U}{dR^2} + \frac{1}{R}\frac{dU}{dR} = -P + \epsilon\theta$$

$$\frac{d^2\theta}{dR^2} + \frac{1}{R}\frac{d\theta}{dR} = -U$$

$$U'(0) = \theta'(0) = 0, \quad U(1) = \theta(1) = 0$$

occurs in a flow of a fluid. Show that

$$U(R) = \frac{P}{4}(1 - R^2) + \epsilon \frac{P}{2304}(R^6 - 9R^4 + 27R^2 - 19)$$

$$-\epsilon^2 \frac{P}{14745600}(R^{10} - 25R^8 + 300R^6 - 1900R^4 + 5275R^2 - 3651) + O(\epsilon^3)$$

$$\theta(R) = \frac{P}{64}(R^4 - 4R^2 + 3) + \epsilon \frac{P}{147456}(R^8 + 16R^6 - 108R^4 + 304R^2 - 211)$$

$$+\epsilon^2 \frac{P}{2123366400}(R^{12} - 36R^{10} + 675R^8 - 7600R^6 + 47475R^4 - 131436R^2$$

$$+ 90921) + O(\epsilon^3).$$

Lecture 18
Singular Perturbations

In many practical problems one often meets cases where the parameter ϵ is involved in the DE in such a way that the methods of regular perturbations cannot be applied. In the literature such problems are known as singular perturbation problems. In this lecture we shall explain the methodology of *singular perturbation technique* with the help of the following examples.

Example 18.1. For the initial value problem

$$\epsilon y'' + (1 + \epsilon)y' + y = 0 \tag{18.1}$$

$$y(0) = c_1, \quad y'(0) = c_2 \tag{18.2}$$

the explicit solution can be written as

$$y(x) = \frac{1}{(\epsilon - 1)} \left[\epsilon(c_1 + c_2)e^{-x/\epsilon} - (c_2\epsilon + c_1)e^{-x} \right].$$

Thus, it follows that

$$y'(x) = \frac{1}{(\epsilon - 1)} \left[-(c_1 + c_2)e^{-x/\epsilon} + (c_2\epsilon + c_1)e^{-x} \right].$$

Hence, as $\epsilon \to 0^+$, $y(x) \to c_1 e^{-x}$, but $y'(x)$ has the following discontinuous behavior:

$$\lim_{\epsilon \to 0^+} y'(x) = \begin{cases} -c_1 e^{-x}, & x > 0 \\ c_2, & x = 0. \end{cases}$$

As a consequence, we find

$$\lim_{x \to 0} \left(\lim_{\epsilon \to 0^+} y'(x) \right) \neq \lim_{\epsilon \to 0^+} \left(\lim_{x \to 0} y'(x) \right).$$

Further, if we set $\epsilon = 0$ in (18.1) then we are left with the first-order DE

$$y' + y = 0. \tag{18.3}$$

Obviously, for the problem (18.3), (18.2) initial conditions are inconsistent unless $c_1 = -c_2$.

R.P. Agarwal, D. O'Regan, *Ordinary and Partial Differential Equations,*
Universitext, DOI 10.1007/978-0-387-79146-3_18,
© Springer Science+Business Media, LLC 2009

If we seek the solution of (18.1), (18.2) in the regular perturbation series form (17.4), then it leads to the system of first-order DEs

$$y_0'(x) + y_0(x) = 0, \quad y_0(0) = c_1, \quad y_0'(0) = c_2$$
$$y_n'(x) + y_n(x) = -(y_{n-1}''(x) + y_{n-1}'(x)), \quad y_n(0) = y_n'(0) = 0, \quad n \geq 1,$$

which can be solved only if the initial conditions are consistent, i.e., $c_1 = -c_2$. Further, in such a case it is easy to obtain $y_0(x) = c_1 e^{-x}$, $y_n(x) = 0$, $n \geq 1$; and hence (17.4) reduces to just $y(x) = c_1 e^{-x}$, which is indeed a solution of (18.1), (18.2).

Example 18.2. For the DE (18.1) together with the boundary conditions

$$y(0) = 0, \quad y(1) = 1 \tag{18.4}$$

an explicit solution can be written as

$$y(x) = \frac{e^{-x} - e^{-x/\epsilon}}{e^{-1} - e^{-1/\epsilon}}, \tag{18.5}$$

which has the following discontinuous behavior

$$\lim_{\epsilon \to 0^+} y(x) = \begin{cases} e^{1-x}, & x > 0 \\ 0, & x = 0. \end{cases}$$

Thus, it follows that

$$\lim_{x \to 0} \left(\lim_{\epsilon \to 0^+} y(x) \right) \neq \lim_{\epsilon \to 0^+} \left(\lim_{x \to 0} y(x) \right).$$

This is due to the fact that the first-order DE (18.3) obtained by substituting $\epsilon = 0$ in (18.1), together with the boundary conditions (18.4) cannot be solved. Hence, we cannot expect the solution of (18.1), (18.4) to have the regular perturbation series form (17.4).

In Figure 18.1 we graph the solution (18.5) for $\epsilon = 0.1$, 0.01 and 0.001, and note that $y(x)$ is slowly varying in the region $\epsilon \ll x \leq 1$. However, in the small interval $0 \leq x \leq O(\epsilon)$ it undergoes an abrupt and rapid change. This small interval is called a *boundary layer*. The boundary layer region is called the *inner region* and the region of slow variation of $y(x)$ is called the *outer region*.

Thus, as illustrated, singular perturbation problems are in general characterized by the nonanalytic dependence of the solution on ϵ. One of the ways of constructing a uniformly valid perturbation solution of such problems is to obtain straight forward solution (called an *outer expansion*) using the original variables, and to obtain a solution (called an *inner expansion*) describing sharp variations using magnifying scales. The outer expansion breaks down in the boundary layer region, whereas the inner expansion breaks down in the outer region. Finally, these two solutions are matched by a procedure known as the method of *inner* and *outer expansions*, or *the method of matched asymptotic expansions*. This technique leads to global approximations to the solutions of singular perturbation problems.

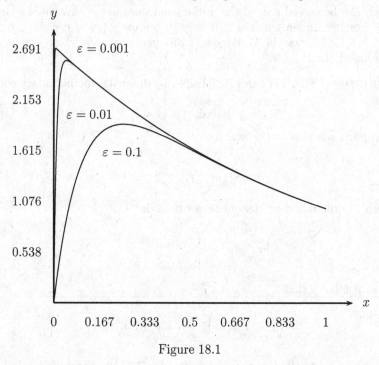

Figure 18.1

To appreciate this method, we reconsider the boundary value problem (18.1), (18.4). Its exact solution (18.5) consists of two parts: e^{-x}, a slowly varying function in $[0, 1]$, and $e^{-x/\epsilon}$, a function of rapid variations in the boundary layer region $0 \leq x \leq O(\epsilon)$. We need to introduce the notion of an *inner* and *outer limits of the solution*. The outer limit of the solution $y(x)$ denoted as $y_{out}(x)$ is obtained by prescribing a fixed x outside the boundary layer, i.e., $O(\epsilon) \ll x \leq 1$ and letting $\epsilon \to 0^+$. We therefore have

$$y_{out}(x) = \lim_{\epsilon \to 0^+} y(x) = e^{1-x}. \tag{18.6}$$

This $y_{out}(x)$ satisfies the first-order DE

$$y'_{out}(x) + y_{out}(x) = 0, \tag{18.7}$$

which is the formal outer limit of the DE (18.1). Since $y_{out}(x)$ satisfies the boundary condition $y(1) = 1$ but not $y(0) = 0$, it is not close to $y(x)$ near $x = 0$.

Next we consider the inner limit of the solution denoted by $y_{in}(x)$ in which $\epsilon \to 0^+$ in the boundary layer region $0 \le x \le O(\epsilon)$. To achieve this we magnify this layer using the stretching transformation $x = \epsilon t$. The variable t is called an inner variable, its introduction is advantageous in the sense that in the boundary layer region the solution given by (18.5) varies rapidly as a function of x but slowly as a function of t. From (18.5) it is clear that

$$y_{in}(x) = z_{in}(t) = \lim_{\epsilon \to 0^+} y(\epsilon t) = e - e^{1-t}. \tag{18.8}$$

Further, defining $y(x) \equiv z(t)$, under the transformation $x = \epsilon t$, (18.1) leads to the DE

$$\frac{1}{\epsilon}\frac{d^2 z}{dt^2} + \left(\frac{1}{\epsilon} + 1\right)\frac{dz}{dt} + z = 0. \tag{18.9}$$

Now for a given t, we let $\epsilon \to 0^+$ to obtain

$$\frac{d^2}{dt^2}z_{in}(t) + \frac{d}{dt}z_{in}(t) = 0. \tag{18.10}$$

The function $z_{in}(t)$ given in (18.8) not only satisfies the DE (18.10), but also $z_{in}(0) = 0$.

The next step is to match $z_{in}(t)$ and $y_{out}(x)$ asymptotically. This matching will take place on an overlapping region described by the intermediate limit $x \to 0$, $t = x/\epsilon \to \infty$, $\epsilon \to 0^+$. From (18.6) and (18.8), we have

$$\lim_{x \to 0} y_{out}(x) = e = \lim_{t \to \infty} z_{in}(t). \tag{18.11}$$

Satisfaction of (18.11) will ensure asymptotic matching. It also provides the second boundary condition $z_{in}(\infty) = e$ for the solution of (18.10) to satisfy. Observe here that although $x \in [0, 1]$, the matching region is unbounded.

We now seek a perturbation expansion of the outer solution of (18.1) in the form

$$y_{out}(x) = \sum_{n=0}^{\infty} \epsilon^n y_n(x) \tag{18.12}$$

satisfying the relevant boundary condition $y(1) = 1$. This leads to the infinite system of initial value problems

$$\begin{aligned} y_0'(x) + y_0(x) &= 0, \quad y_0(1) = 1 \\ y_n'(x) + y_n(x) &= -(y_{n-1}''(x) + y_{n-1}'(x)), \quad y_n(1) = 0, \quad n \ge 1, \end{aligned} \tag{18.13}$$

which can be solved to obtain $y_0(x) = e^{1-x}$, $y_n(x) = 0$, $n \ge 1$. Thus, $y_{out}(x) = e^{1-x}$. (Note that $y_{out}(x)$ in (18.6) is not the same as in (18.12), rather it is $y_0(x)$ in (18.12)).

In order to obtain $z_{in}(t)$, we assume that the DE (18.9) has the expansion

$$z_{in}(t) = \sum_{n=0}^{\infty} \epsilon^n z_n(t) \tag{18.14}$$

satisfying the other boundary condition $z(0) = 0$. This gives the infinite system of initial value problems

$$z_0''(t) + z_0'(t) = 0, \quad z_0(0) = 0$$
$$z_n''(t) + z_n'(t) = -(z_{n-1}'(t) + z_{n-1}(t)), \quad z_n(0) = 0, \quad n \geq 1. \tag{18.15}$$

The system (18.15) can be solved to find

$$z_0(t) = A_0(1 - e^{-t})$$
$$z_n(t) = \int_0^t (A_n e^{-s} - z_{n-1}(s))ds, \quad n \geq 1, \tag{18.16}_n$$

where A_n, $n \geq 0$ are constants of integration. To find these constants, we shall match $y_{out}(x)$ with $z_{in}(t)$. For this, we have

$$y_{out}(x) = e^{1-x} = e\left(1 - \epsilon t + \frac{\epsilon^2 t^2}{2!} - \frac{\epsilon^3 t^3}{3!} + \cdots\right). \tag{18.17}$$

Thus, on comparing $z_0(t)$ obtained in $(18.16)_0$ for large t $(\to \infty)$ with the constant term on the right of (18.17), we have $z_0(t) \sim A_0 = e$. Once $z_0(t)$ is known, we easily obtain $z_1(t) = (A_1 + A_0)(1 - e^{-t}) - et$. Hence, on comparing $z_1(t)$ for large t $(\to \infty)$ with the second term on the right of (18.17), we find $z_1(t) \sim (A_1 + A_0) - et = -et$, i.e., $A_1 + A_0 = 0$, or $A_1 = -A_0 = -e$, so that $z_1(t) = -et$. Proceeding in a similar fashion we arrive at $z_n(t) = (-1)^n(et^n/n!)$, and finally the inner expansion is

$$z_{in}(t) = e\sum_{n=0}^{\infty} \epsilon^n \frac{(-1)^n t^n}{n!} - e^{1-t} = e^{1-\epsilon t} - e^{1-t}. \tag{18.18}$$

Clearly, $z_{in}(t)$ is a valid asymptotic expansion not only for t in the boundary layer region $0 \leq t \leq O(1)$, but also for large t $(t = O(\epsilon^{-\alpha})$, $0 < \alpha < 1)$, while $y_{out}(x)$ is valid for $\epsilon \ll x \leq 1$, and not for $x = O(\epsilon)$ since it does not satisfy the boundary condition $y(0) = 0$. Further, from the above construction we are able to match inner and outer expansions asymptotically in the region $\epsilon \ll x \leq 1$ to all orders in powers of ϵ.

Finally, to construct a uniform approximation $y_{unif}(x)$ of $y(x)$ we may take

$$y_{unif}(x) = y_{in}(x) + y_{out}(x) - y_{match}(x), \tag{18.19}$$

where $y_{match}(x)$ approximates $y(x)$ in the matching region. If we compute $y_{in}(x)$, $y_{out}(x)$ and $y_{match}(x)$ up to nth-order, then $|y(x) - y_{unif}(x)| =$

$O(\epsilon^{n+1})$ ($\epsilon \rightarrow 0^+$, $0 \leq x \leq 1$). Since for the boundary value problem (18.1), (18.4) we can compute $y_{out}(x) = e^{1-x}$, $y_{in}(x) = e^{1-x} - e^{1-t}$ and $y_{match}(x) = e^{1-x}$ it follows that

$$y_{unif}(x) = e^{1-x} - e^{1-x/\epsilon} \qquad (18.20)$$

is an infinite order approximation to $y(x)$.

It is to be noted that $y_{unif}(x) \neq y(x)$, i.e. $y_{unif}(x)$ is only asymptotic to $y(x)$ and one should not expect $y_{unif}(x) \rightarrow y(x)$ as $n \rightarrow \infty$.

We remark that the DE (18.1) is sufficiently simple so that we could perform the preceding perturbation analysis and could obtain uniformly valid approximations to the solution of the problem (18.1), (18.4). However, in general, straightforward computations of $y_{in}(x)$ and $y_{out}(x)$ are practically impossible. Matching criteria are in general very complicated and there is no a priori reason to believe the existence of an overlapping region where both outer and inner expansions remain valid. Finally, to obtain $y_{unif}(x)$ one needs to make necessary modifications, at times one may have to have composite, (e.g., multiplicative) expansions.

A careful analysis of the method of matched asymptotic expansions is too complicated to be included in this elementary discussion on singular perturbations.

Problems

18.1.   For the initial value problem

$$(x + \epsilon y)y' + y = 0, \quad y(1) = 1$$

show that

(i) $y(x) = \dfrac{1}{x} + \epsilon \dfrac{x^2 - 1}{2x^3} - \epsilon^2 \dfrac{x^2 - 1}{2x^5} + O(\epsilon^3)$

(ii)   the exact solution is

$$y(x) = -\frac{x}{\epsilon} + \left(\frac{x^2}{\epsilon^2} + \frac{2}{\epsilon} + 1 \right)^{1/2}$$

(iii) $y(x) = 1 + \dfrac{1}{\epsilon}(1 - x) + \dfrac{1}{\epsilon^2}\dfrac{1}{2}(x^2 - 1) + O\left(\dfrac{1}{\epsilon^3}\right)$, which is the same as the expanded version of the exact solution.

18.2.   For the initial value problem

$$(x + \epsilon y)y' - \frac{1}{2}y = 1 + x^2, \quad y(1) = 1$$

show that

$$y(x) = \frac{1}{3}(2x^2 + 7x^{1/2} - 6) - \epsilon\frac{1}{45}\left(16x^3 - 175x^{3/2} - 240x + 539x^{1/2}\right.$$
$$\left. + 105x^{-1/2} - 245\right) + O(\epsilon^2).$$

18.3. For the initial value problem

$$\frac{d\theta}{d\tau} + \theta^{1+\epsilon} = 0, \quad \theta(0) = 1$$

which occurs in cooling of a lumped system, show that

(i)   the exact solution is $\theta(\tau) = (1 + \epsilon\tau)^{-1/\epsilon}$, $\epsilon \neq 0$

(ii) $\theta(\tau) = e^{-\tau} + \epsilon\frac{1}{2}\tau^2 e^{-\tau} + \epsilon^2\frac{1}{3}\left(\frac{3}{8}\tau^4 - \tau^3\right)e^{-\tau} + O(\epsilon^3).$

18.4. For the boundary value problem

$$\epsilon y'' + y' + y = 0, \quad y(0) = 0, \quad y(1) = 1$$

show that

$$y(x) = e^{(1-x)/2\epsilon}\frac{e^{x\sqrt{1-4\epsilon}/2\epsilon} - e^{-x\sqrt{1-4\epsilon}/2\epsilon}}{e^{\sqrt{1-4\epsilon}/2\epsilon} - e^{-\sqrt{1-4\epsilon}/2\epsilon}}$$
$$\longrightarrow \quad e^{1-x} - e^{1+x-x/\epsilon} \quad \text{as} \quad \epsilon \to 0.$$

18.5. For the boundary value problem

$$y'' - \frac{2}{x^2}y + \frac{2}{x^2} = 0$$
$$y(\epsilon) = 0, \quad y(1) = 1$$

show that

(i)   the exact solution is

$$y(x) = \frac{\epsilon}{1 - \epsilon^3}(x^2 - x^{-1}) + 1$$

(ii) the constant function $z(x) = 1$ satisfies both the DE and the right boundary condition

(iii) $\lim_{\epsilon \to 0} y(x) = z(x)$, $0 < x \leq 1$ and hence the solution has a boundary layer near $x = 0$.

Lecture 19
Sturm–Liouville Problems

In Lecture 14 we have seen that homogeneous boundary value problem (14.3), (14.4) may have nontrivial solutions. If the coefficients of the DE and/or of the boundary conditions depend on a parameter, then one of the pioneer problems of mathematical physics is to determine the value(s) of the parameter for which such nontrivial solutions exist. In this lecture we shall explain some of the essential ideas involved in this vast field, which is continuously growing.

A boundary value problem which consists of the DE

$$(p(x)y')' + q(x)y + \lambda r(x)y = \mathcal{P}[y] + \lambda r(x)y = 0, \quad x \in J = [\alpha, \beta] \quad (19.1)$$

and the boundary conditions

$$\begin{aligned} a_0 y(\alpha) + a_1 y'(\alpha) = 0, & \quad a_0^2 + a_1^2 > 0 \\ d_0 y(\beta) + d_1 y'(\beta) = 0, & \quad d_0^2 + d_1^2 > 0 \end{aligned} \quad (19.2)$$

is called a *Sturm–Liouville problem*. In the DE (19.1), λ is a parameter, and the functions q, $r \in C(J)$, $p \in C^1(J)$ and $p(x) > 0$, $r(x) > 0$ in J.

The problem (19.1), (19.2) satisfying the above conditions is said to be a *regular* Sturm–Liouville problem. Clearly, $y(x) \equiv 0$ is always a solution of (19.1), (19.2). Solving such a problem means finding values of λ called *eigenvalues* and the corresponding nontrivial solutions $\phi_\lambda(x)$ known as *eigenfunctions*. The set of all eigenvalues of a regular problem is called its *spectrum*.

The computation of eigenvalues and eigenfunctions is illustrated in the following examples.

Example 19.1. Consider the boundary value problem

$$y'' + \lambda y = 0 \quad (19.3)$$

$$y(0) = y(\pi) = 0. \quad (19.4)$$

If $\lambda = 0$, then the general solution of (19.3) (reduced to $y'' = 0$) is $y(x) = c_1 + c_2 x$ and this solution satisfies the boundary conditions (19.4) if and only if $c_1 = c_2 = 0$, i.e., $y(x) \equiv 0$ is the only solution of (19.3), (19.4). Hence, $\lambda = 0$ is not an eigenvalue of the problem (19.3), (19.4).

R.P. Agarwal, D. O'Regan, *Ordinary and Partial Differential Equations*,
Universitext, DOI 10.1007/978-0-387-79146-3_19,
© Springer Science+Business Media, LLC 2009

If $\lambda \neq 0$, it is convenient to replace λ by μ^2, where μ is a new parameter and not necessarily real. In this case the general solution of (19.3) is $y(x) = c_1 e^{i\mu x} + c_2 e^{-i\mu x}$, and this solution satisfies the boundary conditions (19.4) if and only if

$$c_1 + c_2 = 0$$
$$c_1 e^{i\mu\pi} + c_2 e^{-i\mu\pi} = 0. \tag{19.5}$$

The system (19.5) has a nontrivial solution if and only if

$$e^{-i\mu\pi} - e^{i\mu\pi} = 0. \tag{19.6}$$

If $\mu = a + ib$, where a and b are real, condition (19.6) reduces to

$$e^{b\pi}(\cos a\pi - i\sin a\pi) - e^{-b\pi}(\cos a\pi + i\sin a\pi) = 0$$

or

$$(e^{b\pi} - e^{-b\pi})\cos a\pi - i(e^{b\pi} + e^{-b\pi})\sin a\pi = 0$$

or

$$2\sinh b\pi \cos a\pi - 2i\cosh b\pi \sin a\pi = 0$$

or

$$\sinh b\pi \cos a\pi = 0 \tag{19.7}$$

and

$$\cosh b\pi \sin a\pi = 0. \tag{19.8}$$

Since $\cosh b\pi > 0$ for all values of b, equation (19.8) requires that $a = n$, where n is an integer. Further, for this choice of a, $\cos a\pi \neq 0$ and equation (19.7) reduces to $\sinh b\pi = 0$, i.e., $b = 0$. However, if $b = 0$, then we cannot have $a = 0$, because then $\mu = 0$, and we have seen it is not an eigenvalue. Hence, $\mu = n$, where n is a nonzero integer. Thus, the eigenvalues of (19.3), (19.4) are $\lambda_n = \mu^2 = n^2$, $n = 1, 2, \cdots$. Further, from (19.5) since $c_2 = -c_1$ for $\lambda_n = n^2$ the corresponding nontrivial solutions of the problem (19.3), (19.4) are

$$\phi_n(x) = c_1(e^{inx} - e^{-inx}) = 2ic_1 \sin nx,$$

or simply $\phi_n(x) = \sin nx$.

Example 19.2. Consider again the DE (19.3) but with the boundary conditions

$$y(0) + y'(0) = 0, \quad y(1) = 0. \tag{19.9}$$

If $\lambda = 0$, then the general solution $y(x) = c_1 + c_2 x$ of (19.3) also satisfies the boundary conditions (19.9) if and only if $c_1 + c_2 = 0$, i.e., $c_2 = -c_1$. Hence, $\lambda = 0$ is an eigenvalue of (19.3), (19.9) and the corresponding eigenfunction is $\phi_0(x) = 1 - x$.

If $\lambda \neq 0$, then once again we replace λ by μ^2 and note that the general solution $y(x) = c_1 e^{i\mu x} + c_2 e^{-i\mu x}$ of (19.3) satisfies the boundary conditions (19.9) if and only if

$$(c_1 + c_2) + i\mu(c_1 - c_2) = 0$$
$$c_1 e^{i\mu} + c_2 e^{-i\mu} = 0. \tag{19.10}$$

The system (19.10) has a nontrivial solution if and only if

$$(1 + i\mu)e^{-i\mu} - (1 - i\mu)e^{i\mu} = 0,$$

which is equivalent to

$$\tan\mu = \mu. \tag{19.11}$$

To find the real roots of (19.11) we graph the curves $y = \mu$ and $y = \tan\mu$ and observe the values of μ where these curves intersect.

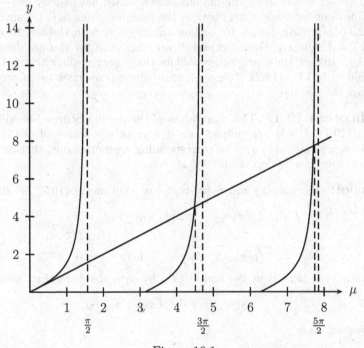

Figure 19.1

From Figure 19.1 it is clear that the equation (19.11) has an infinite number of positive roots μ_n, $n = 1, 2, \cdots$ which are approaching the odd multiples of $\pi/2$, i.e., $\mu_n \simeq (2n+1)\pi/2$. Further, since the equation (19.11) remains unchanged if μ is replaced by $-\mu$, we find that the only nonzero real roots of (19.11) are $\mu_n \simeq \pm(2n+1)\pi/2$, $n = 1, 2, \cdots$.

Thus, the problem (19.3), (19.9) also has an infinite number of eigenvalues, $\lambda_0 = 0$, $\lambda_n \simeq (2n+1)^2\pi^2/4$, $n = 1, 2, \cdots$. Further, from (19.10), since $c_2 = -c_1 e^{2i\mu}$ for these λ_n, $n \geq 1$ the corresponding nontrivial solutions of the problem (19.3), (19.9) are

$$y(x) = c_1 e^{i\sqrt{\lambda_n}x} - c_1 e^{-i\sqrt{\lambda_n}x} e^{2i\sqrt{\lambda_n}} = -2c_1 e^{i\sqrt{\lambda_n}} \sin\sqrt{\lambda_n}(1-x).$$

Hence, the eigenfunctions of (19.3), (19.9) are

$$\phi_0(x) = 1 - x$$
$$\phi_n(x) = \sin\sqrt{\lambda_n}(1-x), \quad n = 1, 2, \cdots.$$

From Example 19.1 it is clear that the problem (19.3), (19.4) has an infinite number of real eigenvalues λ_n, which can be arranged as a monotonic increasing sequence $\lambda_1 < \lambda_2 < \cdots$ such that $\lambda_n \to \infty$ as $n \to \infty$. Also, corresponding to each eigenvalue λ_n of (19.3), (19.4) there exists a one-parameter family of eigenfunctions $\phi_n(x)$, which has exactly $(n-1)$ zeros in the open interval $(0, \pi)$. Further, the eigenfunctions $\phi_n(x) = \sin nx$, $n = 1, 2, \cdots$ of (19.3), (19.4) are orthogonal in $(0, \pi)$ with the weight function $r(x) = 1$. Clearly, these properties are also valid for the problem (19.3), (19.9). In fact, these properties hold for the general regular Sturm–Liouville problem (19.1), (19.2). We shall state these properties as theorems and prove the results.

Theorem 19.1. The eigenvalues of the regular Sturm–Liouville problem (19.1), (19.2) are simple; i.e., if λ is an eigenvalue of (19.1), (19.2) and $\phi_1(x)$ and $\phi_2(x)$ are the corresponding eigenfunctions, then $\phi_1(x)$ and $\phi_2(x)$ are linearly dependent.

Proof. Since $\phi_1(x)$ and $\phi_2(x)$ both are solutions of (19.1), we have

$$(p(x)\phi_1')' + q(x)\phi_1 + \lambda r(x)\phi_1 = 0 \tag{19.12}$$

and

$$(p(x)\phi_2')' + q(x)\phi_2 + \lambda r(x)\phi_2 = 0. \tag{19.13}$$

Multiplying (19.12) by ϕ_2, and (19.13) by ϕ_1 and subtracting, we get

$$\phi_2(p(x)\phi_1')' - (p(x)\phi_2')'\phi_1 = 0. \tag{19.14}$$

However, since

$$[\phi_2(p(x)\phi_1') - (p(x)\phi_2')\phi_1]'$$
$$= \phi_2(p(x)\phi_1')' + \phi_2'(p(x)\phi_1') - (p(x)\phi_2')'\phi_1 - (p(x)\phi_2')\phi_1'$$
$$= \phi_2(p(x)\phi_1')' - (p(x)\phi_2')'\phi_1$$

from (19.14) it follows that

$$[\phi_2(p(x)\phi_1') - (p(x)\phi_2')\phi_1]' = 0$$

and hence

$$p(x)[\phi_2\phi_1' - \phi_2'\phi_1] = \text{constant} = C. \tag{19.15}$$

To find the value of C, we note that ϕ_1 and ϕ_2 satisfy the boundary conditions, and hence

$$a_0\phi_1(\alpha) + a_1\phi_1'(\alpha) = 0$$
$$a_0\phi_2(\alpha) + a_2\phi_2'(\alpha) = 0,$$

which implies $\phi_1(\alpha)\phi_2'(\alpha) - \phi_2(\alpha)\phi_1'(\alpha) = 0$. Thus, from (19.15) it follows that

$$p(x)[\phi_2\phi_1' - \phi_2'\phi_1] = 0 \quad \text{for all} \quad x \in [\alpha, \beta].$$

Since $p(x) > 0$, we must have $\phi_2\phi_1' - \phi_2'\phi_1 = 0$ for all $x \in [\alpha, \beta]$. But, this means that ϕ_1 and ϕ_2 are linearly dependent. ∎

Theorem 19.2. Let λ_n, $n = 1, 2, \cdots$ be the eigenvalues of the regular Sturm–Liouville problem (19.1), (19.2) and $\phi_n(x)$, $n = 1, 2, \cdots$ be the corresponding eigenfunctions. Then, the set $\{\phi_n(x) : n = 1, 2, \cdots\}$ is orthogonal in $[\alpha, \beta]$ with respect to the weight function $r(x)$.

Proof. Let λ_k and λ_ℓ, $(k \neq \ell)$ be eigenvalues, and $\phi_k(x)$ and $\phi_\ell(x)$ be the corresponding eigenfunctions of (19.1), (19.2). Since $\phi_k(x)$ and $\phi_\ell(x)$ are solutions of (19.1), we have

$$(p(x)\phi_k')' + q(x)\phi_k + \lambda_k r(x)\phi_k = 0 \tag{19.16}$$

and

$$(p(x)\phi_\ell')' + q(x)\phi_\ell + \lambda_\ell r(x)\phi_\ell = 0. \tag{19.17}$$

Now following the argument in Theorem 19.1, we get

$$[\phi_\ell(p(x)\phi_k') - (p(x)\phi_\ell')\phi_k]' + (\lambda_k - \lambda_\ell)r(x)\phi_k\phi_\ell = 0,$$

which on integration gives

$$(\lambda_\ell - \lambda_k) \int_\alpha^\beta r(x)\phi_k(x)\phi_\ell(x)dx = p(x)[\phi_\ell(x)\phi_k'(x) - \phi_\ell'(x)\phi_k(x)]\Big|_\alpha^\beta. \tag{19.18}$$

Next since $\phi_k(x)$ and $\phi_\ell(x)$ satisfy the boundary conditions (19.2), i.e.,

$$a_0\phi_k(\alpha) + a_1\phi_k'(\alpha) = 0, \qquad d_0\phi_k(\beta) + d_1\phi_k'(\beta) = 0$$
$$a_0\phi_\ell(\alpha) + a_1\phi_\ell'(\alpha) = 0, \qquad d_0\phi_\ell(\beta) + d_1\phi_\ell'(\beta) = 0$$

it is necessary that

$$\phi_k(\alpha)\phi_\ell'(\alpha) - \phi_k'(\alpha)\phi_\ell(\alpha) = \phi_k(\beta)\phi_\ell'(\beta) - \phi_k'(\beta)\phi_\ell(\beta) = 0.$$

Hence, the identity (19.18) reduces to

$$(\lambda_\ell - \lambda_k) \int_\alpha^\beta r(x)\phi_k(x)\phi_\ell(x)dx = 0. \tag{19.19}$$

However, since $\lambda_\ell \neq \lambda_k$, it follows that $\int_\alpha^\beta r(x)\phi_k(x)\phi_\ell(x)dx = 0$. ∎

Corollary 19.3.   Let λ_1 and λ_2 be two eigenvalues of the regular Sturm–Liouville problem (19.1), (19.2) and $\phi_1(x)$ and $\phi_2(x)$ be the corresponding eigenfunctions. Then, $\phi_1(x)$ and $\phi_2(x)$ are linearly dependent if and only if $\lambda_1 = \lambda_2$.

Proof.   The proof is a direct consequence of equality (19.19).   ∎

Theorem 19.4.   For the regular Sturm–Liouville problem (19.1), (19.2) the eigenvalues are real.

Proof.   Let $\lambda = a + ib$ be a complex eigenvalue and $\phi(x) = \mu(x) + i\nu(x)$ be the corresponding eigenfunction. Then, we have

$$(p(x)(\mu + i\nu)')' + q(x)(\mu + i\nu) + (a + ib)r(x)(\mu + i\nu) = 0$$

and hence

$$(p(x)\mu')' + q(x)\mu + (a\mu - b\nu)r(x) = 0$$

and

$$(p(x)\nu')' + q(x)\nu + (b\mu + a\nu)r(x) = 0.$$

Now following exactly the same argument as in Theorem 19.1, we get

$$0 = p(x)(\nu\mu' - \nu'\mu)\Big|_\alpha^\beta = \int_\alpha^\beta [-(a\mu - b\nu)\nu r(x) + (b\mu + a\nu)\mu r(x)]dx$$

$$= b\int_\alpha^\beta r(x)(\nu^2(x) + \mu^2(x))dx.$$

Hence, it is necessary that $b = 0$, i.e., λ is real.   ∎

Since (19.3), (19.9) is a regular Sturm–Liouville problem, from Theorem 19.4 it is immediate that the equation (19.11) has only real roots.

In the above results we have established several properties of the eigenvalues and eigenfunctions of the regular Sturm–Liouville problem (19.1), (19.2). In all these results the existence of eigenvalues is tacitly assumed. We now state the following very important result whose proof involves some advanced arguments.

Theorem 19.5.   For the regular Sturm–Liouville problem (19.1), (19.2) there exists an infinite number of eigenvalues λ_n, $n = 1, 2, \cdots$. These eigenvalues can be arranged as a monotonically increasing sequence $\lambda_1 < \lambda_2 < \cdots$ such that $\lambda_n \to \infty$ as $n \to \infty$. Further, eigenfunction $\phi_n(x)$ corresponding to the eigenvalue λ_n has exactly $(n - 1)$ zeros in the open interval (α, β).

The following examples show that the above properties for singular Sturm–Liouville problems do not always hold.

Example 19.3. For the singular Sturm–Liouville problem (19.3),

$$y(0) = 0, \quad |y(x)| \leq M < \infty \quad \text{for all} \quad x \in (0, \infty)$$

each $\lambda \in (0, \infty)$ is an eigenvalue and $\sin \sqrt{\lambda} x$ is the corresponding eigenfunction. Thus, in comparison with the regular problems where the spectrum is always discrete, the singular problems may have a continuous spectrum.

Example 19.4. Consider the singular Sturm–Liouville problem (19.3),

$$y(-\pi) = y(\pi), \quad y'(-\pi) = y'(\pi). \tag{19.20}$$

This problem has eigenvalues $\lambda_0 = 0$, $\lambda_n = n^2$, $n = 1, 2, \cdots$. The eigenvalue $\lambda_0 = 0$ is simple and 1 is the corresponding eigenfunction. The eigenvalue $\lambda_n = n^2$, $n \geq 1$ is not simple and two independent eigenfunctions are $\sin nx$ and $\cos nx$. Thus, in contrast with regular problems where the eigenvalues are simple, there may be multiple eigenvalues for singular problems.

Finally, we remark that the properties of the eigenvalues and eigenfunctions of regular Sturm–Liouville problems can be extended under appropriate assumptions to singular problems also in which the function $p(x)$ is zero at α or β, or both, but remains positive in (α, β). Some examples of this type are given in Problems 19.13 – 19.18.

Problems

19.1. The deflection y of a uniform column of length a under a constant axial load p is governed by the boundary value problem

$$EI \frac{d^2 y}{dx^2} + py = 0, \quad y(0) = y(a) = 0$$

here E is Young's modulus, and I is the moment of inertia of the column. Find the values of p for which this problem has nontrivial solutions. The smallest such value of p is the upper limit for the stability of the undeflected equilibrium position of the column. (This problem with different terminology is the same as (15.4)).

19.2. Find the eigenvalues and eigenfunctions of the problem (19.3) with the boundary conditions

(i) $y(0) = 0$, $y'(\beta) = 0$

(ii) $y'(0) = 0$, $y(\beta) = 0$

(iii) $y'(0) = 0$, $y'(\beta) = 0$

(iv) $y(0) = 0$, $y(\beta) + y'(\beta) = 0$

(v) $y(0) - y'(0) = 0$, $y'(\beta) = 0$

(vi) $y(0) - y'(0) = 0$, $y(\beta) + y'(\beta) = 0$.

19.3. Find the eigenvalues and eigenfunctions of each of the following Sturm–Liouville problems:

(i) $y'' + \lambda y = 0$, $y(0) = y(\pi/2) = 0$

(ii) $y'' + (1 + \lambda)y = 0$, $y(0) = y(\pi) = 0$

(iii) $y'' + 2y' + (1 - \lambda)y = 0$, $y(0) = y(1) = 0$

(iv) $(x^2 y')' + \lambda x^{-2} y = 0$, $y(1) = y(2) = 0$

(v) $x^2 y'' + xy' + (\lambda x^2 - (1/4))y = 0$, $y(\pi/2) = y(3\pi/2) = 0$

(vi) $((x^2 + 1)y')' + \lambda(x^2 + 1)^{-1}y = 0$, $y(0) = y(1) = 0$.

19.4. Find the eigenvalues and eigenfunctions of the boundary value problem

$$y'' + \lambda y = 0, \quad y'(0) = 0, \quad hy(\beta) + k\beta y'(\beta) = 0.$$

Further, show that the set of all its eigenfunctions is orthogonal on $[0, \beta]$.

19.5. Consider the boundary value problem

$$x^2 y'' + xy' + \lambda y = 0, \quad 1 < x < e$$
$$y(1) = 0, \quad y(e) = 0. \tag{19.21}$$

(i) Show that (19.21) is equivalent to the Sturm–Liouville problem

$$(xy')' + \frac{\lambda}{x}y = 0, \quad 1 < x < e$$
$$y(1) = 0, \quad y(e) = 0. \tag{19.22}$$

(ii) Verify that for (19.22) the eigenvalues are $\lambda_n = n^2\pi^2$, $n = 1, 2, \cdots$ and the corresponding eigenfunctions are $\phi_n(x) = \sin(n\pi \ln x)$.

(iii) Show that

$$\int_1^e \frac{1}{x}\phi_m(x)\phi_n(x)dx = \begin{cases} 0, & m \neq n \\ 1/2, & m = n. \end{cases}$$

19.6. Verify that for the Sturm–Liouville problem

$$(xy')' + \frac{\lambda}{x}y = 0, \quad 1 < x < e^{2\pi}$$
$$y'(1) = 0, \quad y'(e^{2\pi}) = 0$$

the eigenvalues are $\lambda_n = n^2/4$, $n = 0, 1, \cdots$ and the corresponding eigenfunctions are $\phi_n(x) = \cos\left(\frac{n \ln x}{2}\right)$. Show that

$$\int_1^{e^{2\pi}} \frac{1}{x}\phi_m(x)\phi_n(x)dx = 0, \quad m \neq n.$$

19.7.   Consider the DE in Problem 6.4 with $n = 4$, i.e.,

$$x^4 y'' + k^2 y = 0. \tag{19.23}$$

(i)   Verify that the general solution of (19.23) is

$$y(x) = x \left(A \cos \frac{k}{x} + B \sin \frac{k}{x} \right).$$

(ii)   Find the eigenvalues and eigenfunctions of the Sturm–Liouville problem (19.23), $y(\alpha) = y(\beta) = 0$, $0 < \alpha < \beta$.

19.8.   Show that the problem

$$y'' - 4\lambda y' + 4\lambda^2 y = 0, \quad y(0) = 0, \quad y(1) + y'(1) = 0$$

has only one eigenvalue, and find the corresponding eigenfunction.

19.9.   Show that for the singular Sturm–Liouville problem (19.1), (14.9) with $p(\alpha) = p(\beta)$ eigenfunctions corresponding to different eigenvalues are orthogonal in $[\alpha, \beta]$ with respect to the weight function $r(x)$.

19.10.   Solve the following singular Sturm–Liouville problems:

(i) $y'' + \lambda y = 0, \quad y'(0) = 0, \quad |y(x)| < \infty \quad$ for all $\quad x \in (0, \infty)$

(ii) $y'' + \lambda y = 0, \quad |y(x)| < \infty \quad$ for all $\quad x \in (-\infty, \infty)$.

19.11.   Find the eigenvalues and eigenfunctions of the problem

$$y^{(4)} - \lambda y = 0 \tag{19.24}$$

with the boundary conditions

(i) $y(0) = y''(0) = y(\beta) = y''(\beta) = 0$ $\hspace{3.5cm}$ (19.25)

(ii) $y(0) = y'(0) = y''(\beta) = y'''(\beta) = 0$ $\hspace{3cm}$ (19.26)

(iii) $y(0) = y'(0) = y(\beta) = y'(\beta) = 0.$ $\hspace{3.3cm}$ (19.27)

19.12.   Find the eigenvalues and eigenfunctions of problem

$$y^{(4)} + \lambda y'' = 0 \tag{19.28}$$

with the boundary conditions (19.25). Further, show that (19.28), (19.26) is not an eigenvalue problem.

19.13.   Consider the singular Sturm–Liouville problem

$$(1 - x^2) y'' - 2xy' + \lambda y = ((1 - x^2) y')' + \lambda y = 0 \tag{19.29}$$

$$\lim_{x \to -1} y(x) < \infty, \quad \lim_{x \to 1} y(x) < \infty. \tag{19.30}$$

Show that the eigenvalues of this problem are $\lambda_n = n(n+1),\cdot n = 0, 1, 2, \cdots$ and the corresponding eigenfunctions are the Legendre polynomials $P_n(x)$.

19.14. Consider the singular Sturm–Liouville problem (19.29),

$$y'(0) = 0, \quad \lim_{x \to 1} y(x) < \infty. \tag{19.31}$$

Show that the eigenvalues of this problem are $\lambda_n = 2n(2n + 1)$, $n = 0, 1, 2, \cdots$ and the corresponding eigenfunctions are the even Legendre polynomials $P_{2n}(x)$.

19.15. Consider the singular Sturm–Liouville problem (19.29),

$$y(0) = 0, \quad \lim_{x \to 1} y(x) < \infty. \tag{19.32}$$

Show that the eigenvalues of this problem are $\lambda_n = (2n + 1)(2n + 2)$, $n = 0, 1, 2, \cdots$ and the corresponding eigenfunctions are the odd Legendre polynomials $P_{2n+1}(x)$.

19.16. Consider the singular Sturm–Liouville problem

$$y'' - 2xy' + \lambda y = 0 = \left(e^{-x^2}y'\right)' + \lambda e^{-x^2}y \tag{19.33}$$

$$\lim_{x \to -\infty} \frac{y(x)}{|x|^k} < \infty, \quad \lim_{x \to \infty} \frac{y(x)}{x^k} < \infty \quad \text{for some positive integer} \quad k. \tag{19.34}$$

Show that the eigenvalues of this problem are $\lambda_n = 2n$, $n = 0, 1, 2, \cdots$ and the corresponding eigenfunctions are the Hermite polynomials $H_n(x)$.

19.17. Consider the singular Sturm–Liouville problem

$$xy'' + (1 - x)y' + \lambda y = 0 = \left(xe^{-x}y'\right)' + \lambda e^{-x}y \tag{19.35}$$

$$\lim_{x \to 0} |y(x)| < \infty, \quad \lim_{x \to \infty} \frac{y(x)}{x^k} < \infty \quad \text{for some positive integer} \quad k. \tag{19.36}$$

Show that the eigenvalues of this problem are $\lambda_n = n$, $n = 0, 1, 2, \cdots$ and the corresponding eigenfunctions are the Laguerre polynomials $L_n(x)$.

19.18. Let $a \geq 0$ be fixed, and b_n, $n = 0, 1, 2, \cdots$ be the zeros of the Bessel function $J_a(x)$. Show that the singular Sturm–Liouville problem

$$x^2 y'' + xy' + (\lambda x^2 - a^2)y = 0 = (xy')' + \left(\lambda x - \frac{a^2}{x}\right)y \tag{19.37}$$

$$\lim_{x \to 0} y(x) < \infty, \quad y(1) = 0 \tag{19.38}$$

has the eigenvalues $\lambda_n = b_n^2$, $n = 0, 1, 2, \cdots$ and the corresponding eigenfunctions are $J_a(b_n x)$.

19.19. Let λ_k be an eigenvalue of the regular Sturm–Liouville problem (19.1), (19.2) and $\phi_k(x)$ be the corresponding eigenfunction. Show that

$$\lambda_k = \frac{-p(x)\phi_k(x)\phi_k'(x)|_\alpha^\beta + \int_\alpha^\beta \left[p(x)(\phi_k'(x))^2 - q(x)\phi_k^2(x) \right] dx}{\int_\alpha^\beta r(x)\phi_k^2(x)dx}.$$

This expression is called *Rayleigh quotient*. From this quotient it follows that

(i) $\lambda_k \geq 0$ if $-p(x)\phi_k(x)\phi_k'(x)|_\alpha^\beta \geq 0$ and $q(x) \leq 0$, $x \in [\alpha, \beta]$

(ii) the minimum value of the Rayleigh quotient over all continuous functions satisfying the conditions (19.2) is the smallest eigenvalue λ_1.

Answers or Hints

19.1. $n^2\pi^2 EI/a^2$, $n = 1, 2, \cdots$.

19.2. (i) $\left(\frac{2n-1}{2\beta}\right)^2 \pi^2$, $\sin\left(\frac{2n-1}{2\beta}\right)\pi x$ (ii) $\left(\frac{2n-1}{2\beta}\right)^2 \pi^2$, $\cos\left(\frac{2n-1}{2\beta}\right)\pi x$
(iii) $\left(\frac{n-1}{\beta}\right)^2 \pi^2$, $\cos\left(\frac{n-1}{\beta}\right)\pi x$ (iv) λ_n^2, where $\lambda = \lambda_n$ is a solution of $\tan \lambda\beta + \lambda = 0$, $\sin \lambda_n x$ (v) λ_n^2, where $\lambda = \lambda_n$ is a solution of $\cot \lambda\beta = \lambda$, $\sin \lambda_n x + \lambda_n \cos \lambda_n x$ (vi) λ_n^2, where $\lambda = \lambda_n$ is a solution of $\tan \lambda\beta = 2\lambda/(\lambda^2 - 1)$, $\sin \lambda_n x + \lambda_n \cos \lambda_n x$.

19.3. (i) $4n^2$, $\sin 2nx$ (ii) $n^2 - 1$, $\sin nx$ (iii) $-n^2\pi^2$, $e^{-x} \sin n\pi x$ (iv) $4n^2 \times \pi^2$, $\sin 2n\pi \left(1 - \frac{1}{x}\right)$ (v) n^2, $\frac{1}{\sqrt{x}} \sin n\left(x - \frac{\pi}{2}\right)$ (vi) $16n^2$, $\sin(4n \tan^{-1} x)$.

19.4. $\lambda_n = \alpha_n^2/\beta^2$ where α_n is a root of $\alpha \tan \alpha = h/k$, $\cos \frac{\alpha_n x}{\beta}$.

19.5. Verify directly.

19.6. Verify directly.

19.7. (i) Verify directly (ii) $k_n = \frac{n\pi\alpha\beta}{\beta-\alpha}$, $x \sin\left[\frac{n\pi\beta(x-\alpha)}{x(\beta-\alpha)}\right]$.

19.8. -1, xe^{-2x}.

19.9. Use (19.18).

19.10. (i) $\lambda \geq 0$, $\phi(x) = \cos\sqrt{\lambda}x$ (ii) $\lambda \geq 0$, $\phi(x) = c_1 \cos\sqrt{\lambda}x + c_2 \sin\sqrt{\lambda}x$.

19.11.(i) $\lambda_n = \frac{n^4\pi^4}{\beta^4}$, $\phi_n(x) = \sin\frac{n\pi x}{\beta}$

(ii) $\lambda_n = (\mu_n/\beta)^4$ where μ_n is the n–th root of the equation $\cosh\mu\cos\mu + 1 = 0$, $\left[\frac{\cosh(\mu_n x/\beta)-\cos(\mu_n x/\beta)}{\cosh\mu_n+\cos\mu_n} - \frac{\sinh(\mu_n x/\beta)-\sin(\mu_n x/\beta)}{\sinh\mu_n+\sin\mu_n}\right]$

(iii) $\lambda_n = (\mu_n/\beta)^4$ where μ_n is the n–th root of the equation $\cosh\mu\cos\mu - 1 = 0$, $\left[\frac{\cos(\mu_n x/\beta)-\cosh(\mu_n x/\beta)}{\cos\mu_n-\cosh\mu_n} - \frac{\sin(\mu_n x/\beta)-\sinh(\mu_n x/\beta)}{\sin\mu_n-\sinh\mu_n}\right]$.

19.12. $\lambda_n = \frac{n^2\pi^2}{\beta^2}$, $\phi_n(x) = \sin\frac{n\pi x}{\beta}$.

19.13. See Lecture 7.

19.14. See Lecture 7.

19.15. See Lecture 7.

19.16. See Lecture 8.

19.17. See Lecture 8.

19.18. See Lectures 9 and 13.

19.19. Multiply (19.16) by $\phi_k(x)$ and integrate over $[\alpha, \beta]$.

Lecture 20
Eigenfunction Expansions

Often we need to expand a given function in terms of other functions with specified exactness so as to be able to compute it in practice. In this and the next lecture we shall show that the sets of orthogonal polynomials and functions we have provided in earlier lectures can be used effectively as the basis in the expansions of general functions.

The basis $\{e^1, \cdots, e^n\}$ (e^k is the unit vector) of $\mathbb{R}^n$ has an important characteristic—namely, for every $u \in \mathbb{R}^n$ there is a unique choice of constants $\alpha_1, \cdots, \alpha_n$ for which $u = \sum_{i=1}^{n} \alpha_i e^i$. Further, from the orthonormality of the vectors e^i, $1 \le i \le n$ we can determine α_i, $1 \le i \le n$ as follows:

$$<u, e^j> = \left\langle \sum_{i=1}^{n} \alpha_i e^i, e^j \right\rangle = \sum_{i=1}^{n} \alpha_i < e^i, e^j > = \alpha_j, \quad 1 \le j \le n.$$

Thus, the vector u has a unique representation $u = \sum_{i=1}^{n} <u, e^i> e^i$.

A natural generalization of this result which is widely applicable and has led to a vast amount of advanced mathematics can be stated as follows: Let $\{\phi_n(x), \ n = 0, 1, 2, \cdots\}$ be an orthogonal set of functions in the interval $[\alpha, \beta]$ with respect to the weight function $r(x)$. Then, an arbitrary function $f(x)$ can be expressed as an infinite series involving orthogonal functions $\phi_n(x), \ n = 0, 1, 2, \cdots$ as

$$f(x) = \sum_{n=0}^{\infty} c_n \phi_n(x). \tag{20.1}$$

It is natural to ask the meaning of equality in (20.1), i.e., the type of convergence, if any, of the infinite series on the right so that we will have some idea as to how well this represents $f(x)$. We shall also determine the constant coefficients c_n, $n = 0, 1, 2, \cdots$ in (20.1).

Let us first proceed formally without considering the question of convergence. We multiply (20.1) by $r(x)\phi_m(x)$ and integrate from α to β, to obtain

$$\int_{\alpha}^{\beta} r(x)\phi_m(x)f(x)dx = \int_{\alpha}^{\beta} \sum_{n=0}^{\infty} c_n r(x)\phi_n(x)\phi_m(x)dx$$

$$= c_m \int_{\alpha}^{\beta} r(x)\phi_m^2(x)dx = c_m \|\phi_m\|^2.$$

R.P. Agarwal, D. O'Regan, *Ordinary and Partial Differential Equations*,
Universitext, DOI 10.1007/978-0-387-79146-3_20,
© Springer Science+Business Media, LLC 2009

Thus, under suitable convergence conditions the constant coefficients c_n, $n = 0, 1, 2, \cdots$ are given by the formula

$$c_n = \int_\alpha^\beta r(x)\phi_n(x)f(x)dx \bigg/ \|\phi_n\|^2. \tag{20.2}$$

However, if the set $\{\phi_n(x)\}$ is orthonormal, so that $\|\phi_n\| = 1$, then we have

$$c_n = \int_\alpha^\beta r(x)\phi_n(x)f(x)dx. \tag{20.3}$$

If the series $\sum_{n=0}^\infty c_n\phi_n(x)$ converges uniformly to $f(x)$ in $[\alpha, \beta]$, then the above formal procedure is justified, and then the coefficients c_n are given by (20.2).

The coefficients c_n obtained in (20.2) are called the *Fourier coefficients* of the function $f(x)$ with respect to the orthogonal set $\{\phi_n(x)\}$ and the series $\sum_{n=0}^\infty c_n\phi_n(x)$ with coefficients (20.2) is called the *Fourier series* of $f(x)$.

We shall write

$$f(x) \sim \sum_{n=0}^\infty c_n\phi_n(x)$$

which, in general, is just a correspondence, i.e., often $f(x) \neq \sum_{n=0}^\infty c_n\phi_n(x)$, unless otherwise is proved.

Fourier cosine series. In Lecture 12, we have seen that the set

$$\left\{ \phi_0(x) = \frac{1}{\sqrt{\pi}}, \ \phi_n(x) = \sqrt{\frac{2}{\pi}} \cos nx, \ n = 1, 2, \cdots \right\}$$

is orthonormal on $0 < x < \pi$. Thus, for any piecewise continuous function $f(x)$ on $0 < x < \pi$,

$$f(x) \sim c_0\phi_0(x) + \sum_{n=1}^\infty c_n\phi_n(x) = \frac{c_0}{\sqrt{\pi}} + \sum_{n=1}^\infty c_n\sqrt{\frac{2}{\pi}} \cos nx,$$

where

$$c_0 = \int_0^\pi f(t)\frac{1}{\sqrt{\pi}}dt, \quad c_n = \int_0^\pi f(t)\sqrt{\frac{2}{\pi}} \cos nt\, dt, \quad n \geq 1.$$

Hence,

$$f(x) \sim \frac{1}{2} \times \frac{2}{\pi} \int_0^\pi f(t)dt + \sum_{n=1}^\infty \left(\frac{2}{\pi} \int_0^\pi f(t) \cos nt\, dt \right) \cos nx,$$

which can be written as

$$f(x) \sim \frac{a_0}{2} + \sum_{n=1}^{\infty} a_n \cos nx, \qquad (20.4)$$

where

$$a_n = \frac{2}{\pi} \int_0^{\pi} f(t) \cos nt\, dt, \quad n \geq 0. \qquad (20.5)$$

The series (20.4) is known as the *Fourier cosine series*.

Example 20.1. We shall find the Fourier cosine series of the function $f(x) = x$, $0 < x < \pi$. Clearly,

$$a_0 = \frac{2}{\pi} \int_0^{\pi} t \cos 0t\, dt = \frac{2}{\pi} \int_0^{\pi} t\, dt = \pi,$$

$$
\begin{aligned}
a_n = \frac{2}{\pi} \int_0^{\pi} t \cos nt\, dt &= \frac{2}{\pi} \left[t \frac{\sin nt}{n} \Big|_0^{\pi} - \int_0^{\pi} \frac{\sin nt}{n}\, dt \right] \\
&= \frac{2}{\pi} \left[\frac{0-0}{n} + \frac{\cos n\pi - 1}{n^2} \right] = \frac{2}{\pi} \frac{(-1)^n - 1}{n^2}, \quad n \geq 1.
\end{aligned}
$$

Thus, from (20.4), we have

$$x \sim \frac{\pi}{2} + \sum_{n=1}^{\infty} \frac{2}{\pi} \frac{(-1)^n - 1}{n^2} \cos nx, \quad 0 < x < \pi.$$

Fourier sine series. Recall that the set

$$\left\{ \phi_n(x) = \sqrt{\frac{2}{\pi}} \sin nx, \quad n = 1, 2, \cdots \right\}$$

is orthonormal on $0 < x < \pi$. Thus, for any piecewise continuous function $f(x)$ on $0 < x < \pi$,

$$f(x) \sim \sum_{i=1}^{n} c_n \sqrt{\frac{2}{\pi}} \sin nx,$$

where

$$c_n = \int_0^{\pi} f(t) \sqrt{\frac{2}{\pi}} \sin nt\, dt, \quad n \geq 1.$$

Again we can rewrite it as

$$f(x) \sim \sum_{n=1}^{\infty} b_n \sin nx, \qquad (20.6)$$

where

$$b_n = \frac{2}{\pi} \int_0^\pi f(t) \sin nt\, dt, \quad n \geq 1. \tag{20.7}$$

This representation of $f(x)$ is called the *Fourier sine series*.

Example 20.2. We shall find the Fourier sine series of the function

$$f(x) = \begin{cases} 2, & 0 < x < \pi/2 \\ 0, & \pi/2 < x < \pi. \end{cases}$$

Clearly, for $n \geq 1$

$$b_n = \frac{2}{\pi} \int_0^\pi f(t) \sin nt\, dt = \frac{2}{\pi} \int_0^{\pi/2} 2 \sin nt\, dt = \frac{4}{\pi} \left[\frac{-\cos(n\pi/2) + 1}{n} \right].$$

Thus,

$$f(x) \sim \sum_{n=1}^\infty \frac{4}{\pi} \left[\frac{1 - \cos(n\pi/2)}{n} \right] \sin nx, \quad 0 < x < \pi.$$

Fourier trigonometric series. In Lecture 12, we have verified that the set

$$\left\{ \phi_0(x) = \frac{1}{\sqrt{2\pi}}, \ \phi_{2n-1}(x) = \frac{1}{\sqrt{\pi}} \cos nx, \ \phi_{2n}(x) = \frac{1}{\sqrt{\pi}} \sin nx, \ n = 1, 2, \cdots \right\}$$

is orthonormal on $-\pi < x < \pi$. Thus, for any piecewise continuous function $f(x)$ on $-\pi < x < \pi$,

$$f(x) \sim \frac{c_0}{\sqrt{2\pi}} + \sum_{n=1}^\infty \left[c_{2n-1} \frac{\cos nx}{\sqrt{\pi}} + c_{2n} \frac{\sin nx}{\sqrt{\pi}} \right],$$

where

$$c_0 = \int_{-\pi}^\pi f(t) \frac{1}{\sqrt{2\pi}} dt$$

$$c_{2n-1} = \int_{-\pi}^\pi f(t) \frac{1}{\sqrt{\pi}} \cos nt\, dt, \quad n \geq 1$$

$$c_{2n} = \int_{-\pi}^\pi f(t) \frac{1}{\sqrt{\pi}} \sin nt\, dt, \quad n \geq 1.$$

On rewriting the above relation, we have the *Fourier trigonometric series*

$$f(x) \sim \frac{a_0}{2} + \sum_{n=1}^\infty (a_n \cos nx + b_n \sin nx), \tag{20.8}$$

where

$$a_n = \frac{1}{\pi} \int_{-\pi}^{\pi} f(t) \cos ntdt, \quad n \ge 0$$

$$b_n = \frac{1}{\pi} \int_{-\pi}^{\pi} f(t) \sin ntdt, \quad n \ge 1.$$

(20.9)

For our later use we note the following relations:

$$c_0 = \sqrt{\frac{\pi}{2}} a_0, \quad c_{2n-1} = \sqrt{\pi} a_n, \quad c_{2n} = \sqrt{\pi} b_n.$$

(20.10)

Example 20.3.   We shall find the Fourier trigonometric series of the function

$$f(x) = \begin{cases} 1, & -\pi < x < 0 \\ x, & 0 < x < \pi. \end{cases}$$

From (20.9), we have

$$a_n = \frac{1}{\pi} \int_{-\pi}^{\pi} f(t) \cos ntdt = \frac{1}{\pi} \int_{-\pi}^{0} 1 \cdot \cos ntdt + \frac{1}{\pi} \int_{0}^{\pi} t \cos ntdt$$

$$= 0 + \frac{1}{\pi} \left[\frac{t \sin nt}{n} + \frac{\cos nt}{n^2} \right]\Big|_0^{\pi}$$

$$= \frac{1}{\pi} \left(\frac{(-1)^n - 1}{n^2} \right), \quad n \ge 1$$

$$a_0 = \frac{1}{\pi} \int_{-\pi}^{0} 1 \cdot 1 dt + \frac{1}{\pi} \int_{0}^{t} t \cdot 1 dt = 1 + \frac{\pi}{2},$$

$$b_n = \frac{1}{\pi} \int_{-\pi}^{0} 1 \cdot \sin ntdt + \frac{1}{\pi} \int_{0}^{\pi} t \sin ntdt = \frac{-1 + (1 - \pi)(-1)^n}{n\pi}, \quad n \ge 1.$$

Hence, we have

$$f(x) \sim \frac{1 + \frac{\pi}{2}}{2} + \sum_{n=1}^{\infty} \left[\frac{(-1)^n - 1}{\pi n^2} \cos nx + \frac{-1 + (1 - \pi)(-1)^n}{n\pi} \sin nx \right],$$

$$-\pi < x < \pi.$$

It is clear that the constants a_n, b_n for the Fourier trigonometric series are different from those for the Fourier cosine and Fourier sine series. However, if the function $f(x)$ is odd, then since

$$a_n = \frac{1}{\pi} \int_{-\pi}^{\pi} f(t) \cos ntdt = 0, \quad n \ge 0$$

$$b_n = \frac{1}{\pi} \int_{-\pi}^{\pi} f(t) \sin ntdt = \frac{2}{\pi} \int_{0}^{\pi} f(t) \sin ntdt, \quad n \ge 1$$

the Fourier trigonometric series reduces to the Fourier sine series. Thus, we conclude that if $f(x)$ is odd, or defined only on $(0,\pi)$ and we make its odd extension then the Fourier sine series (20.6) holds on $(-\pi, \pi)$. Exactly, in the same way if $f(x)$ is even, or defined only on $(0, \pi)$ and we make its even extension then since

$$a_n = \frac{1}{\pi} \int_{-\pi}^{\pi} f(t) \cos ntdt = \frac{2}{\pi} \int_0^{\pi} f(t) \cos ntdt, \quad n \geq 0$$

$$b_n = \frac{1}{\pi} \int_{-\pi}^{\pi} f(t) \sin ntdt = 0, \quad n \geq 1$$

the Fourier trigonometric series reduces to the Fourier cosine series.

Finally, we remark that if the piecewise continuous function $f(x)$ is defined on $-a < x < a$, then its Fourier trigonometric series (20.8) takes the form

$$f(x) \sim \frac{a_0}{2} + \sum_{n=1}^{\infty} \left(a_n \cos \frac{n\pi x}{a} + b_n \sin \frac{n\pi x}{a} \right), \qquad (20.11)$$

where

$$a_n = \frac{1}{a} \int_{-a}^{a} f(t) \cos \frac{n\pi t}{a} dt, \quad n \geq 0$$

$$b_n = \frac{1}{a} \int_{-a}^{a} f(t) \sin \frac{n\pi t}{a} dt, \quad n \geq 1. \qquad (20.12)$$

Lecture 21
Eigenfunction Expansions (Cont'd.)

In this lecture also, we shall expand a given function in terms of orthogonal polynomials and functions we have provided in Lectures 12 and 13.

Fourier–Legendre series. We have proved that the set of Legendre polynomials $\{\phi_n(x) = P_n(x), \ n = 0, 1, \cdots\}$ is orthogonal on $[-1, 1]$ with $r(x) = 1$. Also,

$$\|P_n\|^2 = \int_{-1}^{1} P_n^2(x)dx = \frac{2}{2n+1}.$$

Thus, for any piecewise continuous function $f(x)$ on $-1 < x < 1$ the Fourier–Legendre series appears as

$$f(x) \sim \sum_{n=0}^{\infty} c_n P_n(x), \tag{21.1}$$

where

$$c_n = \frac{2n+1}{2} \int_{-1}^{1} P_n(x)f(x)dx, \quad n \geq 0. \tag{21.2}$$

Example 21.1. For the function $f(x) = \cos(\pi x/2)$ (recall explicit form of $P_0(x)$, $P_1(x)$, $P_2(x)$, $P_3(x)$ and $P_4(x)$) it is easy to find from (21.2) that

$$c_0 = \frac{2}{\pi}, \quad c_1 = 0, \quad c_2 = -\frac{10}{\pi^3}(12 - \pi^2)$$

$$c_3 = 0, \quad c_4 = \frac{18}{\pi^5}(\pi^4 - 180\pi^2 + 1680), \quad c_5 = 0.$$

Hence, the Fourier–Legendre series of the function $f(x) = \cos(\pi x/2)$ up to $P_4(x)$ in the interval $[-1, 1]$ is

$$\frac{2}{\pi}P_0(x) - \frac{10}{\pi^3}(12 - \pi^2)P_2(x) + \frac{18}{\pi^5}(\pi^4 - 180\pi^2 + 1680)P_4(x).$$

Note that $P_{2m}(x)$ is an even function, whereas $P_{2m+1}(x)$ is odd. So, if $f(x)$ is even or defined on $(0, 1)$ and we make its even extension, then the

R.P. Agarwal, D. O'Regan, *Ordinary and Partial Differential Equations*,
Universitext, DOI 10.1007/978-0-387-79146-3_21,
© Springer Science+Business Media, LLC 2009

Fourier–Legendre series (21.1) reduces to

$$f(x) \sim \sum_{m=0}^{\infty} c_{2m} P_{2m}(x), \qquad (21.3)$$

where

$$c_{2m} = (4m+1) \int_0^1 P_{2m}(x) f(x) dx, \quad m \ge 0. \qquad (21.4)$$

In our previous example the function $f(x) = \cos(\pi x/2)$ is even, and hence all odd c_n's are zero.

Similarly, if $f(x)$ is odd or defined on $(0,1)$ and we make its odd extension, then the Fourier–Legendre series (21.1) reduces to

$$f(x) \sim \sum_{m=0}^{\infty} c_{2m+1} P_{2m+1}(x), \qquad (21.5)$$

where

$$c_{2m+1} = (4m+3) \int_0^1 P_{2m+1}(x) f(x) dx, \quad m \ge 0. \qquad (21.6)$$

Example 21.2. For the function

$$g(x) = \begin{cases} 0, & -1 < x < 0 \\ x, & 0 \le x < 1 \end{cases}$$

denote the Fourier–Legendre series by $\sum_{n=0}^{\infty} c_n P_n(x)$. Then, since the function

$$f(x) = g(x) - \frac{P_1(x)}{2} = g(x) - \frac{x}{2} = \begin{cases} -x/2, & -1 < x < 0 \\ x/2, & 0 \le x < 1 \end{cases} = \frac{|x|}{2}$$

is even, in the expansion

$$f(x) \sim \sum_{n=0}^{\infty} c_n P_n(x) - \frac{P_1(x)}{2}$$

the odd Fourier coefficients are zero, i.e.,

$$c_1 - \frac{1}{2} = 0, \quad c_3 = 0, \quad c_5 = 0, \cdots.$$

Hence, the expansion of the function $g(x)$ reduces to

$$c_0 P_0(x) + \frac{1}{2} P_1(x) + \sum_{m=1}^{\infty} c_{2m} P_{2m}(x). \qquad (21.7)$$

Fourier–Chebyshev series.   We have shown that the set of Chebyshev polynomials $\{\phi_n(x) = T_n(x), \quad n = 0, 1, \cdots\}$ is orthogonal in $[-1, 1]$ with $r(x) = (1 - x^2)^{-1/2}$. Also,

$$\|T_n\|^2 = \int_{-1}^{1} (1 - x^2)^{-1/2} T_n^2(x)dx = \begin{cases} \pi/2, & n \neq 0 \\ \pi, & n = 0. \end{cases}$$

Thus, for any piecewise continuous function $f(x)$ on $-1 < x < 1$ the Fourier–Chebyshev series appears as

$$f(x) \sim \sum_{n=0}^{\infty} c_n T_n(x), \tag{21.8}$$

where

$$c_n = \frac{2d_n}{\pi} \int_{-1}^{1} \frac{f(x)T_n(x)}{\sqrt{1 - x^2}} dx, \quad d_0 = \frac{1}{2}, \quad d_n = 1, \ n \geq 1. \tag{21.9}$$

Fourier–Hermite series.   Following the argument in the earlier cases this series over the interval $(-\infty, \infty)$ appears as

$$f(x) \sim \sum_{n=0}^{\infty} c_n H_n(x), \tag{21.10}$$

where

$$c_n = \frac{1}{2^n \, n! \, \sqrt{\pi}} \int_{-\infty}^{\infty} e^{-x^2} f(x) H_n(x)dx, \quad n \geq 0. \tag{21.11}$$

Fourier–Bessel series.   Let $m \geq 0$ be a fixed integer and let $\{b_n, \ n = 0, 1, \cdots\}$ be the zeros of the Bessel function $J_m(x)$. In Lecture 13 we have seen that the set $\{J_m(b_n x), \ n = 0, 1, \cdots\}$ is orthogonal on $[0, 1]$ with respect to the weight function x, i.e.,

$$\int_{0}^{1} x J_m(b_p x) J_m(b_q x)dx = \begin{cases} 0, & p \neq q \\ \frac{1}{2} J_{m+1}^2(b_p), & q = p. \end{cases}$$

Now following the argument in the earlier cases the Fourier–Bessel series can be written as

$$f(x) \sim \sum_{n=0}^{\infty} c_n J_m(b_n x), \quad x \in [0, 1] \tag{21.12}$$

where

$$c_n = \frac{2}{J_{m+1}^2(b_n)} \int_{0}^{1} x J_m(b_n x) f(x)dx, \quad n \geq 0. \tag{21.13}$$

We also remark that with little adjustment the above series can be written over the interval $[0, c]$ instead of $[0, 1]$. Indeed, if $\{a_n,\ n = 0, 1, \cdots\}$ are the positive roots of the equation $J_m(ac) = 0$, then (21.12) can be written as

$$f(x) \sim \sum_{n=0}^{\infty} c_n J_m(a_n x), \quad x \in [0, c] \tag{21.14}$$

where

$$c_n = \frac{2}{c^2 J_{m+1}^2(a_n c)} \int_0^c x J_m(a_n x) f(x) dx, \quad n \geq 0. \tag{21.15}$$

Our last two examples illustrate how the eigenfunctions of a regular Sturm–Liouville problem can be used to expand a given function.

Example 21.3. To obtain the Fourier series of the function $f(x) = 1$ in the interval $[0, \pi]$ in terms of the eigenfunctions $\phi_n(x) = \sin nx$, $n = 1, 2, \cdots$ of the eigenvalue problem $y'' + \lambda y = 0$, $y(0) = y(\pi) = 0$ (see Example 19.1) we recall that

$$\|\phi_n\|^2 = \int_0^\pi \sin^2 nx\, dx = \frac{\pi}{2}.$$

Thus, it follows that

$$c_n = \frac{1}{\|\phi_n\|^2} \int_0^\pi f(x)\, \sin nx\, dx = \frac{2}{\pi} \int_0^\pi \sin nx\, dx = \frac{2}{n\pi}\left(1 - (-1)^n\right).$$

Hence, we have

$$1 \sim \frac{4}{\pi} \sum_{n=1}^{\infty} \frac{1}{(2n-1)} \sin(2n-1)x = F_1(x), \quad \text{say.} \tag{21.16}$$

Example 21.4. We shall obtain the Fourier series of the function $f(x) = x - x^2$, $x \in [0, 1]$ in terms of the eigenfunctions $\phi_0(x) = 1 - x$, $\phi_n(x) = \sin\sqrt{\lambda_n}(1 - x)$, $n = 1, 2, \cdots$ of the eigenvalue problem $y'' + \lambda y = 0$, $y(0) + y'(0) = 0$, $y(1) = 0$ (see Example 19.2). For this, we note that

$$\|\phi_0\|^2 = \int_0^1 (1-x)^2 dx = \frac{1}{3}$$

$$\|\phi_n\|^2 = \int_0^1 \sin^2\sqrt{\lambda_n}(1-x) dx = \frac{1}{2}\int_0^1 (1 - \cos 2\sqrt{\lambda_n}(1-x)) dx$$

$$= \frac{1}{2}\left[x + \frac{1}{2\sqrt{\lambda_n}}\sin 2\sqrt{\lambda_n}(1-x)\right]\Big|_0^1 = \frac{1}{2}\left[1 - \frac{1}{2\sqrt{\lambda_n}}\sin 2\sqrt{\lambda_n}\right]$$

$$= \frac{1}{2}\left[1 - \frac{1}{2\sqrt{\lambda_n}}2\sin\sqrt{\lambda_n}\cos\sqrt{\lambda_n}\right] = \frac{1}{2}[1 - \cos^2\sqrt{\lambda_n}]$$

$$= \frac{1}{2}\sin^2\sqrt{\lambda_n},$$

where we have used the fact that $\tan \sqrt{\lambda_n} = \sqrt{\lambda_n}$.

Thus, it follows that

$$c_0 = 3 \int_0^1 (1-x)(x-x^2)dx = \frac{1}{4}$$

and for $n \geq 1$,

$$
\begin{aligned}
c_n &= \frac{2}{\sin^2 \sqrt{\lambda_n}} \int_0^1 (x-x^2) \sin \sqrt{\lambda_n}(1-x)dx \\
&= \frac{2}{\sin^2 \sqrt{\lambda_n}} \left[(x-x^2)\frac{\cos \sqrt{\lambda_n}(1-x)}{\sqrt{\lambda_n}} \Big|_0^1 - \int_0^1 (1-2x)\frac{\cos \sqrt{\lambda_n}(1-x)}{\sqrt{\lambda_n}}dx \right] \\
&= \frac{-2}{\sqrt{\lambda_n}\sin^2 \sqrt{\lambda_n}} \left[(1-2x)\frac{\sin \sqrt{\lambda_n}(1-x)}{-\sqrt{\lambda_n}} \Big|_0^1 - \int_0^1 -2\frac{\sin \sqrt{\lambda_n}(1-x)}{-\sqrt{\lambda_n}}dx \right] \\
&= \frac{-2}{\sqrt{\lambda_n}\sin^2 \sqrt{\lambda_n}} \left[\frac{\sin \sqrt{\lambda_n}}{\sqrt{\lambda_n}} - \frac{2}{\sqrt{\lambda_n}} \frac{\cos \sqrt{\lambda_n}(1-x)}{\sqrt{\lambda_n}} \Big|_0^1 \right] \\
&= \frac{-2}{\lambda_n^{3/2}\sin^2 \sqrt{\lambda_n}} \left[\sqrt{\lambda_n} \sin \sqrt{\lambda_n} - 2 + 2\cos \sqrt{\lambda_n} \right] \\
&= \frac{-2}{\lambda_n^{3/2}\sin^2 \sqrt{\lambda_n}} \left[\lambda_n \cos \sqrt{\lambda_n} - 2 + 2\cos \sqrt{\lambda_n} \right] \\
&= \frac{2}{\lambda_n^{3/2}\sin^2 \sqrt{\lambda_n}} \left[2 - (2+\lambda_n)\cos \sqrt{\lambda_n} \right].
\end{aligned}
$$

Hence, we have

$$x - x^2 \sim \frac{1}{4}(1-x) + \sum_{n=1}^{\infty} \frac{2}{\lambda_n^{3/2}\sin^2 \sqrt{\lambda_n}}(2-(2+\lambda_n)\cos \sqrt{\lambda_n})\sin \sqrt{\lambda_n}(1-x)$$

$$= F_2(x), \quad \text{say.} \quad (21.17)$$

Problems

21.1. Find the Fourier cosine series on the interval $0 < x < \pi$ of each of the following functions:

(i) $f(x) = x^2$ (ii) $f(x) = \begin{cases} 0, & 0 < x \leq \pi/2 \\ 1, & \pi/2 < x < \pi \end{cases}$ (iii) $f(x) = \cos x$.

.**21.2.** Find the Fourier sine series on the interval $0 < x < \pi$ of each of the following functions:

(i) $f(x) = 1$ (ii) $f(x) = \pi - x$.

21.3. Find the Fourier sine series on the interval $0 < x < \pi$ of each of the functions (i)–(iii) given in Problem 21.1.

21.4. Find the Fourier trigonometric series of each of the following functions:

(i) $f(x) = \begin{cases} 0, & -\pi < x < 0 \\ 1, & 0 < x < \pi \end{cases}$ (ii) $f(x) = \begin{cases} 1, & -\pi < x < 0 \\ 2, & 0 < x < \pi \end{cases}$

(iii) $f(x) = x - \pi, \quad -\pi < x < \pi$ (iv) $f(x) = |x|, \quad -\pi < x < \pi$

(v) $f(x) = x^2, \quad -\pi < x < \pi$ (vi) $f(x) = \begin{cases} x, & -\pi < x < 0 \\ 2, & x = 0 \\ e^{-x}, & 0 < x < \pi \end{cases}$

(vii) $f(x) = x^4, \quad -\pi < x < \pi.$

Further, for each series sum a few terms graphically, and compare with the function expanded.

21.5. Find the Fourier trigonometric series of each of the following functions:

(i) $f(x) = x^2, \quad 0 < x < 2\pi$ (ii) $f(x) = \begin{cases} 0, & -2 < x < -1 \\ 1, & -1 < x < 1 \\ 0, & 1 < x < 2. \end{cases}$

21.6. Establish the identities $\sin^3 x = (3/4)\sin x - (1/4)\sin 3x$ and $\cos^3 x = (3/4)\cos x + (1/4)\cos 3x$. Prove that in each case expression on the right-hand side is the Fourier series for the function on the left-hand side.

21.7. Let $f(x)$ be a periodic function with period 2π such that its Fourier coefficients exist. Suppose further that $f(\pi - x) = f(x)$ for all x. Show that $a_n = 0$ when n is odd and $b_n = 0$ when n is even.

21.8. In the expansion (21.7) find c_0, c_2 and c_4.

21.9. Find the Fourier–Legendre series of the function

$$f(x) = \begin{cases} 0, & -1 < x < 0 \\ 1, & 0 < x < 1. \end{cases}$$

21.10. Show that the Fourier series expansion of the function $f(x) = \pi x - x^2$, $0 \le x \le \pi$ in terms of the orthonormal functions $\phi_n(x) = \sqrt{\frac{2}{\pi}}\sin nx$, $0 \le x \le \pi$, $n = 1, 2, \cdots$ can be written as

$$\sum_{n=1}^{\infty} \frac{8}{\pi(2n-1)^3}\sin(2n-1)x.$$

21.11.   Expand the function $f(x) = 1$ when $0 < x < c$ in series of functions $J_0(\lambda_j x)$, where λ_j are the positive roots of the equation $J_0(\lambda c) = 0$.

21.12.   Expand the function

$$f(x) = \begin{cases} 1, & 0 < x < 1 \\ 1/2, & x = 1 \\ 0, & 1 < x < 2 \end{cases}$$

in series of $J_0(\lambda_j x)$ where λ_j are the roots of $J_0(2\lambda) = 0$.

21.13.   Expand the function $f(x) = x$, $0 < x < 1$ in series of $J_1(\lambda_j x)$ where λ_j are the positive roots of $J_1(\lambda) = 0$. Also find the function represented by the series in the interval $-1 < x \leq 0$.

21.14.   Show that

$$x^2 = \sum_{j=1}^{\infty} \frac{2(\lambda_j^2 - 4)J_0(\lambda_j x)}{\lambda_j^3 J_1(\lambda_j)}, \quad -1 < x < 1$$

where λ_j are the positive roots of $J_0(\lambda) = 0$.

Answers or Hints

21.1.   (i) $\frac{\pi^2}{3} + \sum_{n=1}^{\infty} \frac{4(-1)^n}{n^2} \cos nx$ (ii) $\frac{1}{2} + \sum_{n=1}^{\infty} \frac{-2\sin(n\pi/2)}{n\pi} \cos nx$
(iii) $\cos x$.

21.2.   (i) $\frac{4}{\pi} \sum_{n=1}^{\infty} \frac{\sin(2n-1)x}{2n-1}$ (ii) $2 \sum_{n=1}^{\infty} \frac{\sin nx}{n}$.

21.3.   (i) $\sum_{n=1}^{\infty} \left(\frac{2\pi(-1)^{n+1}}{n} + 4\frac{-1+(-1)^n}{\pi\, n^3} \right) \sin nx$
(ii) $\sum_{n=1}^{\infty} \frac{2}{n\pi} \left((-1)^{n+1} + \cos\frac{n\pi}{2} \right) \sin nx$ (iii) $\sum_{n=2}^{\infty} \frac{2n}{\pi} \frac{1+(-1)^n}{n^2-1} \sin nx$.

21.4.   (i) $\frac{1}{2} + \frac{2}{\pi} \left(\sin x + \frac{1}{3}\sin 3x + \frac{1}{5}\sin 5x + \cdots \right)$
(ii) $\frac{3}{2} + \frac{2}{\pi} \left(\sin x + \frac{1}{3}\sin 3x + \frac{1}{5}\sin 5x + \cdots \right)$ (iii) $-\pi + \sum_{n=1}^{\infty} \frac{2(-1)^{n+1}}{n} \sin nx$
(iv) $\frac{\pi}{2} + \sum_{n=1}^{\infty} \frac{2}{\pi n^2}((-1)^n - 1) \cos nx$ (v) $\frac{\pi^2}{3} + 4\sum_{n=1}^{\infty} \frac{(-1)^n}{n^2} \cos nx$
(vi) $-\left(\frac{e^{-\pi}-1}{2\pi} + \frac{\pi}{4} \right) + \frac{1}{\pi} \sum_{n=1}^{\infty} \left[\left(\frac{1+(-1)^{n+1}}{n^2} + \frac{1+(-1)^{n+1}e^{-\pi}}{1+n^2} \right) \cos nx \right.$
$\left. + \left(\frac{n}{1+n^2} \left(1 + (-1)^{n+1}e^{-\pi} \right) + \frac{\pi(-1)^{n+1}}{n} \right) \sin nx \right]$
(vii) $\frac{\pi^4}{5} + 8\sum_{n=1}^{\infty} \left(\frac{\pi^2}{n^2} - \frac{6}{n^4} \right) (-1)^n \cos nx$.

21.5.   (i) $\frac{4\pi^2}{3} + \sum_{n=1}^{\infty} \left(\frac{4}{n^2} \cos nx - \frac{4\pi}{n} \sin nx \right)$

(ii) $\frac{1}{2} + \frac{2}{\pi} \sum_{n=1}^{\infty} \frac{1}{n} \sin \frac{n\pi}{2} \cos \frac{n\pi}{2} x.$

21.6. Note that

$$a_n = \frac{1}{\pi} \frac{3}{4} \int_{-\pi}^{\pi} \sin x \cos nx \, dx - \frac{1}{\pi} \frac{1}{4} \int_{-\pi}^{\pi} \sin 3x \cos nx \, dx,$$

$$b_n = \frac{1}{\pi} \frac{3}{4} \int_{-\pi}^{\pi} \sin x \sin nx \, dx - \frac{1}{\pi} \frac{1}{4} \int_{-\pi}^{\pi} \sin 3x \sin nx \, dx$$

and hence $a_n = 0$, $n \geq 0$, $b_1 = \frac{3}{4\pi}\pi$, $b_2 = 0$, $b_3 = -\frac{1}{4\pi}\pi$, $b_n = 0$, $n \geq 4$.

21.7. For n odd, successively, we have

$$
\begin{aligned}
a_n &= \frac{1}{\pi} \int_{-\pi}^{\pi} f(x) \cos nx \, dx = \frac{1}{\pi} \int_{2\pi}^{0} f(\pi - y) \cos n(\pi - y)(-1) dy \\
&= \frac{1}{\pi} \int_{0}^{2\pi} f(y)(-1)^n \cos ny \, dy = -\frac{1}{\pi} \int_{0}^{2\pi} f(y) \cos ny \, dy \\
&= -\frac{1}{\pi} \int_{-\pi}^{\pi} f(y) \cos ny \, dy = -a_n,
\end{aligned}
$$

where we have used the fact that $f(y) \cos ny$ is periodic. Hence, $a_n = 0$.

21.8. $c_0 = 1/4$, $c_2 = 5/16$, $c_4 = -3/32$.

21.9. From (21.2), $c_0 = 1/2$, $c_1 = 3/4$, use Problems 7.8(ii) and 7.9(i) to get $c_n = \frac{1}{2}(P_{n-1}(0) - P_{n+1}(0))$ which in view of Problem 7.2 gives $c_{2n} = 0$, $c_{2n+1} = \frac{1}{2}(-1)^n \frac{1 \cdot 3 \cdots (2n-1)}{2 \cdot 4 \cdots 2n} \frac{4n+3}{2n+2}.$

21.10. See Example 21.3.

21.11. $1 = \frac{2}{c} \sum_{j=1}^{\infty} \frac{J_0(\lambda_j x)}{\lambda_j J_1(\lambda_j c)}$, $0 < x < c.$

21.12. $f(x) = \frac{1}{2} \sum_{j=1}^{\infty} \frac{J_1(\lambda_j)}{\lambda_j [J_1(2\lambda_j)]^2} J_0(\lambda_j x)$, $0 < x < 2.$

21.13. $x = 2 \sum_{j=1}^{\infty} \frac{J_1(\lambda_j x)}{\lambda_j J_2(\lambda_j)}$, $-1 < x < 1.$

Lecture 22
Convergence of the Fourier Series

In this and the next lectures we shall examine the convergence of the Fourier series of the function $f(x)$ to $f(x)$. For this, to make the analysis widely applicable, we assume that the functions $\phi_n(x)$, $n = 0, 1, \cdots$ and $f(x)$ are only piecewise continuous on $[\alpha, \beta]$. Let the sum of first $N + 1$ terms $\sum_{n=0}^{N} c_n \phi_n(x)$ be denoted by $S_N(x)$. We consider the difference $|S_N(x) - f(x)|$ for various values of N and x. If for an arbitrary $\epsilon > 0$ there is an integer $N(\epsilon) > 0$ such that $|S_N(x) - f(x)| < \epsilon$, then the Fourier series *converges uniformly* to $f(x)$ for all x in $[\alpha, \beta]$. On the other hand, if N depends on x and ϵ both, then the Fourier series *converges pointwise* to $f(x)$. However, for the moment both of these types of convergence are too demanding, and we will settle for something less. To this end, we need the following definition.

Definition 22.1. Let each of the functions $\psi_n(x)$, $n \geq 0$ and $\psi(x)$ be piecewise continuous in $[\alpha, \beta]$. We say that the sequence $\{\psi_n(x)\}$ *converges in the mean* to $\psi(x)$ (with respect to the weight function $r(x)$) in the interval $[\alpha, \beta]$ if

$$\lim_{n \to \infty} \|\psi_n - \psi\|^2 = \lim_{n \to \infty} \int_\alpha^\beta r(x)(\psi_n(x) - \psi(x))^2 dx = 0. \qquad (22.1)$$

Thus, the Fourier series converges in the mean to $f(x)$ provided

$$\lim_{N \to \infty} \int_\alpha^\beta r(x)(S_N(x) - f(x))^2 dx = 0. \qquad (22.2)$$

Before we prove the convergence of the Fourier series, let us consider the possibility of representing $f(x)$ by a series of the form $\sum_{n=0}^{\infty} d_n \phi_n(x)$, where the coefficients d_n are not necessarily the Fourier coefficients. Let

$$T_N(x; d_0, d_1, \cdots, d_N) = \sum_{n=0}^{N} d_n \phi_n(x)$$

and let e_N be the quantity $\|T_N - f\|$. Then, from the orthogonality of the

R.P. Agarwal, D. O'Regan, *Ordinary and Partial Differential Equations*,
Universitext, DOI 10.1007/978-0-387-79146-3_22,
© Springer Science+Business Media, LLC 2009

functions $\phi_n(x)$ it is clear that

$$
\begin{aligned}
e_N^2 &= \|T_N - f\|^2 = \int_\alpha^\beta r(x) \left(\sum_{n=0}^N d_n \phi_n(x) - f(x) \right)^2 dx \\
&= \sum_{n=0}^N d_n^2 \int_\alpha^\beta r(x) \phi_n^2(x) dx - 2 \sum_{n=0}^N d_n \int_\alpha^\beta r(x) \phi_n(x) f(x) dx \\
&\quad + \int_\alpha^\beta r(x) f^2(x) dx \\
&= \sum_{n=0}^N d_n^2 \|\phi_n\|^2 - 2 \sum_{n=0}^N d_n c_n \|\phi_n\|^2 + \|f\|^2 \\
&= \sum_{n=0}^N \|\phi_n\|^2 (d_n - c_n)^2 - \sum_{n=0}^N \|\phi_n\|^2 c_n^2 + \|f\|^2.
\end{aligned}
$$

(22.3)

Thus, the quantity e_N is least when $d_n = c_n$ for $n = 0, 1, \cdots, N$. Therefore, we have established the following theorem.

Theorem 22.1. For any nonnegative integer N, the best approximation in the mean to a function $f(x)$ by an expression of the form $\sum_{n=0}^N d_n \phi_n(x)$ is obtained when the coefficients d_n are the Fourier coefficients of $f(x)$.

Now in (22.3) let $d_n = c_n$, $n = 0, 1, \cdots, N$ to obtain

$$
\|S_N - f\|^2 = \|f\|^2 - \sum_{n=0}^N \|\phi_n\|^2 c_n^2. \tag{22.4}
$$

Thus, it follows that

$$
\|T_n - f\|^2 = \sum_{n=0}^N \|\phi_n\|^2 (d_n - c_n)^2 + \|S_N - f\|^2. \tag{22.5}
$$

Hence, we find

$$
0 \le \|S_N - f\| \le \|T_N - f\|. \tag{22.6}
$$

If the series $\sum_{n=0}^\infty d_n \phi_n(x)$ converges in the mean to $f(x)$, i.e., if $\lim_{N \to \infty} \|T_N - f\| = 0$, then from (22.6) it is clear that the Fourier series converges in the mean to $f(x)$, i.e., $\lim_{N \to \infty} \|S_N - f\| = 0$. However, then (22.5) implies that

$$
\lim_{N \to \infty} \sum_{n=0}^N \|\phi_n\|^2 (d_n - c_n)^2 = 0.
$$

But this is possible only if $d_n = c_n$, $n = 0, 1, \cdots$. Thus, we have proved the following result.

Theorem 22.2. If a series of the form $\sum_{n=0}^{\infty} d_n \phi_n(x)$ converges in the mean to $f(x)$, then the coefficients d_n must be the Fourier coefficients of $f(x)$.

Now from the equality (22.4) we note that

$$0 \leq \|S_{N+1} - f\| \leq \|S_N - f\|.$$

Thus, the sequence $\{\|S_N - f\|, \ N = 0, 1, \cdots\}$ is nonincreasing and bounded below by zero, and therefore it must converge. If it converges to zero, then the Fourier series of $f(x)$ converges in the mean to $f(x)$. Further, from (22.4) we have the inequality

$$\sum_{n=0}^{N} \|\phi_n\|^2 c_n^2 \leq \|f\|^2.$$

Since the sequence $\{C_N, \ N = 0, 1, \cdots\}$ where $C_N = \sum_{n=0}^{N} \|\phi_n\|^2 c_n^2$ is nondecreasing and bounded above by $\|f\|^2$, it must converge. Therefore, we have

$$\sum_{n=0}^{\infty} \|\phi_n\|^2 c_n^2 \leq \|f\|^2. \tag{22.7}$$

Hence, from (22.4) we see that the Fourier series of $f(x)$ converges in the mean to $f(x)$ if and only if

$$\|f\|^2 = \sum_{n=0}^{\infty} \|\phi_n\|^2 c_n^2. \tag{22.8}$$

For the case when $\phi_n(x), \ n = 0, 1, 2, \cdots$ are orthonormal, (22.7) reduces to *Bessel's inequality*

$$\sum_{n=0}^{\infty} c_n^2 \leq \|f\|^2 \tag{22.9}$$

and (22.8) becomes the well-known *Parseval's equality*

$$\|f\|^2 = \sum_{n=0}^{\infty} c_n^2. \tag{22.10}$$

We summarize the above considerations in the following theorem.

Theorem 22.3. Let $\{\phi_n(x), \ n = 0, 1, 2, \cdots\}$ be an orthonormal set, and let c_n be the Fourier coefficients of $f(x)$ given in (20.3). Then, the following hold:

(i)   the series $\sum_{n=0}^{\infty} c_n^2$ converges, and therefore

$$\lim_{n\to\infty} c_n = \lim_{n\to\infty} \int_\alpha^\beta r(x)\phi_n(x)f(x)dx = 0,$$

(ii)   the Bessel inequality (22.9) holds,

(iii) the Fourier series of $f(x)$ converges in the mean to $f(x)$ if and only if Parseval's equality (22.10) holds.

Now let $C_p[\alpha, \beta]$ be the space of all piecewise continuous functions in $[\alpha, \beta]$. The orthogonal set $\{\phi_n(x), \; n = 0, 1, \cdots\}$ is said to be *complete* in $C_p[\alpha, \beta]$ if for every function $f(x)$ of $C_p[\alpha, \beta]$ its Fourier series converges in the mean to $f(x)$. Clearly, if $\{\phi_n(x), \; n = 0, 1, \cdots\}$ is orthonormal then it is complete if and only if Parseval's equality holds for every function in $C_p[\alpha, \beta]$. The following property of an orthogonal set is fundamental.

Theorem 22.4. If an orthogonal set $\{\phi_n(x), \; n = 0, 1, \cdots\}$ is complete in $C_p[\alpha, \beta]$, then any function of $C_p[\alpha, \beta]$ that is orthogonal to every $\phi_n(x)$ must be zero except possibly at a finite number of points in $[\alpha, \beta]$.

Proof.   Without loss of generality, let the set $\{\phi_n(x), \; n = 0, 1, \cdots\}$ be orthonormal. If $f(x)$ is orthogonal to every $\phi_n(x)$, then from (20.3) all Fourier coefficients c_n of $f(x)$ are zero. But, then from the Parseval equality (22.10) the function $f(x)$ must be zero except possibly at a finite number of points in $[\alpha, \beta]$. ∎

The importance of this result lies in the fact that if we delete even one member from an orthogonal set, then the remaining functions cannot be a complete set. For example, the sets $\{\cos nx, \; n = 1, 2, \cdots\}$ and $\{\sin n\pi x, \; n = 1, 2, \cdots\}$ are orthogonal in $[0, \pi]$ and $[-1, 1]$, respectively, with respect to the weight function $r(x) = 1$, but not complete.

Unfortunately, there is no single procedure for establishing the completeness of a given orthogonal set. However, the following results are known.

Theorem 22.5. The orthogonal set $\{\phi_n(x), \; n = 0, 1, \cdots\}$ in the interval $[\alpha, \beta]$ with respect to the weight function $r(x)$ is complete in $C_p[\alpha, \beta]$ if $\phi_n(x)$ is a polynomial of degree n.

As a consequence of this result, it is clear that the Fourier–Legendre series of a piecewise continuous function $f(x)$ in $[-1, 1]$ converges in the mean to $f(x)$. The same conclusion holds for other series also.

Theorem 22.6. The set of all eigenfunctions $\{\phi_n(x), \; n = 1, 2, \cdots\}$ of a regular Sturm–Liouville problem is complete in the space $C_p[\alpha, \beta]$.

Example 22.1. In view of Theorem 22.6 the expansion $F_1(x)$ of the function 1 obtained in (21.16) converges in the mean to 1 on the interval $[0, \pi]$. Similarly, the expansion $F_2(x)$ of $x - x^2$ given in (21.17) converges in the mean to the function $x - x^2$ on the interval $[0, 1]$.

Theorem 22.6 can be extended to encompass the periodic eigenvalue problem (19.1), (19.20). Thus, in particular, the set $\{1, \cos nx, \sin nx, n \geq 1\}$ is complete in $C_p[-\pi, \pi]$, and therefore, the Fourier trigonometric series of any function $f(x)$ in $C_p[-\pi, \pi]$ converges in the mean to $f(x)$. Similarly, Fourier sine and cosine series of any function $f(x)$ in $C_p[0, \pi]$ converge in the mean to $f(x)$.

In view of the above remark and the relations (20.10), for the Fourier trigonometric series Parseval's equality can be written as

$$\left(\sqrt{\frac{\pi}{2}}a_0\right)^2 + \sum_{n=1}^{\infty}((\sqrt{\pi}a_n)^2 + (\sqrt{\pi}b_n)^2) = \int_{-\pi}^{\pi} f^2(x)dx,$$

or

$$\frac{a_0^2}{2} + \sum_{n=1}^{\infty}(a_n^2 + b_n^2) = \frac{1}{\pi}\int_{-\pi}^{\pi} f^2(x)dx. \tag{22.11}$$

Example 22.2. The Fourier trigonometric expansion of the function $|x|, \ -\pi < x < \pi$ is

$$|x| \sim \frac{\pi}{2} + \sum_{n=1}^{\infty} \frac{-4}{\pi(2n-1)^2} \cos(2n-1)x, \quad -\pi < x < \pi. \tag{22.12}$$

Clearly, $|x| \in C_p(-\pi, \pi)$. Further, comparing (22.12) with the Fourier trigonometric series, we have

$$\frac{a_0}{2} = \frac{\pi}{2}, \quad a_{2n} = 0, \quad a_{2n-1} = -\frac{4}{\pi(2n-1)^2}, \quad b_n = 0.$$

Thus, equality (22.11) gives

$$\frac{\pi^2}{2} + \sum_{n=1}^{\infty}\left[-\frac{4}{\pi(2n-1)^2}\right]^2 = \frac{1}{\pi}\int_{-\pi}^{\pi} |x|^2 dx,$$

which is the same as

$$\frac{\pi^2}{2} + \frac{16}{\pi^2}\sum_{n=1}^{\infty}\frac{1}{(2n-1)^4} = \frac{2\pi^2}{3}$$

and hence

$$\sum_{n=1}^{\infty}\frac{1}{(2n-1)^4} = \frac{\pi^4}{96}.$$

Lecture 23
Convergence of the
Fourier Series (Cont'd.)

The analytic discussion of uniform and pointwise convergence of the Fourier series of the function $f(x)$ to $f(x)$ is difficult. Therefore, in this lecture we shall state several results without proofs. These results are easily applicable to concrete problems.

Theorem 23.1. Let $\{\phi_n(x), \ n = 1, 2, \cdots\}$ be the set of all eigenfunctions of a regular Sturm–Liouville problem. Then, the following hold:

(i)   the Fourier series of $f(x)$ converges to $[f(x+)+f(x-)]/2$ at each point in the open interval (α, β) provided $f(x)$ and $f'(x)$ are piecewise continuous in $[\alpha, \beta]$,

(ii)   the Fourier series of $f(x)$ converges uniformly and absolutely to $f(x)$ in $[\alpha, \beta]$ provided $f(x)$ is continuous having a piecewise continuous derivative $f'(x)$ in $[\alpha, \beta]$, and is such that $f(\alpha) = 0$ if $\phi_n(\alpha) = 0$ and $f(\beta) = 0$ if $\phi_n(\beta) = 0$.

Example 23.1.   For the expansion $F_1(x)$ of the function $f(x) = 1$ obtained in (21.16), Theorem 23.1(i) ensures that $F_1(x) = 1$ at each point of the open interval $(0, \pi)$. However, $F_1(x) \neq 1$ at $x = 0$ and π, i.e., at the end points of the interval. We also note that since $\phi_n(0) = \phi_n(\pi) = 0$, but $f(0) = f(\pi) = 1$, Theorem 23.1(ii) cannot be applied.

Example 23.2.   For the expansion $F_2(x)$ of the function $f(x) = x - x^2$ given in (21.17), Theorem 23.1(ii) ensures that $F_2(x) = x - x^2$ uniformly in $[0, 1]$. In fact, here $\phi_n(1) = 0$ so does $f(1) = 0$, however, $\phi_n(0) \neq 0$, but $f(0) = 0$ (see *if* in Theorem 23.1(ii)).

Theorem 23.2.   Let $f(x)$ and $f'(x)$ be piecewise continuous in the interval $[-1, 1]$. Then, the Fourier–Legendre series of $f(x)$ converges to $[f(x+)+f(x-)]/2$ at each point in the open interval $(-1, 1)$, and at $x = -1$ the series converges to $f(-1+)$ and at $x = 1$ it converges to $f(1-)$.

Theorem 23.3.   Let $f(x)$ and $f'(x)$ be piecewise continuous in the interval $[0, c]$. Then, the Fourier–Bessel series of $f(x)$ converges to $[f(x+)+f(x-)]/2$ at each point in the open interval $(0, c)$.

Theorem 23.4.   Let $f(x)$ and $f'(x)$ be piecewise continuous in the

R.P. Agarwal, D. O'Regan, *Ordinary and Partial Differential Equations*,
Universitext, DOI 10.1007/978-0-387-79146-3_23,
© Springer Science+Business Media, LLC 2009

interval $[-\pi, \pi]$. Then, the Fourier trigonometric series of $f(x)$ converges to $[f(x+)+f(x-)]/2$ at each point in the open interval $(-\pi, \pi)$ and at $x = \pm\pi$ the series converges to $[f(-\pi+) + f(\pi-)]/2$.

Example 23.3.   Consider the function $f(x) = \begin{cases} 0, & x \in [-\pi, 0) \\ 1, & x \in [0, \pi]. \end{cases}$
Clearly, $f(x) \in C_p^1(-\pi, \pi)$ and has a single jump discontinuity at 0. For this function (see Problem 21.4(i)), the Fourier trigonometric coefficients are $a_0 = 1$, $a_n = 0$, $b_n = (1 - (-1)^n)/n\pi$. Thus, we have

$$f(x) \sim \frac{1}{2} + \frac{2}{\pi} \sum_{n=1}^{\infty} \frac{1}{(2n-1)} \sin(2n-1)x = F(x), \quad \text{say.} \qquad (23.1)$$

From Theorem 23.4 in (23.1) the equality $F(x) = f(x)$ holds at each point in the open intervals $(-\pi, 0)$ and $(0, \pi)$, whereas at $x = 0$ the right–hand side is $1/2$ which is the same as $[f(0+) + f(0-)]/2$. Also, at $x = \pm\pi$ the right–hand side is again $1/2$ which is the same as $[f(-\pi+) + f(\pi-)]/2$.

Example 23.4.   We shall use the Fourier trigonometric series

$$x + \frac{1}{4}x^2 \sim \frac{\pi^2}{12} + \sum_{n=1}^{\infty} \frac{(-1)^n}{n^2}[\cos nx - 2n \sin nx], \quad -\pi < x < \pi$$

to show that $\sum_{n=1}^{\infty}(1/n^2) = \pi^2/6$.

Since $f(x) = x + x^2/4 \in C_p^1(-\pi, \pi)$, by Theorem 23.4, we have

$$\frac{\pi^2}{12} + \sum_{n=1}^{\infty} \frac{(-1)^n}{n^2}[\cos nx - 2n \sin nx] = \begin{cases} [f(x+) + f(x-)]/2, & -\pi < x < \pi \\ [f(\pi-) + f(-\pi+)]/2, & x = \pm\pi. \end{cases}$$

Thus, at $x = \pi$, we find

$$\frac{\pi^2}{12} + \sum_{n=1}^{\infty} \frac{(-1)^n}{n^2}[\cos n\pi - 2n \sin n\pi] = \frac{f(\pi-) + f(-\pi+)}{2}$$

and hence

$$\frac{\pi^2}{12} + \sum_{n=1}^{\infty} \frac{(-1)^n}{n^2}[(-1)^n - 0] = \frac{(\pi + \pi^2/4) + (-\pi + \pi^2/4)}{2} = \frac{\pi^2}{4},$$

or

$$\sum_{n=1}^{\infty} \frac{1}{n^2} = \frac{\pi^2}{4} - \frac{\pi^2}{12} = \frac{\pi^2}{6}.$$

Example 23.5.   We shall find the Fourier trigonometric series of the function $\cos ax$ on $[-\pi, \pi]$ where $a \neq 0, \pm 1, \pm 2, \cdots$. For this, since $\cos ax$ is an even function, $b_n = 0$, $n \geq 1$ and

$$
\begin{aligned}
a_n &= \frac{2}{\pi} \int_0^\pi \cos ax \, \cos nx dx = \frac{1}{\pi} \int_0^\pi [\cos(nx - ax) + \cos(nx + ax)] dx \\
&= \frac{1}{\pi} \left[\frac{\sin(nx - ax)}{n - a} + \frac{\sin(nx + ax)}{n + a} \right] \Bigg|_0^\pi \\
&= \frac{1}{\pi} (-1)^{n+1} \left[\frac{\sin a\pi}{n - a} - \frac{\sin a\pi}{n + a} \right] = \frac{1}{\pi} (-a)^{n+1} \sin a\pi \, \frac{2a}{n^2 - a^2}.
\end{aligned}
$$

Hence, it follows that

$$
f(x) \sim \frac{2a}{\pi} \sin a\pi \left[\frac{1}{2a^2} + \sum_{n=1}^\infty \frac{(-1)^{n+1}}{n^2 - a^2} \cos nx \right].
$$

Now in view of Theorem 23.4 at $x = 0$, we have

$$
\frac{2a}{\pi} \sin a\pi \left[\frac{1}{2a^2} + \sum_{n=1}^\infty \frac{(-1)^{n+1}}{n^2 - a^2} \right] = \cos 0 = 1
$$

and hence

$$
\frac{a\pi}{\sin a\pi} = 1 + 2a^2 \sum_{n=1}^\infty \frac{(-1)^{n+1}}{n^2 - a^2}.
$$

Example 23.6. We shall use the Fourier trigonometric series

$$
|\sin x| \sim \frac{2}{\pi} - \frac{4}{\pi} \sum_{n=1}^\infty \frac{\cos 2nx}{4n^2 - 1}, \qquad -\pi < x < \pi
$$

to find $\sum_{n=1}^\infty (-1)^n / (16n^2 - 1)$.

Clearly, the function $f(x) = |\sin x| \in C_p^1[-\pi, \pi]$, and hence from Theorem 23.4 we have

$$
\frac{2}{\pi} - \frac{4}{\pi} \sum_{n=1}^\infty \frac{\cos 2nx}{4n^2 - 1} = \begin{cases} [f(x+) + f(x-)]/2, & -\pi < x < \pi \\ [f(\pi-) + f(-\pi+)]/2, & x = \pm\pi. \end{cases}
$$

At $x = \pi/4$, we find

$$
\frac{2}{\pi} - \frac{4}{\pi} \sum_{n=1}^\infty \frac{\cos(2n\pi/4)}{4n^2 - 1} = \frac{f(\pi/4+) + f(\pi/4-)}{2} = \frac{1}{\sqrt{2}}
$$

and hence

$$
\frac{2}{\pi} - \frac{4}{\pi} \left(\sum_{k=1}^\infty \frac{\cos(2k\pi/2)}{4(2k)^2 - 1} + \sum_{k=1}^\infty \frac{\cos(2k - 1)\pi/2}{4(2k - 1)^2 - 1} \right) = \frac{1}{\sqrt{2}},
$$

or

$$\frac{2}{\pi} - \frac{4}{\pi} \sum_{k=1}^{\infty} \frac{(-1)^k}{16k^2 - 1} = \frac{1}{\sqrt{2}},$$

which gives

$$\sum_{k=1}^{\infty} \frac{(-1)^k}{16k^2 - 1} = \frac{\pi}{4} \left(\frac{2}{\pi} - \frac{1}{\sqrt{2}} \right) = \frac{1}{2} - \frac{\pi}{4\sqrt{2}}.$$

Theorem 23.5. Let $f(x)$ and $f'(x)$ be piecewise continuous in the interval $[0, \pi]$. Then, the Fourier cosine series of $f(x)$ converges to $[f(x+) + f(x-)]/2$ at each point in the open interval $(0, \pi)$ and at $x = 0$ and π the series converges, respectively, to $f(0+)$ and $f(\pi-)$.

Example 23.7. Since

$$x^2 \sim \frac{\pi^2}{3} + 4 \sum_{n=1}^{\infty} \frac{(-1)^n}{n^2} \cos nx, \quad 0 < x < \pi$$

from Theorem 23.5 it follows that

$$\frac{\pi^2}{3} + 4 \sum_{n=1}^{\infty} \frac{(-1)^n}{n^2} \cos nx = \begin{cases} f(x) = x^2, & 0 < x < \pi \\ f(0+) = 0, & x = 0 \\ f(\pi-) = \pi^2, & x = \pi. \end{cases}$$

Thus, at $x = \pi$, we have

$$\frac{\pi^2}{3} + 4 \sum_{n=1}^{\infty} \frac{(-1)^n}{n^2} \cos n\pi = \pi^2$$

and hence

$$\frac{\pi^2}{3} + 4 \sum_{n=1}^{\infty} \frac{1}{n^2} = \pi^2,$$

which gives $\sum_{n=1}^{\infty} (1/n^2) = \pi^2/6$.

Theorem 23.6. Let $f(x)$ and $f'(x)$ be piecewise continuous in the interval $[0, \pi]$. Then, the Fourier sine series of $f(x)$ converges to $[f(x+) + f(x-)]/2$ at each point in the open interval $(0, \pi)$ and at $x = 0$ and π the series converges to 0.

Example 23.8. Since

$$\pi - x \sim 2 \sum_{n=1}^{\infty} \frac{\sin nx}{n}, \quad 0 < x < \pi$$

from Theorem 23.6 it follows that

$$2 \sum_{n=1}^{\infty} \frac{\sin nx}{n} = \begin{cases} f(x) = \pi - x, & 0 < x < \pi \\ 0, & x = 0, \pi. \end{cases}$$

Thus, at $x = \pi/2$ we have

$$2 \sum_{n=1}^{\infty} \frac{\sin n\pi/2}{n} = \pi - \frac{\pi}{2}$$

and hence

$$1 - \frac{1}{3} + \frac{1}{5} - \frac{1}{7} + \cdots = \frac{\pi}{4}. \tag{23.2}$$

Example 23.9. For the function $f(x) = \begin{cases} 0, & -3 < x < 0 \\ 1, & 0 < x < 3 \end{cases}$ in view of (20.12) we have

$$a_0 = \frac{1}{3} \int_{-3}^{3} f(x)dx = 1, \quad a_n = \frac{1}{3} \int_{-3}^{3} f(x) \cos \frac{n\pi x}{3} dx = 0$$

and

$$b_n = \frac{1}{3} \int_{-3}^{3} f(x) \sin \frac{n\pi x}{3} dx = \frac{1 - (-1)^n}{n\pi}, \quad n \geq 1$$

and hence the Fourier trigonometric series (20.11) for this function is

$$f(x) \sim \frac{1}{2} + \sum_{n=1}^{\infty} \frac{1 - (-1)^n}{n\pi} \sin \frac{n\pi x}{3} = F(x), \text{ say.}$$

Now from Theorem 23.4 (over the interval $[-3, 3]$) it follows that

$$F(x) = \begin{cases} f(x), & x \in (-3, 0) \cup (0, 3) \\ 1/2, & x = 0, -3, 3. \end{cases}$$

For the differentiation and integration of Fourier trigonometric series we have the following results.

Theorem 23.7. Suppose $f(x)$ is continuous in $(-\pi, \pi)$, $f(-\pi) = f(\pi)$ and $f'(x)$ is piecewise continuous. Then,

$$f'(x) = \sum_{n=1}^{\infty} (-na_n \sin nx + nb_n \cos nx)$$

at each point $x \in (-\pi, \pi)$ where $f''(x)$ exists.

Theorem 23.8. Suppose $f(x)$ is piecewise continuous in $(-\pi, \pi)$ and

$$f(x) \sim \frac{a_0}{2} + \sum_{n=1}^{\infty} (a_n \cos nx + b_n \sin nx), \quad -\pi < x < \pi.$$

Then, for $x \in [-\pi, \pi]$

$$\int_{-\pi}^{x} f(t)dt = \int_{-\pi}^{x} \frac{a_0}{2} dt + \sum_{n=1}^{\infty} \int_{-\pi}^{x} (a_n \cos nt + b_n \sin nt)dt$$

$$= \frac{a_0}{2}(x+\pi) + \sum_{n=1}^{\infty} \frac{1}{n} [a_n \sin nx - b_n (\cos nx - (-1)^n)].$$

Example 23.10. For the function $f(x) = |x|$, $x \in [-\pi, \pi]$ we have (see Problem 21.4(iv))

$$f(x) \sim \frac{\pi}{2} - \frac{4}{\pi} \sum_{n=1}^{\infty} \frac{\cos(2n-1)x}{(2n-1)^2}, \quad -\pi < x < \pi.$$

Clearly, $f(x)$ is continuous on $[-\pi, \pi]$, $f(-\pi) = f(\pi) = \pi$, $f'(x) = \begin{cases} -1, & -\pi < x < 0 \\ 1, & 0 < x < \pi \end{cases}$, $f'(0)$ does not exist, $f'(x) \in C_p(-\pi, \pi)$, $f''(x) = 0$, $x \in (-\pi, 0) \cup (0, \pi)$ and $f''(0)$ does not exist. Thus, Theorem 23.7 is applicable and we have

$$\begin{cases} -1, & -\pi < x < 0 \\ 1, & 0 < x < \pi \end{cases} = -\frac{4}{\pi} \sum_{n=1}^{\infty} \frac{-(2n-1)\sin(2n-1)x}{(2n-1)^2} = \frac{4}{\pi} \sum_{n=1}^{\infty} \frac{\sin(2n-1)x}{2n-1}.$$

$$(23.3)$$

At $x = \pi/2$ this series immediately gives (23.2). Also, note that at $x = 0$ the left-hand side of (23.3) is not defined, but its right-hand side is zero.

Example 23.11. We shall use the relation

$$x^2 \sim \frac{\pi^2}{3} + 4 \sum_{n=1}^{\infty} \frac{(-1)^n}{n^2} \cos nx, \quad -\pi < x < \pi$$

to find a polynomial $p(x)$ such that

$$p(x) = \sum_{n=1}^{\infty} \frac{(-1)^n}{n^3} \sin nx, \quad -\pi \leq x \leq \pi.$$

Clearly, for the function $f(x) = x^2$ all the conditions of Theorem 23.8 are satisfied. Thus, on integrating the above relation from 0 to x, we get

$$\int_0^x t^2 dt = \int_0^x \frac{\pi^2}{3} dt + 4 \sum_{n=1}^{\infty} \int_0^x \frac{(-1)^n}{n^2} \cos nt dt,$$

or

$$\frac{x^3}{3} = \frac{\pi^2 x}{3} + 4 \sum_{n=1}^{\infty} \frac{(-1)^n}{n^3} \sin nx$$

and hence

$$\sum_{n=1}^{\infty} \frac{(-1)^n}{n^3} \sin nx = \frac{x^3 - \pi^2 x}{12} = p(x), \quad -\pi \le x \le \pi.$$

Problems

23.1. Functions $\phi_1(x) = 1$, $\phi_2(x) = \sqrt{3}(2x - 1)$ are orthonormal on $0 < x < 1$. Define

$$F(a, b) = \int_0^1 \left(\sqrt{3}x^2 - a\phi_1(x) - b\phi_2(x) \right)^2 dx.$$

Find a, b such that $F(a, b)$ attains its minimum. Find also the minimum value of $F(a, b)$.

23.2. Consider the function $f(x) = \begin{cases} 1, & 0 < x \le \pi/2 \\ 0, & \pi/2 < x < \pi. \end{cases}$ It is given

that $\phi_n(x) = \sqrt{\frac{2}{\pi}} \sin nx$, $n = 1, 2, 3$ are orthonormal on the interval $0 < x < \pi$. Find the values of a, b at which the integral

$$F(a, b) = \int_0^{\pi} (f(x) - a\phi_1(x) - b(\phi_2(x) + \phi_3(x)))^2 dx$$

attains its minimum.

23.3. Let $f(x) = \begin{cases} 0, & x \in [-1, 0) \\ 1, & x \in [0, 1]. \end{cases}$ Show that

$$\int_{-1}^1 \left(f(x) - \frac{1}{2} - \frac{3}{4}x \right)^2 dx \le \int_{-1}^1 (f(x) - d_0 - d_1 x - d_2 x^2)^2 dx$$

for any set of constants d_0, d_1 and d_2.

23.4. Show that the following cannot be the Fourier series representation for any piecewise continuous function

(i) $\displaystyle\sum_{n=1}^{\infty} n^{1/n} \phi_n(x)$ (ii) $\displaystyle\sum_{n=1}^{\infty} \frac{1}{\sqrt{n}} \phi_n(x)$.

23.5. Find Parseval's equality for the function $f(x) = 1$, $x \in [0, c]$ with respect to the orthonormal set $\left\{ \sqrt{\frac{2}{c}} \sin \frac{n\pi x}{c}, \ n = 1, 2, \cdots \right\}$.

23.6. Consider the function $f(x) = \begin{cases} 0, & -\pi < x < 0 \\ \sqrt{x}, & 0 < x < \pi. \end{cases}$ Use Parseval's equality (22.11) to show that

$$\sum_{n=1}^{\infty} \left(\int_{-\pi}^{\pi} f(x) \cos nx dx \right) \left(\int_{-\pi}^{\pi} f(x) \sin nx dx \right) \leq \frac{5\pi^3}{36}.$$

23.7. Assume that Parseval's equality holds for piecewise continuous functions. If $\{a_0, a_1, b_1, \cdots\}$ and $\{\alpha_0, \alpha_1, \beta_1, \cdots\}$ are the sets of Fourier coefficients of piecewise continuous functions f and g, respectively, on $[-\pi, \pi]$. Show that

$$\frac{1}{2}a_0\alpha_0 + \sum_{n=1}^{\infty}(a_n\alpha_n + b_n\beta_n) = \frac{1}{\pi} \int_{-\pi}^{\pi} f(x)g(x)dx.$$

23.8. Let $f(x)$ and $g(x)$ be piecewise continuous in the interval $[\alpha, \beta]$ and have the same Fourier coefficients with respect to a complete orthonormal set. Show that $f(x) = g(x)$ at each point of $[\alpha, \beta]$ where both functions are continuous.

23.9. (i) Let $\psi_n(x) = n\sqrt{x}e^{-nx^2/2}$, $x \in [0, 1]$. Show that $\psi_n(x) \to 0$ as $n \to \infty$ for each x in $[0, 1]$. Further, show that

$$e_n^2 = \int_0^1 (\psi_n(x) - 0)^2 dx = \frac{n}{2}(1 - e^{-n}),$$

and hence $e_n \to \infty$ as $n \to \infty$. Thus, pointwise convergence does not imply convergence in the mean.

(ii) Let $\psi_n(x) = x^n$, $x \in [0, 1]$, and $f(x) = 0$ in $[0, 1]$. Show that

$$e_n^2 = \int_0^1 (\psi_n(x) - f(x))^2 dx = \frac{1}{2n + 1},$$

and hence $\psi_n(x)$ converges in the mean to $f(x)$. Further, show that $\psi_n(x)$ does not converge to $f(x)$ pointwise in $[0, 1]$. Thus, mean convergence does not imply pointwise convergence.

23.10. Show that the sequence $\{x/(x + n)\}$ converges pointwise on $[0, \infty)$ and uniformly on $[0, a]$, $a > 0$.

23.11. Consider the function $f(x) = \begin{cases} 0, & -\pi \leq x \leq 0 \\ \sin x, & 0 < x \leq \pi. \end{cases}$ The

Fourier trigonometric series of $f(x)$ is given by

$$f(x) \sim \frac{1}{\pi} + \frac{1}{2}\sin x - \frac{2}{\pi}\sum_{n=1}^{\infty}\frac{\cos 2nx}{4n^2 - 1}.$$

(i)   Show that the above series converges to $f(x)$ everywhere in the interval $[-\pi, \pi]$.

(ii)   Use Part (i) to show that

$$\sum_{n=1}^{\infty}\frac{(-1)^n}{4n^2 - 1} = \frac{1}{2} - \frac{\pi}{4}.$$

(iii)   Find the sum of the series $\sum_{n=1}^{\infty} 1/(4n^2 - 1)$.

23.12.   The Fourier series of e^x on the interval $-\pi < x < \pi$ is given by

$$e^x \sim \frac{2\sinh\pi}{\pi}\left[\frac{1}{2} + \sum_{n=1}^{\infty}\frac{(-1)^n}{n^2 + 1}(\cos nx - n\sin nx)\right].$$

(i)   Use the above correspondence to show that

$$\sum_{n=1}^{\infty}\frac{1}{n^2 + 1} = \frac{\pi}{2}\coth\pi - \frac{1}{2}.$$

(ii)   Find the sum of the series $\displaystyle\sum_{n=1}^{\infty}\frac{(-1)^n}{n^2 + 1}$. Justify your answer.

23.13.   Let $f(x)$ be a twice continuously differentiable, periodic function with a period 2π. Show that

(i)   the Fourier trigonometric coefficients a_n and b_n of $f(x)$ satisfy

$$|a_n| \leq \frac{M}{n^2} \quad\text{and}\quad |b_n| \leq \frac{M}{n^2}, \quad n = 1, 2, \cdots$$

where $M = \frac{1}{\pi}\int_{-\pi}^{\pi}|f''(x)|dx$

(ii)   the Fourier trigonometric series of $f(x)$ converges uniformly to $f(x)$ on $[-\pi, \pi]$.

23.14.   Suppose that $f(x)$ is continuous in $[-\pi, \pi]$, $f(-\pi) = f(\pi)$, and $f'(x)$ is piecewise continuous in $[-\pi, \pi]$. Show that

(i) $\sum_{n=1}^{\infty}|a_n|$ and $\sum_{n=1}^{\infty}|b_n|$ converge,

(ii) $na_n \to 0$ and $nb_n \to 0$,

(iii) the Fourier trigonometric series of $f(x)$ converges to $f(x)$ absolutely and uniformly in $[-\pi, \pi]$.

23.15. The proof of Theorem 23.4 requires the following steps:

(i) Establish Lagrange's trigonometric identity

$$2 \sum_{n=1}^{N} \cos n\theta = -1 + \frac{\sin \left(N + \frac{1}{2}\right) \theta}{\sin \frac{1}{2}\theta}.$$

(ii) If $f(x)$ is piecewise continuous in the interval $[a, b]$, then show that

$$\lim_{k \to \infty} \int_a^b f(x) \sin kx\, dx = 0.$$

(iii) If $f(x)$ is piecewise continuous in the interval $[0, b]$ and has a right-hand derivative at $x = 0$, then show that

$$\lim_{k \to \infty} \int_0^b f(x) \frac{\sin kx}{x}\, dx = \frac{\pi}{2} f(+0).$$

(iv) Show that

$$S_n(x) = \frac{1}{2}a_0 + \sum_{m=1}^{n} (a_m \cos mx + b_m \sin mx)$$

$$= \frac{1}{\pi} \int_{-\pi}^{\pi} f(t) \frac{\sin \left((n + \frac{1}{2})(t - x)\right)}{2 \sin \left(\frac{1}{2}(t - x)\right)}\, dt.$$

(v) If $f(x)$ is piecewise continuous in the interval (a, b) and has derivatives from the right and left at a point $x = x_0$ where $a < x_0 < b$, then show that

$$\lim_{k \to \infty} \int_a^b f(x) \frac{\sin k(x - x_0)}{x - x_0}\, dx = \frac{\pi}{2}[f(x_0 + 0) + f(x_0 - 0)].$$

Answers or Hints

23.1. In view of Theorem 22.1, $F(a, b)$ is minimum when $a = <\sqrt{3}x^2, \phi_1> = 1/\sqrt{3}$, $b = <\sqrt{3}x^2, \phi_2> = 1/2$. Now from (22.4), $\min F(a, b) = \|\sqrt{3}x^2\|^2 - a^2 - b^2 = 1/60$.

23.2. Let $\psi(x) = (\phi_2(x) + \phi_3(x))/\sqrt{2}$. Then, $\phi_1(x), \psi(x)$ are orthonormal on $0 < x < \pi$. Thus, $F(a, b)$ is the same as $F(a, b) = \int_0^\pi (f(x) - a\phi_1(x) - (\sqrt{2}b)\psi(x))^2\, dx$. Now $F(a, b)$ is minimum when $a = \sqrt{2/\pi}$, $\sqrt{2}b = 4/(3\sqrt{\pi})$.

23.3. For the given function Fourier–Legendre coefficients are $c_0 = 1/2$, $c_1 = 3/4$, $c_2 = 0$.

23.4. Use Theorem 22.3(i).

23.5. $\|f\|^2 = c$, $c_n^2 = \begin{cases} 0, & n \text{ even} \\ 8c/(n^2\pi^2), & n \text{ odd}. \end{cases}$ Thus, $c = \frac{8c}{\pi^2} \sum_{n=1}^{\infty} \frac{1}{(2n-1)^2}$.

23.6. Left-hand side $= \sum_{n=1}^{\infty}(\pi a_n)(\pi b_n) \le \frac{\pi^2}{2}\sum_{n=1}^{\infty}(a_n^2 + b_n^2)$.

23.7. Add and subtract Fourier trigonometric series of f and g, write Parseval's equalities for $(f+g)$ and $(f-g)$, use $(u+v)^2 - (u-v)^2 = 4uv$.

23.8. Let $h(x) = f(x) - g(x)$ and $\{\phi_n(x)\}$ be a complete orthonormal set on the interval $[\alpha, \beta]$ with respect to the weight function $r(x)$. Note that for the function $h(x)$ Fourier coefficients $c_n = 0$, $n \ge 0$. Now apply Theorem 22.4.

23.9. (i) Use l'Hospital's rule to show that $\psi_n(x) \to 0$ as $n \to \infty$, $x \in [0,1]$ (ii) $\psi_n(x) \to 0$ as $n \to \infty$, $x \in [0,1)$, and $\psi_n(1) = 1$, $n \ge 0$.

23.10. Use definition.

23.11. (i) Use Theorem 23.4 (ii) Let $x = \pi/2$ (iii) $1/2$.

23.12. (i) Use Theorem 23.4 and let $x = \pi$ (ii) Let $x = 0$, $\sum_{n=1}^{\infty} \frac{(-1)^n}{n^2+1} = \frac{\pi}{2\sinh\pi} - \frac{1}{2}$.

23.13. (i) $a_n = \frac{1}{\pi}\int_{-\pi}^{\pi} f(x)\cos nx\, dx = -\frac{1}{n^2\pi}\int_{-\pi}^{\pi} f''(x)\cos nx\, dx$ (ii) $\left|\frac{1}{2}a_0 + \sum_{n=1}^{\infty}(a_n\cos nx + b_n\sin nx)\right| \le \frac{1}{2}|a_0| + \sum_{n=1}^{\infty}(|a_n| + |b_n|)$.

23.14. (i) If $f'(x) \sim (\alpha_0/2) + \sum_{n=1}^{\infty}(\alpha_n\cos nx + \beta_n\sin nx)$, then $\alpha_0 = 0$, $\alpha_n = nb_n$, $\beta_n = -na_n$. Now Bessel's inequality applied to $f'(x)$ implies that $\sum_{n=1}^{\infty}\beta_n^2$ converges. Since $\rho_n = \sum_{k=1}^{n}\frac{|\beta_k|}{k} \le \sqrt{\sum_{k=1}^{n}\beta_k^2}\sqrt{\sum_{k=1}^{n}\frac{1}{k^2}}$, for each n, ρ_n is bounded. Further, since ρ_n is increasing, the sequence $\{\rho_n\}$ converges. Thus, $\rho_n = \sum_{k=1}^{n}\frac{|\beta_k|}{k} = \sum_{k=1}^{n}|a_k|$ converges (ii) The convergence of $\sum_{n=1}^{\infty}\beta_n^2$ implies that $\beta_n = -na_n \to 0$ (iii) From Theorem 23.4, $f(x) = (a_0/2) + \sum_{n=1}^{\infty}(a_n\cos nx + b_n\cos nx)$, $x \in [-\pi, \pi]$. Finally, note that $|(a_0/2) + \sum_{n=1}^{\infty}(a_n\cos nx + b_n\cos nx)| \le (|a_0|/2) + \sum_{n=1}^{\infty}(|a_n| + |b_n|)$ and use (i)

Lecture 24
Fourier Series Solutions of Ordinary Differential Equations

In this lecture we shall use Fourier series expansions to find periodic particular solutions of nonhomogeneous DEs, and solutions of nonhomogeneous self-adjoint DEs satisfying homogeneous boundary conditions.

We begin with the second-order nonhomogeneous DE

$$y'' + ay' + by = f(x), \quad x \in [-\pi, \pi], \tag{24.1}$$

where a, b are constants, and the function $f(x)$ is periodic of period 2π, and satisfies the conditions of Theorem 23.4 so that it can be expanded in a Fourier trigonometric series (20.8). Our interest here is to find a *periodic* particular solution of (24.1), which we shall denote by $y_p(x)$ and call it a *steady periodic solution*. We assume that

$$y_p(x) = \frac{A_0}{2} + \sum_{n=1}^{\infty} (A_n \cos nx + B_n \sin nx) \tag{24.2}$$

so that the term by term differentiation gives

$$y_p'(x) = \sum_{n=1}^{\infty} (-nA_n \sin nx + nB_n \cos nx) \tag{24.3}$$

and

$$y_p''(x) = \sum_{n=1}^{\infty} (-n^2 A_n \cos nx - n^2 B_n \sin nx). \tag{24.4}$$

Substituting (24.2) – (24.4) and (20.8) in (24.1), we get

$$b\frac{A_0}{2} + \sum_{n=1}^{\infty} ([-n^2 A_n + anB_n + bA_n] \cos nx + [-n^2 B_n - anA_n + bB_n] \sin nx)$$

$$= \frac{a_0}{2} + \sum_{n=1}^{\infty} (a_n \cos nx + b_n \sin nx).$$

Now matching the coefficients, we find

$$bA_0 = a_0$$
$$(b - n^2)A_n + anB_n = a_n$$
$$-anA_n + (b - n^2)B_n = b_n.$$

R.P. Agarwal, D. O'Regan, *Ordinary and Partial Differential Equations*,
Universitext, DOI 10.1007/978-0-387-79146-3_24,
© Springer Science+Business Media, LLC 2009

Solving this system, we obtain

$$A_0 = \frac{a_0}{b}, \quad A_n = \frac{(b - n^2)a_n - anb_n}{(b - n^2)^2 + a^2n^2}, \quad B_n = \frac{(b - n^2)b_n + ana_n}{(b - n^2)^2 + a^2n^2}. \quad (24.5)$$

Since for a given function $f(x)$ the coefficients a_n, $n \geq 0$ and b_n, $n \geq 1$ are known from (20.9), the unknowns A_n, $n \geq 0$ and B_n, $n \geq 1$ can be determined explicitly from (24.5).

Example 24.1.   We shall find the steady periodic solution of the nonhomogeneous DE

$$y'' + 3y = f(x) = \begin{cases} 1, & -\pi < x < 0 \\ 2, & 0 < x < \pi. \end{cases}$$

From Problem 21.4 (ii), we have

$$f(x) = \frac{3}{2} + \frac{2}{\pi}\left(\sin x + \frac{1}{3}\sin 3x + \frac{1}{5}\sin 5x + \cdots\right)$$

and hence $a_n = 0$, $n = 1, 2, \cdots$, and

$$a_0 = 3, \quad b_{2n-1} = \frac{2}{\pi(2n - 1)}, \quad b_{2n} = 0, \quad n = 1, 2, \cdots.$$

Now since $a = 0$, $b = 3$ from (24.5), it follows that $A_n = 0$, $n = 1, 2, \cdots$ and

$$A_0 = 1, \quad B_{2n-1} = \frac{2}{\pi(2n - 1)[3 - (2n - 1)^2]}, \quad B_{2n} = 0, \quad n = 1, 2, \cdots.$$

Hence, the required particular solution is

$$y_p(x) = \frac{1}{2} + \sum_{n=1}^{\infty} \frac{2}{\pi(2n - 1)[3 - (2n - 1)^2]} \sin(2n - 1)x.$$

Using the same procedure we can obtain a steady periodic solution of (24.1) over any interval $[-L, L]$, $L > 0$. For this, we need to use the Fourier series (20.11) of $f(x)$ with the coefficients a_n, b_n given in (20.12), and let

$$y_p(x) = \frac{A_0}{2} + \sum_{n=1}^{\infty}\left(A_n \cos\frac{n\pi x}{L} + B_n \sin\frac{n\pi x}{L}\right). \quad (24.6)$$

The unknown A_n, $n \geq 0$ and B_n, $n \geq 1$ are then determined from the relations

$$A_0 = \frac{a_0}{b}, \quad A_n = \frac{\left(b - \frac{n^2\pi^2}{L^2}\right)a_n - a\frac{n\pi}{L}b_n}{\left(b - \frac{n^2\pi^2}{L^2}\right)^2 + a^2\frac{n^2\pi^2}{L^2}}, \quad B_n = \frac{\left(b - \frac{n^2\pi^2}{L^2}\right)b_n + a\frac{n\pi}{L}a_n}{\left(b - \frac{n^2\pi^2}{L^2}\right)^2 + a^2\frac{n^2\pi^2}{L^2}}.$$

$$(24.7)$$

Example 24.2.   We shall find the steady periodic solution of the nonhomogeneous DE

$$y'' + 5y = x, \quad -2 < x < 2.$$

Since

$$x = \frac{4}{\pi} \sum_{n=1}^{\infty} \frac{(-1)^{n+1}}{n} \sin \frac{n\pi}{2} x$$

we have $a_0 = 0$, $a_n = 0$, $b_n = 4(-1)^{n+1}/(n\pi)$, and since $a = 0$, $b = 5$ from (24.7) we obtain $A_0 = 0$, $A_n = 0$ and

$$B_n = \frac{b_n}{\left(5 - \frac{n^2\pi^2}{4}\right)} = \frac{16(-1)^{n+1}}{n\pi(20 - n^2\pi^2)}.$$

Thus, the required particular solution is

$$y_p(x) = \frac{16}{\pi} \sum_{n=1}^{\infty} \frac{(-1)^{n+1}}{\pi(20 - n^2\pi^2)} \sin \frac{n\pi}{2} x.$$

Next we shall consider the nonhomogeneous self-adjoint DE

$$(p(x)y')' + q(x)y + \mu r(x)y = P_2[y] + \mu r(x)y = f(x) \tag{24.8}$$

together with the homogeneous boundary conditions (19.2). In (24.8) functions $p(x)$, $q(x)$ and $r(x)$ are assumed to satisfy the same restrictions as in (19.1), μ is a given constant and $f(x)$ is a given function in $[\alpha, \beta]$. For the nonhomogeneous boundary value problem (24.8), (19.2) we shall assume that the solution $y(x)$ can be expanded in terms of eigenfunctions $\phi_n(x)$, $n = 1, 2, \cdots$ of the corresponding homogeneous Sturm–Liouville problem (19.1), (19.2), i.e., $y(x) = \sum_{n=1}^{\infty} c_n \phi_n(x)$. To compute the coefficients c_n in this expansion first we note that the infinite series $\sum_{n=1}^{\infty} c_n \phi_n(x)$ does satisfy the boundary conditions (19.2) since each $\phi_n(x)$ does so. Next consider the DE (24.8) that $y(x)$ must satisfy. For this, we have

$$P_2\left[\sum_{n=1}^{\infty} c_n \phi_n(x)\right] + \mu r(x) \sum_{n=1}^{\infty} c_n \phi_n(x) = f(x).$$

Thus, if we can interchange the operations of summation and differentiation, then

$$\sum_{n=1}^{\infty} c_n P_2[\phi_n(x)] + \mu r(x) \sum_{n=1}^{\infty} c_n \phi_n(x) = f(x).$$

Since $P_2[\phi_n(x)] = -\lambda_n r(x)\phi_n(x)$, this relation is the same as

$$\sum_{n=1}^{\infty} (\mu - \lambda_n) c_n \phi_n(x) = \frac{f(x)}{r(x)}. \tag{24.9}$$

Now we assume that the function $f(x)/r(x)$ satisfies the conditions of Theorem 23.1, so that it can be written as

$$\frac{f(x)}{r(x)} = \sum_{n=1}^{\infty} d_n \phi_n(x),$$

where from (20.2) the coefficients d_n are given by

$$d_n = \frac{1}{\|\phi_n\|^2} \int_{\alpha}^{\beta} r(x)\phi_n(x)\frac{f(x)}{r(x)}dx = \frac{1}{\|\phi_n\|^2} \int_{\alpha}^{\beta} \phi_n(x)f(x)dx. \qquad (24.10)$$

With this assumption (24.9) takes the form

$$\sum_{n=1}^{\infty} [(\mu - \lambda_n)c_n - d_n]\phi_n(x) = 0.$$

Since this equation holds for each x in $[\alpha, \beta]$, it is necessary that

$$(\mu - \lambda_n)c_n - d_n = 0, \quad n = 1, 2, \cdots. \qquad (24.11)$$

Thus, if μ is not equal to any eigenvalue of the corresponding homogeneous Sturm–Liouville problem (19.1), (19.2), i.e., $\mu \neq \lambda_n$, $n = 1, 2, \cdots$, then

$$c_n = \frac{d_n}{\mu - \lambda_n}, \quad n = 1, 2, \cdots. \qquad (24.12)$$

Hence, the solution $y(x)$ of the nonhomogeneous problem (24.8), (19.2) can be written as

$$y(x) = \sum_{n=1}^{\infty} \frac{d_n}{\mu - \lambda_n}\phi_n(x). \qquad (24.13)$$

Of course, the convergence of (24.13) is yet to be established.

If $\mu = \lambda_m$, then for $n = m$ equation (24.11) is of the form $0 \cdot c_m - d_m = 0$. Thus, if $d_m \neq 0$ then it is impossible to solve (24.11) for c_m, and hence the nonhomogeneous problem (24.8), (19.2) has no solution. Further, if $d_m = 0$ then (24.11) is satisfied for any arbitrary value of c_m, and hence the nonhomogeneous problem (24.8), (19.2) has an infinite number of solutions. From (24.10), $d_m = 0$ if and only if

$$\int_{\alpha}^{\beta} \phi_m(x)f(x)dx = 0,$$

i.e., $f(x)$ in (24.8) is orthogonal to the eigenfunction $\phi_m(x)$.

This formal discussion for the problem (24.8), (19.2) is summarized in the following theorem.

Theorem 24.1. Let $f(x)$ be continuous in the interval $[\alpha, \beta]$. Then, the nonhomogeneous boundary value problem (24.8), (19.2) has a unique solution provided μ is different from all eigenvalues of the corresponding homogeneous Sturm–Liouville problem (19.1), (19.2). This solution $y(x)$ is given by (24.13), and the series converges for each x in $[\alpha, \beta]$. If μ is equal to an eigenvalue λ_m of the corresponding homogeneous Sturm–Liouville problem (19.1), (19.2) then the nonhomogeneous problem (24.8), (19.2) has no solution unless $f(x)$ is orthogonal to $\phi_m(x)$, i.e., unless

$$\int_\alpha^\beta \phi_m(x)f(x)dx = 0.$$

Further, in this case the solution is not unique.

Alternatively, this result can be stated as follows:

Theorem 24.2 (Fredholm's Alternative). For a given constant μ and a continuous function $f(x)$ in $[\alpha, \beta]$ the nonhomogeneous problem (24.8), (19.2) has a unique solution, or else the corresponding homogeneous problem (19.1), (19.2) has a nontrivial solution.

Example 24.3. Consider the nonhomogeneous boundary value problem

$$y'' + \pi^2 y = x - x^2$$
$$y(0) + y'(0) = 0 = y(1). \tag{24.14}$$

This problem can be solved directly to obtain the unique solution

$$y(x) = \frac{2}{\pi^4} \cos \pi x - \frac{1}{\pi^3}\left(1 + \frac{4}{\pi^2}\right)\sin \pi x + \frac{1}{\pi^2}\left(x - x^2 + \frac{2}{\pi^2}\right). \tag{24.15}$$

From Example 19.2 we know that π^2 is not an eigenvalue of the Sturm–Liouville problem (19.3), (19.9). Thus, from Theorem 24.1 the nonhomogeneous problem (24.14) has a unique solution. To find this solution in terms of the eigenvalues λ_n and eigenfunctions $\phi_n(x)$ of (19.3), (19.9) we note that the function $f(x) = x - x^2$ has been expanded in Example 21.4, and hence from (24.9) and (24.11), we have

$$d_0 = \frac{1}{4}, \quad d_n = \frac{2}{\lambda_n^{3/2} \sin^2 \sqrt{\lambda_n}}(2 - (2 + \lambda_n)\cos\sqrt{\lambda_n}), \quad n \geq 1.$$

Thus, from (24.13) we find that the solution $y(x)$ of (24.14) has the expansion

$$y(x) = \frac{1}{4\pi^2}(1-x) + \sum_{n=1}^{\infty} \frac{2}{(\pi^2 - \lambda_n)\lambda_n^{3/2}\sin^2\sqrt{\lambda_n}}$$
$$\times (2 - (2+\lambda_n)\cos\sqrt{\lambda_n})\sin\sqrt{\lambda_n}(1-x). \tag{24.16}$$

Problems

24.1. Find periodic particular solutions of the following nonhomogeneous DEs:

(i) $\quad y'' + 3y = \begin{cases} x, & -\pi < x < 0 \\ 0, & 0 < x < \pi \end{cases}$

(ii) $\quad y'' + y = \begin{cases} -x, & -1 < x < 0 \\ x, & 0 < x < 1 \end{cases}$

(iii) $\quad y'' + 2y' + 3y = e^x, \quad -\pi < x < \pi$

(iv) $\quad y'' + 3y' + 7y = \begin{cases} 1, & -\pi < x < 0 \\ x, & 0 < x < \pi. \end{cases}$

24.2. Use Fourier trigonometric series to solve the following initial value problems:

(i) $\quad y' + y = \dfrac{4}{\pi} \displaystyle\sum_{n=1}^{\infty} \dfrac{\sin(2n-1)\pi x}{2n-1}, \quad y(0) = 0$

(ii) $\quad y' + y = \displaystyle\sum_{n=1}^{\infty} \dfrac{\cos nx}{n!}, \quad y(0) = 0$

(iii) $\quad y'' + 4y = \dfrac{4}{\pi} \displaystyle\sum_{n=1}^{\infty} \dfrac{\sin(2n-1)x}{2n-1}, \quad y(0) = y'(0) = 0$

(iv) $\quad y'' + y = \dfrac{4}{\pi} \displaystyle\sum_{n=1}^{\infty} \dfrac{\sin(2n-1)x}{2n-1}, \quad y(0) = y'(0) = 0.$

24.3. Solve the following nonhomogeneous boundary value problems by means of an eigenfunction expansion:

(i) $\quad y'' + 4y = 4x, \quad y(0) = 0 = y(1)$

(ii) $\quad y'' + 11y = x^2, \quad y(0) = 0 = y(1)$

(iii) $\quad y'' + 3y = e^x, \quad y(0) = 0 = y(1)$

(iv) $\quad y'' - x = x, \quad y(0) = 0 = y'(\pi)$

(v) $\quad y'' + 2y = -x, \quad y'(0) = 0 = y(1) + y'(1)$

(vi) $\quad y'' + 2y = x, \quad y'(0) = 0 = y'(\pi)$

(vii) $(xy')' + \dfrac{3}{x}y = \dfrac{1}{x}\sin(\ln x), \quad y(1) = y(2) = 0.$

24.4. A simply supported beam of length a has a constant load q_0/a distributed over its length. The small deflections of the beam are governed by the boundary value problem

$$EIy^{(4)} = \frac{q_0}{a}, \quad y(0) = y''(0) = 0, \quad y(a) = y''(a) = 0.$$

Show that the deflections are given by the series

$$y(x) = \frac{4q_0 a^4}{EI\pi^5} \sum_{n=0}^{\infty} \frac{1}{(2n+1)^5} \sin\frac{(2n+1)\pi x}{a}.$$

Answers or Hints

24.1. (i) $-\frac{\pi}{12} - \frac{2}{\pi} \sum_{n=1}^{\infty} \frac{1}{(2n-1)^2[(2n-1)^2-3]} \cos(2n-1)x$

$+ \sum_{n=1}^{\infty} \frac{(-1)^n}{n(n^2-3)} \sin nx$ (ii) $\frac{1}{2} + \frac{4}{\pi^2} \sum_{n=1}^{\infty} \frac{1}{(2n-1)^2[(2n-1)^2\pi^2-1]} \cos(2n-1)n\pi x$

(iii) $a = 2$, $b = 3$, $a_0 = \sinh\pi$, $a_n = \frac{(-1)^n}{1+n^2}\sinh\pi$, $b_n = \frac{(-1)^{n-1}n}{1+n^2}\sinh\pi$

(iv) $a = 3$, $b = 7$, $a_0 = 1 + \frac{\pi}{2}$, $a_n = \frac{(-1)^n-1}{\pi n^2}$, $b_n = \frac{(-1)^n(1-\pi)-1}{\pi n}$.

24.2. (i) $4 \sum_{n=1}^{\infty} \frac{1}{1+(2n-1)^2\pi^2} \left(e^{-x} + \frac{1}{\pi(2n-1)} \sin(2n-1)\pi x - \cos(2n-1)\pi x \right)$

(ii) $\sum_{n=1}^{\infty} \frac{1}{n!(n^2+1)} \left(-e^{-x} + \cos nx + n\sin nx \right)$ (iii) $\frac{2}{\pi} \sum_{n=1}^{\infty} \frac{1}{4-(2n-1)^2} \times$

$\left(-\sin 2x + \frac{2}{2n-1}\sin(2n-1)x \right)$ (iv) $\frac{2}{\pi}(\sin x - x\cos x)$

$+ \frac{4}{\pi} \sum_{n=2}^{\infty} \frac{1}{(2n-1)^2-1} \left(\sin x - \frac{1}{2n-1}\sin(2n-1)x \right)$.

24.3. (i) $\frac{8}{\pi} \sum_{n=1}^{\infty} \frac{(-1)^{n+1}\sin n\pi x}{n(4-n^2\pi^2)}$ (ii) $2\sum_{n=1}^{\infty} \left[\frac{(-1)^{n+1}}{n\pi} - \frac{2\{1-(-1)^n\}}{n^3\pi^3} \right]$

$\times \frac{1}{(11-n^2\pi^2)} \sin n\pi x$ (iii) $2\sum_{n=1}^{\infty} \frac{n\pi(1+e(-1)^{n+1})\sin n\pi x}{(1+n^2\pi^2)(3-n^2\pi^2)}$

(iv) $\frac{32}{\pi} \sum_{n\,\text{odd}}^{\infty} \frac{(-1)^{(n+1)/2}}{n^2(n^2+4)} \sin\frac{nx}{2}$ (v) $2\sum_{n=1}^{\infty} \frac{(2\cos\sqrt{\lambda_n}-1)\cos\sqrt{\lambda_n}x}{\lambda_n(\lambda_n-2)(1+\sin^2\sqrt{\lambda_n})}$, where

$\cot\sqrt{\lambda_n} = \sqrt{\lambda_n}$ (vi) $\frac{\pi}{2} + \frac{2}{\pi} \sum_{n=1}^{\infty} \frac{1-(-1)^n}{n^2(n^2-2)} \cos nx$ (vii) $2(\ln 2)^2 \sin(\ln 2)\pi$

$\times \sum_{n=1}^{\infty} \frac{(-1)^n n}{[3(\ln 2)^2-(n\pi)^2][(\ln 2)^2-(n\pi)^2]} \sin\left(\frac{n\pi}{\ln 2}\ln x \right)$.

Lecture 25
Partial Differential Equations

Partial differential equations arise in many branches of science and technology, for example, in electromagnetic theory, elasticity, fluid mechanics, heat transfer, acoustics, quantum mechanics, and so on. In this lecture we shall introduce partial differential equations, and explain several concepts through elementary examples. Here we shall also provide the most fundamental classification of second-order linear equations in two independent variables.

A partial differential equation (DE) is a relation that involves partial derivatives of an unknown function. Let the unknown function be u, and x, y, z, $\cdots$ be independent variables, i.e., $u = u(x, y, z, \cdots)$. Often, one of these variables represents the time. Thus, a partial DE is an equation of the form

$$F(x, y, z, u, u_x, u_y, u_z, u_{xx}, u_{xy}, \cdots, u_{xxx}, \cdots) = 0. \qquad (25.1)$$

In (25.1) we have used the subscript notation for the partial differentiation, i.e.,

$$u_x = \frac{\partial u}{\partial x}, \quad u_{xy} = \frac{\partial^2 u}{\partial x \partial y}, \quad \text{and so on.}$$

We will always assume that the unknown function u is *sufficiently well behaved* so that all necessary partial derivatives exist and corresponding mixed partial derivatives are equal, e.g.,

$$u_{xy} = u_{yx}, \quad u_{xzx} = u_{xxz}, \quad \text{and so on.}$$

As in the case of ordinary DEs, we define the *order* of the partial DE (25.1) to be the highest order partial derivative appearing in the equation. Furthermore, we say that the partial DE (25.1) is *linear* if F is linear as a function of the variables u, u_x, u_y, u_z, $u_{xx}, \cdots$, i.e., F is a linear combination of the unknown function and its derivatives. Equation (25.1) is said to be *quasilinear* if F is linear as a function of the highest-order derivatives.

The following are examples of partial DEs:

$$
\begin{array}{ll}
u_x + u_y = 3u_z - 2x^2 - 5z & \text{(first-order linear)} \\
u_{xx} + u = 4x^2 & \text{(second-order linear)} \\
5xyu_{xy} - 3zu_y + 2u = 0 & \text{(second-order linear)} \\
u_{xz} + 2uu_y - 4z = 0. & \text{(second-order quasilinear)}
\end{array}
$$

R.P. Agarwal, D. O'Regan, *Ordinary and Partial Differential Equations*,
Universitext, DOI 10.1007/978-0-387-79146-3_25,
© Springer Science+Business Media, LLC 2009

With a very few exceptions, we will limit our discussion to only first and second-order partial DEs. Thus, our most general partial DE in three independent variables can be written as

$$a_1 u_{xx} + a_2 u_{yy} + a_3 u_{zz} + a_4 u_{xy} + a_5 u_{yz} + a_6 u_{zx} + a_7 u_x + a_8 u_y + a_9 u_z + a_{10} u = f, \tag{25.2}$$

where $u = u(x, y, z)$, $f = f(x, y, z)$ and $a_i = a_i(x, y, z)$, $i = 1, \cdots, 10$.

In (25.2) it is understood that the function f and the coefficients a_i are known, and u is unknown. By a *solution* of (25.2) we mean a continuous function $u = u(x, y, z)$, with continuous first and second-order partial derivatives, which, when substituted in (25.2), reduces equation (25.2) to an identity.

Example 25.1. For the first-order partial DE

$$u_x + u_y = 0, \quad u = u(x, y)$$

we shall show that $u = \phi(x - y)$, where ϕ is any function having continuous first-order partial derivatives is a solution. Indeed, since

$$u_x = \phi'(x - y) \times (1) \quad \text{and} \quad u_y = \phi'(x - y) \times (-1)$$

it immediately follows that $u_x + u_y = \phi'(x - y) - \phi'(x - y) = 0$.

If $f(x, y, z) \equiv 0$ the partial DE (25.2) is called *homogeneous*; otherwise it is called *nonhomogeneous*. If each coefficient a_i is constant, then (25.2) is called a partial DE with *constant coefficients*. If at least one of the a_i is not a constant, equation (25.2) is called a partial DE with *variable coefficients*.

Example 25.2. Consider the first-order partial DE

$$u_x = x + y, \quad u = u(x, y). \tag{25.3}$$

Integrating (25.3) partially with respect to x, i.e., treating y as a constant, we obtain

$$u = \frac{1}{2} x^2 + xy + c. \tag{25.4}$$

We note that the constant of integration is denoted by c. In order to verify that u as given in (25.4) is a solution of (25.3), we need only to substitute this expression for u in (25.3). When verifying this, notice that even when $c = c(y)$, u as given in (25.4) still is a solution of (25.3), since $\partial c(y)/\partial x = 0$. Thus, the general solution of (25.3) is

$$u = \frac{1}{2} x^2 + xy + c(y), \tag{25.5}$$

where c is an arbitrary function of y.

Hence, in contrast with ordinary DEs the solution (25.5) of the partial DE (25.3) contains an arbitrary function rather than an arbitrary constant.

Example 25.3. Consider the partial DE

$$u_{xy} = z + x, \quad u = u(x, y, z). \tag{25.6}$$

First we integrate (25.6) partially with respect to y (treating x and z as constants), to obtain

$$u_x = yz + xy + f_1(x, z),$$

where f_1 is an arbitrary function of the variables x and z. Next we integrate partially with respect to x (treating y and z as constants), to get

$$u = xyz + \frac{1}{2}x^2y + \int^x f_1(s, z)ds + g(y, z),$$

where $g(y, z)$ is an arbitrary function of the variables y and z. If we set

$$f(x, z) = \int^x f_1(s, z)ds,$$

then our solution takes the form

$$u = xyz + \frac{1}{2}x^2y + f(x, z) + g(y, z), \tag{25.7}$$

where f is an arbitrary function of x and z, and g is an arbitrary function of y and z. Functions f and g must have continuous first-order partial derivatives with respect to their arguments.

The *general solution* of a given partial DE of order n in k independent variables usually involves n arbitrary functions of $k-1$ variables. Thus, the solutions (25.5) and (25.7) are the general solutions of equations (25.3) and (25.6), respectively. Each specific assignment of the arbitrary function(s) in the general solution gives rise to a *particular solution* of the corresponding partial DE Thus,

$$u = \frac{1}{2}x^2 + xy + e^y \sin y$$

is a particular solution of equation (25.3), and

$$u = xyz + \frac{1}{2}x^2y + z^2 \cos x + ye^z$$

is a particular solution of equation (25.6).

Now let

$$\mathcal{A}[u] = a_1(x, y, z)u_{xx} + \cdots + a_{10}(x, y, z)u$$

so that the partial DE (25.2) can be written as

$$\mathcal{A}[u] = f. \tag{25.8}$$

The *principle of superposition*, which plays a fundamental role in ordinary DEs, for partial DEs can be stated as follows: Let $f_i = f_i(x, y, z)$, $1 \leq i \leq m$ be given functions and let c_i, $1 \leq i \leq m$ be arbitrary constants. If $u_i = u_i(x, y, z)$, $1 \leq i \leq m$ are respective solutions of the partial DEs $\mathcal{A}[u_i] = f_i$, $1 \leq i \leq m$ then $u = c_1 u_1 + \cdots + c_m u_m$ is a solution of the partial DE $\mathcal{A}[u] = c_1 f_1 + \cdots + c_m f_m$.

Two important consequences of the principle of superposition are as follows:

(i) If $u_1, \cdots, u_m$ are solutions of $\mathcal{A}[u] = 0$ and $c_1, \cdots, c_m$ are any constants, then $\sum_{i=1}^{m} c_i u_i$ is also a solution of $\mathcal{A}[u] = 0$, i.e., any linear combination of solutions is also a solution. Also a series built from an infinite number of solutions $\sum_{i=1}^{\infty} c_i u_i$ is a solution in a region D, provided the series and the various derivative series required for substitution into the partial DE converge uniformly in D.

(ii) If u_h is a general solution of $\mathcal{A}[u] = 0$ and if u_p is a particular solution of $\mathcal{A}[u] = f$, then $u = u_h + u_p$ is a solution of $\mathcal{A}[u] = f$, i.e., the sum of a general solution of the homogeneous equation and a particular solution is also a solution.

If $u(x, y, z, \mu)$ is a solution of $\mathcal{A}[u] = 0$ containing a parameter μ, then by the principle of superposition

$$\frac{u(x, y, z, \mu + \Delta\mu) - u(x, y, z, \mu)}{\Delta\mu}$$

is also a solution. Generally, the limit $u_\mu(x, y, z, \mu)$ of this difference quotient is also a solution. For all problems that we shall consider, for all choices of integrable functions ϕ, the integral

$$\int_{\mu_0}^{\mu} \phi(\mu) u(x, y, z, \mu) d\mu$$

is another solution. Since the integral is the limit of a sum, this is a further extension of the superposition theorem. We also note that if all coefficients of the equation $\mathcal{A}[u] = 0$ are constants, then the various derivatives u_x, u_y, u_{xx}, $\cdots$ etc., are also solutions. In conclusion, the general solution often does not represent the collection of all possible solutions.

Example 25.4.   Consider again the partial DE (25.6). Clearly, $u_h = f(x, z) + g(y, z)$ is the general solution of $u_{xy} = 0$, where f and g are arbitrary functions; $u_1 = xyz$ is a particular solution of $u_{xy} = z$; and

$u_2 = x^2 y/2$ is a particular solution of $u_{xy} = x$. Thus, the general solution of (25.6) can be written as (25.7).

Now we shall discuss the most crucial classification of second-order linear partial DEs in two independent variables

$$\begin{aligned}
\mathcal{L}[u] \;=\; & A(x,y)u_{xx} + B(x,y)u_{xy} + C(x,y)u_{yy} \\
& + D(x,y)u_x + E(x,y)u_y + F(x,y)u = G(x,y), \quad u = u(x,y),
\end{aligned}$$
(25.9)

where the functions $A(x,t), \cdots, G(x,t)$ are continuous in some open set $\Omega \subseteq \mathbb{R}^2$. In (25.9), often $y = t$ represents the time, and sometimes for convenience x and y are interchanged. Following the analogy of the quadratic equation

$$ax^2 + bxy + cy^2 + dx + ey + f = 0$$

that it represents a hyperbola, parabola, or ellipse according as $b^2 - 4ac$ is positive, zero, or negative, the operator $\mathcal{L}$ (and so the partial DE (25.9)) is said to be hyperbolic, parabolic, or elliptic at a point $(x_0, y_0) \in \Omega$ according as

$$B^2(x_0, y_0) - 4A(x_0, y_0)C(x_0, y_0)$$
(25.10)

is positive, zero, or negative. It is said to be hyperbolic, parabolic, or elliptic in a domain, if it has the required property at each point of the domain.

Example 25.5. The *wave equation in one dimension*

$$\frac{\partial^2 u}{\partial t^2} = c^2 \frac{\partial^2 u}{\partial x^2}, \quad c > 0$$
(25.11)

occurs in the study of processes of transversal vibrations of a string, the longitudinal vibrations of rods, electric oscillations in conductors, the torsional oscillations of shafts, gas vibrations, and so on. Clearly, for (25.11), $A = -c^2$, $B = 0$, $C = 1$ so that $B^2 - 4AC = 4c^2 > 0$, and hence it is hyperbolic in any domain.

Example 25.6. The *one-dimensional heat equation*

$$\frac{\partial u}{\partial t} = a \frac{\partial^2 u}{\partial x^2}, \quad a > 0$$
(25.12)

arises in the study of processes of the propagation of heat, the filtration of liquids and gases in a porous medium, e.g., the filtration of oil and gas in subterranean sandstones, some problems in probability theory, and so on. Clearly, for (25.12), $A = -a$, $B = 0$, $C = 0$ so that $B^2 - 4AC = 0$, and hence it is parabolic in any domain.

Example 25.7. The *potential, or Laplace equation in two dimensions,*

$$\frac{\partial^2 u}{\partial x^2} + \frac{\partial^2 u}{\partial y^2} = 0$$
(25.13)

is invoked in the study of problems dealing with electric and magnetic fields, stationary states, problems in hydrodynamics, diffusion, and so on. Clearly, for (25.13), $A = 1$, $B = 0$, $C = 1$, so that $B^2 - 4AC = -4 < 0$; and hence it is elliptic in any domain. Solutions of (25.13) are often called *potential functions* as well as *harmonic functions*.

In the literature equations (25.11)–(25.13) are known as the classical equations of mathematical physics, as these equations keep on popping up in various other applications. Further, these equations are important examples of the three major types of linear partial differential equations.

Example 25.8. The *Tricomi equation*

$$yu_{xx} + u_{yy} = 0 \qquad\qquad (25.14)$$

occurs in the study of aerodynamics. This equation is elliptic for $y > 0$, parabolic for $y = 0$, and hyperbolic for $y < 0$. The elliptic region corresponds to smooth subsonic flow, the parabolic region to a sonic barrier, and the hyperbolic region to supersonic propagation of shock waves.

Problems

25.1. Show that the given function satisfies the corresponding partial DE

(i) $u = f(x^2 + y^2)$, $yu_x = xu_y$
(ii) $u = f(xy)$, $xu_x - yu_y = 0$
(iii) $u = e^y f(x - y)$, $u = u_x + u_y$
(iv) $u = ax + by + ab$, $u = xu_x + yu_y + u_xu_y$
(v) $(u + a^2)^3 = (x + ay + b)^2$, $uu_x^2 + u_y^2 = 4/9$.

25.2. Find where the following operators are hyperbolic, parabolic, or elliptic:

(i) $u_{yy} + yu_{xy} + xu_{xx}$
(ii) $x^2 u_{yy} - u_{xx} + u$
(iii) $yu_{yy} + 2u_{xy} + xu_{xx} + u_x$.

25.3. Show that for the one-dimensional heat equation (25.12) with $a = 1$ the function

$$\phi(x,t) = A + B \int_0^{x/\sqrt{t}} e^{-s^2/4} ds$$

as well as its partial derivatives $\phi_x(x,t)$ and $\phi_t(x,t)$ are solutions; here A and B are arbitrary constants.

25.4. Let $\psi(\tau)$ be an arbitrary function such that

$$\phi(x,t) = \int_0^\infty \psi(\tau) \frac{1}{(t+\tau)^{1/2}} \exp\left(-\frac{x^2}{4(t+\tau)}\right) d\tau$$

is convergent. Show that $\phi(x,t)$ is a solution of (25.12) with $a = 1$.

25.5. Show that if $u = u(x,y)$ and $v = v(x,y)$ satisfy Cauchy–Riemann equations

$$u_x = v_y, \quad u_y = -v_x$$

then each is a solution of Laplace's equation (25.13).

25.6. Show that $u = \ln 1/r$, where $r = \sqrt{(x-a)^2 + (y-b)^2}$ is a solution of Laplace's equation (25.13).

25.7. A linear approximation of one-dimensional isentropic flow of an ideal gas (a gas in which the only stress across any element of area is normal to it) is given by

$$u_t + \rho_x = 0$$
$$u_x + \alpha^2 \rho_t = 0,$$

where $u = u(x,t)$ is the velocity, and $\rho = \rho(x,t)$ is the density of the gas. Show that u and ρ both satisfy the wave equation (25.11).

25.8. A nonlinear partial DE that arises in shallow-water theory is the *Korteweg–de Vries equation*

$$u_t + (\alpha + \epsilon u)u_x + \beta u_{xxx} = 0,$$

where α, β, ϵ are constants. Show that its one solution (the solitary wave, or *soliton*) is given by

$$u = A\,\mathrm{sech}^2[(\epsilon A/12\beta)^{1/2}(x - Vt)], \quad V = \alpha + (1/3)\epsilon A.$$

Answers or Hints

25.1. Verify directly.

25.2. (i) Hyperbolic when $y^2 > 4x$, parabolic on the parabola $y^2 = 4x$, elliptic when $y^2 < 4x$ (ii) Hyperbolic when $x \neq 0$, parabolic when $x = 0$ (iii) Hyperbolic when $xy < 1$, parabolic on the hyperbola $xy = 1$, elliptic when $xy > 1$.

25.3. Verify directly.

25.4. Verify directly.

25.5. $u_{xy} = u_{yx}$.

25.6. Verify directly.

25.7. Differentiate the first equation with respect to x, and the second equation with respect to t, and subtract.

25.8. Verify directly.

Lecture 26
First-Order Partial Differential Equations

We begin this lecture with simultaneous DEs, which play an important role in the theory of partial DEs. Then we shall consider quasilinear partial DEs of the Lagrange type, and show that such equations can be solved rather easily provided we can find solutions of related simultaneous DEs. Finally, in this lecture we shall explain a general method to find solutions of nonlinear first-order partial DEs which is due to Charpit.

Simultaneous DEs. To solve simultaneous DEs of the form

$$\frac{dx}{P} = \frac{dy}{Q} = \frac{du}{R}, \tag{26.1}$$

where P, Q, and R are functions of x, y, u, several different techniques are known. We shall present here only the following two methods.

The method of grouping. If in $dx/P = du/R$ the variable y can be canceled or absent, leaving the equation in x and u only, then an integration of this equation gives

$$\phi(x, u) = c_1. \tag{26.2}$$

Again, if one variable, say, x, is absent or can be removed from $dy/Q = du/R$, then an integration of this equation leads to

$$\psi(y, u) = c_2. \tag{26.3}$$

The general solution of (26.1) is then the solutions (26.2) and (26.3) taken together.

Example 26.1. For the DE

$$\frac{dx}{u^2 y} = \frac{dy}{u^2 x} = \frac{du}{y^2 x} \tag{26.4}$$

we take first two fractions and cancel out u^2, to get $dx/y = dy/x$ or $xdx - ydy = 0$, which can be integrated to obtain

$$x^2 - y^2 = c_1. \tag{26.5}$$

Again, we take second and third fractions and cancel out x, to have $dy/u^2 = du/y^2$ or $y^2 dy - u^2 du = 0$, which on integration yields

$$y^3 - u^3 = c_2. \tag{26.6}$$

R.P. Agarwal, D. O'Regan, *Ordinary and Partial Differential Equations*,
Universitext, DOI 10.1007/978-0-387-79146-3_26,
© Springer Science+Business Media, LLC 2009

Equations (26.5) and (26.6) taken together gives the solution of (26.4).

The method of multipliers. By a proper choice of multipliers ℓ, m, n which are not necessarily constants, we write

$$\frac{dx}{P} = \frac{dy}{Q} = \frac{du}{R} = \frac{\ell dx + mdy + ndu}{\ell P + mQ + nR}$$

so that $\ell P + mQ + nR = 0$. Then $\ell dx + mdy + ndu = 0$, and this can be solved to get the integral

$$\phi(x, y, u) = c_1. \tag{26.7}$$

Again we search for another set of multipliers λ, μ, ν so that $\lambda P + \mu Q + \nu R = 0$ giving $\lambda dx + \mu dy + \nu du = 0$, which on integration yields

$$\psi(x, y, u) = c_2. \tag{26.8}$$

These two integrals (26.7) and (26.8) taken together give the required solution of (26.1).

Example 26.2. For the DE

$$\frac{dx}{x(y^2 - u^2)} = \frac{dy}{-y(u^2 + x^2)} = \frac{du}{u(x^2 + y^2)} \tag{26.9}$$

we use the multipliers x, y, u so that each fraction is the same as

$$\frac{xdx + ydy + udu}{x^2(y^2 - u^2) - y^2(u^2 + x^2) + u^2(x^2 + y^2)} = \frac{xdx + ydy + udu}{0},$$

and hence $xdx + ydy + udu = 0$, which on integration gives

$$x^2 + y^2 + u^2 = c_1. \tag{26.10}$$

Again using the multipliers $1/x$, $-1/y$, $-1/u$ we obtain

$$\frac{1}{x}dx - \frac{1}{y}dy - \frac{1}{u}du = 0,$$

which can be solved to get the integral

$$yu = c_2 x. \tag{26.11}$$

Hence, the solution of (26.9) is (26.10) and (26.11) taken together.

Lagrange's first-order linear partial DE   Now we shall study the first-order quasilinear partial DE

$$Pu_x + Qu_y = R, \quad u = u(x, y) \tag{26.12}$$

where P, Q, and R are functions of x, y, u. Such an equation is obtained by eliminating an arbitrary function from the relation

$$\phi(\mu, \nu) = 0, \qquad (26.13)$$

where μ, ν are some functions of x, y, u. Indeed differentiating (26.13) partially with respect to x and y, we get

$$\phi_\mu(\mu_x + \mu_u u_x) + \phi_\nu(\nu_x + \nu_u u_x) = 0$$

and

$$\phi_\mu(\mu_y + \mu_u u_y) + \phi_\nu(\nu_y + \nu_u u_y) = 0.$$

Eliminating ϕ_μ and ϕ_ν from these relations, we get

$$\begin{vmatrix} \mu_x + \mu_u u_x & \nu_x + \nu_u u_x \\ \mu_y + \mu_u u_y & \nu_y + \nu_u u_y \end{vmatrix} = 0,$$

which simplifies to

$$(\mu_y \nu_u - \mu_u \nu_y) u_x + (\mu_u \nu_x - \mu_x \nu_u) u_y = \mu_x \nu_y - \mu_y \nu_x. \qquad (26.14)$$

Clearly, equation (26.14) is of the form (26.12).

Now suppose that $\mu = a$ and $\nu = b$, where a and b are constants, so that

$$\mu_x dx + \mu_y dy + \mu_u du = d\mu = 0$$

and

$$\nu_x dx + \nu_y dy + \nu_u du = d\nu = 0.$$

By cross-multiplication, we have

$$\frac{dx}{\mu_y \nu_u - \nu_y \mu_u} = \frac{dy}{\mu_u \nu_x - \mu_x \nu_u} = \frac{du}{\mu_x \nu_y - \mu_y \nu_x},$$

which in view of (26.14) and (26.12) is of the same form as (26.1). Now since the solutions of these equations are $\mu = a$ and $\nu = b$ the required solution of (26.12) can be written as $\phi(\mu, \nu) = 0$. Thus, to solve (26.12) first we need to form the subsidiary equations (26.1), and need to solve these to obtain $\mu = a$ and $\nu = b$, and finally write the solution as $\phi(\mu, \nu) = 0$ or $\mu = f(\nu)$.

We also note that any integral of (26.12), $u = f(x, y)$ represents a surface. We call it an *integral surface* of (26.12) and denote it by S. Consider any point $M(x_0, y_0, u_0)$ on S. If u_x and u_y are evaluated at M, then $u_x, u_y, -1$ are direction numbers of the normal to S and M. Thus, DE (26.12) expresses the geometric fact that this normal is perpendicular to the line through M whose direction numbers are the values of P, Q and R evaluated at M. This line, which is tangent to S at M, has the equations

$$\frac{x - x_0}{P} = \frac{y - y_0}{Q} = \frac{u - u_0}{R}.$$

Hence, at each point of S there is defined a direction whose direction numbers dx, dy, du satisfy the equations (26.1).

Although our above discussion is only for two independent variables, it can be extended rather easily to the case of n independent variables. In fact, analytic methods can be used to prove the following general theorem.

Theorem 26.1. If $\phi_i(x_1, x_2, \cdots, x_n, u) = c_i$, $1 \leq i \leq n$ are independent solutions of the equations

$$\frac{dx_1}{P_1} = \frac{dx_2}{P_2} = \cdots = \frac{dx_n}{P_n} = \frac{du}{R}, \tag{26.15}$$

then the relation $\Psi(\phi_1, \phi_2, \cdots, \phi_n) = 0$, where the function Ψ is arbitrary, is a general solution of the linear partial DE

$$P_1 u_{x_1} + P_2 u_{x_2} + \cdots + P_n u_{x_n} = R. \tag{26.16}$$

Example 26.3. For the partial DE

$$(mu - ny)u_x + (nx - \ell u)u_y = \ell y - mx \tag{26.17}$$

the subsidiary equations are

$$\frac{dx}{mu - ny} = \frac{dy}{nx - \ell u} = \frac{du}{\ell y - mx}.$$

Using multipliers x, y, u we find that each fraction is the same as $(xdx + ydy + udu)/0$ and hence $xdx + ydy + udu = 0$, which on integration gives

$$x^2 + y^2 + u^2 = a. \tag{26.18}$$

Again using multipliers ℓ, m, n we have each fraction equal to $(\ell dx + mdy + ndu)/0$ and thus $\ell dx + mdy + ndu = 0$, which on integration yields

$$\ell x + my + nu = b. \tag{26.19}$$

Thus, from (26.18) and (26.19) the solution of (26.17) is $x^2 + y^2 + u^2 = f(\ell x + my + nu)$.

Charpit's method. We shall now explain a general method for finding the solutions of first-order nonlinear partial DEs which is due to Charpit. Consider the equation

$$f(x, y, u, p, q) = 0, \tag{26.20}$$

where for simplicity $p = u_x$ and $q = u_y$. Since u depends on x and y, we have

$$du = u_x dx + u_y dy = p dx + q dy. \tag{26.21}$$

If we can find another relation involving x, y, u, p, q such as

$$\phi(x, y, u, p, q) = 0, \tag{26.22}$$

then we can solve (26.20) and (26.22) for p and q and substitute in (26.21). This will give the solution provided the resulting equation is integrable.

To determine ϕ, we differentiate (26.20) and (26.22) with respect to x and y, to obtain

$$f_x + f_u p + f_p p_x + f_q q_x = 0 \tag{26.23}$$

$$\phi_x + \phi_u p + \phi_p p_x + \phi_q q_x = 0 \tag{26.24}$$

$$f_y + f_u q + f_p p_y + f_q q_y = 0 \tag{26.25}$$

$$\phi_y + \phi_u q + \phi_p p_y + \phi_q q_y = 0. \tag{26.26}$$

Eliminating p_x between the equations (26.23) and (26.24), we get

$$(f_x \phi_p - \phi_x f_p) + (f_u \phi_p - \phi_u f_p)p + (f_q \phi_p - \phi_q f_p)q_x = 0. \tag{26.27}$$

Also eliminating q_y between the equations (26.25) and (26.26), we obtain

$$(f_y \phi_q - \phi_y f_q) + (f_u \phi_q - \phi_u f_q)q + (f_p \phi_q - \phi_p f_q)p_y = 0. \tag{26.28}$$

Adding (26.27) and (26.28) and using $q_x = u_{xy} = p_y$, we find that the last terms in both cancel and the other terms, on rearrangement, give

$$(-f_p)\phi_x + (-f_q)\phi_y + (-pf_p - qf_q)\phi_u + (f_x + pf_u)\phi_p + (f_y + qf_u)\phi_q = 0. \tag{26.29}$$

This is Lagrange's DE (26.16) with x, y, u, p, q as independent variables and ϕ as the dependent variable. Its solution will depend on the solution of the subsidiary equations

$$\frac{dx}{-f_p} = \frac{dy}{-f_q} = \frac{du}{-pf_p - qf_q} = \frac{dp}{f_x + pf_u} = \frac{dq}{f_y + qf_u} = \frac{d\phi}{0}. \tag{26.30}$$

An integral of these equations involving p or q or both, can be taken as the required relation (26.22).

Example 26.4. For the partial DE

$$(p^2 + q^2)y = qu, \tag{26.31}$$

we have

$$f(x, y, u, p, q) = (p^2 + q^2)y - qu = 0. \tag{26.32}$$

Thus, the subsidiary equations are

$$\frac{dx}{-2py} = \frac{dy}{u - 2qy} = \frac{du}{-qu} = \frac{dp}{-pq} = \frac{dq}{p^2}.$$

The last two equations give $pdp + qdq = 0$, and hence

$$p^2 + q^2 = c^2. \tag{26.33}$$

Now we solve (26.32) and (26.33), to get $p = (c/u)\sqrt{u^2 - c^2y^2}$, $q = yc^2/u$.
Thus,

$$du = pdx + qdy = \frac{c}{u}\sqrt{u^2 - c^2y^2}dx + \frac{c^2y}{u}du,$$

which is the same as

$$\frac{1}{2}\frac{d(u^2 - c^2y^2)}{\sqrt{u^2 - c^2y^2}} = cdx,$$

and hence on integration and squaring, we get

$$u^2 = (a + cx)^2 + c^2y^2,$$

which is the required solution of (26.31).

Problems

26.1. Solve the following simultaneous DEs:

(i) $\quad \dfrac{dx}{x^2} = \dfrac{dy}{y^2} = \dfrac{du}{nxy}$

(ii) $\quad \dfrac{dx}{mu - ny} = \dfrac{dy}{nx - \ell u} = \dfrac{du}{\ell y - mx}$

(iii) $\quad \dfrac{dx}{x^2 - yu} = \dfrac{dy}{y^2 - ux} = \dfrac{du}{u^2 - xy}$

(iv) $\quad \dfrac{dx}{u(x + y)} = \dfrac{dy}{u(x - y)} = \dfrac{du}{x^2 + y^2}$

(v) $\quad \dfrac{dx}{x(y^2 - u^2)} = \dfrac{dy}{y(u^2 - x^2)} = \dfrac{du}{u(x^2 - y^2)}$

(vi) $\quad \dfrac{dx}{x^2 - y^2 - u^2} = \dfrac{dy}{2xy} = \dfrac{du}{2xu}.$

26.2. Solve the following first-order linear partial DEs:

(i) $\quad (x^2 - y^2 - u^2)u_x + 2xyu_y = 2xu$

(ii) $\quad (y - u)u_x + (x - y)u_y = u - x$

(iii) $(x^2 - yu)u_x + (y^2 - ux)u_y = u^2 - xy$

(iv) $y^2 u_x - xyu_y = x(u - 2y)$

(v) $x^2 u_x + y^2 u_y = (x + y)u.$

26.3. Show that the conditions for exactness of the ordinary DE

$$\mu(x, y)M(x, y)dx + \mu(x, y)N(x, y)dy = 0$$

is a linear partial DE of the form (26.12).

26.4. Show that the general solution of the partial DE

$$u_x + (a(x)y + b(x)u)u_y = c(x)y + d(x)u$$

is of the form

$$\phi\left(\frac{yv_1(x) - uw_1(x)}{Z(x)}, \ \frac{yv_2(x) - uw_2(x)}{Z(x)}\right) = 0,$$

where $W(x) = c_1 w_1(x) + c_2 w_2(x)$, $V(x) = c_1 v_1(x) + c_2 v_2(x)$ is the general solution of the system of equations

$$\frac{DW}{dx} = aW + bV, \qquad \frac{dV}{dx} = cW + dV$$

and $Z = w_1 v_2 - w_2 v_1$. Hence, solve the following partial DEs:

(i) $u_x + (-y + 2u)u_y = 4y + u$

(ii) $u_x + \dfrac{2y + u}{x}u_y = \dfrac{4y + 2u}{x}.$

26.5. Solve the following first-order nonlinear partial DEs:

(i) $u = p^2 x + q^2 y$

(ii) $u^2 = xypq$

(iii) $1 + p^2 = qu$

(iv) $pxy + pq + qy = yu$

(v) $u^2(p^2 + q^2) = x^2 + y^2.$

Answers or Hints

26.1. (i) $\frac{1}{x} = \frac{1}{y} + c_1$, $u = c_2 + \frac{nxy}{y-x}\ln\frac{x}{y}$ (ii) $\ell x + my + nu = c_1$, $x^2 + y^2 + u^2 = c_2$ (iii) $\frac{x-y}{y-u} = c_1$, $xy + yu + xu = c_2$ (iv) $x^2 - y^2 - 2xy = c_1$, $x^2 - y^2 - u^2 = c_2$ (v) $xyu = c_1$, $x^2 + y^2 + u^2 = c_2$ (vi) $y = c_1 u$, $x^2 + y^2 + u^2 = c_2 u$.

26.2. (i) $x^2 + y^2 + u^2 = uf(y/u)$ (ii) $\frac{1}{2}x^2 + yu = \phi(x + y + u)$

(iii) $\phi\left(\frac{x-y}{y-u}, xy + yu + xu\right) = 0$ (iv) $x^2 + y^2 = f(2\ln y + u/y)$

(v) $\phi\left(\frac{xy}{u}, \frac{x-y}{u}\right) = 0$.

26.3. $N\mu_x - M\mu_y = \mu(M_y - N_x)$.

26.4. (i) $\phi(e^{-3x}(y+u), e^{3x}(u-2y)) = 0$ (ii) $f((2y+u)/x^4, 2y-u) = 0$.

26.5. (i) $u = (\sqrt{a} + \sqrt{x})^2 + (b + \sqrt{y})^2$ (ii) $u = ax^b y^{1/b}$ (iii) $\frac{u^2}{2} \pm \left\{\frac{u}{2}\sqrt{u^2 - 4a^2} - 2a^2 \ln(u + \sqrt{u^2 - 4a^2})\right\} = 2ax + 2y + b$ (iv) $\ln(u - ax) = y - a\ln(a+y) + b$ (v) Put $u^2 = U, u^2 = b + x\sqrt{(x^2 + a^2)} + a\ln\{x + \sqrt{(x^2 + a^2)}\} + y\sqrt{(y^2 - a^2)} + a\ln\{y + \sqrt{(y^2 - a^2)}\}$.

Lecture 27
Solvable Partial Differential Equations

In this lecture we shall show that like ordinary DEs, partial DEs with constant coefficients can be solved explicitly. We shall begin with homogeneous second-order DEs involving only second-order terms, and then show how the operator method can be used to solve some particular nonhomogeneous DEs. Then, we shall extend the method to general second and higher order partial DEs.

For partial DEs of the form

$$u_{xx} + k_1 u_{xy} + k_2 u_{yy} = 0, \qquad (27.1)$$

where k_1 and k_2 are constants, we write $D_1^r = \partial^r/\partial x^r$ and $D_2^r = \partial^r/\partial y^r$, so that in symbolic form it can be written as

$$(D_1^2 + k_1 D_1 D_2 + k_2 D_2^2)u = 0. \qquad (27.2)$$

Its symbolic operator equated to zero, i.e.,

$$D_1^2 + k_1 D_1 D_2 + k_2 D_2^2 = 0 \qquad (27.3)$$

is called the *auxiliary equation*. Let its roots be $D_1/D_2 = m_1,\ m_2$.

Case I. If the roots are distinct, then (27.2) is equivalent to

$$(D_1 - m_1 D_2)(D_1 - m_2 D_2)u = 0. \qquad (27.4)$$

It will be satisfied by the solution of $(D_1 - m_2 D_2)u = 0$, i.e., $u_x - m_2 u_y = 0$. This is a Lagrange equation and its subsidiary equations are

$$\frac{dx}{1} = \frac{dy}{-m_2} = \frac{du}{0}$$

and hence $y + m_2 x = a$ and $u = b$. Therefore, its solution is $u = \phi(y + m_2 x)$.

Similarly, equation (27.4) will also be satisfied by the solution of $(D_1 - m_1 D_2)u = 0$, i.e., $u = f(y + m_1 x)$. Hence, in this case the general solution of (27.1) is

$$u = f(y + m_1 x) + \phi(y + m_2 x).$$

R.P. Agarwal, D. O'Regan, *Ordinary and Partial Differential Equations*,
Universitext, DOI 10.1007/978-0-387-79146-3_27,
© Springer Science+Business Media, LLC 2009

Case II. If the roots are equal, i.e., $m_1 = m_2$, then (27.4) is equivalent to

$$(D_1 - m_1 D_2)^2 u = 0. \tag{27.5}$$

Putting $(D_1 - m_1 D_2)u = v$, it becomes $(D_1 - m_1 D_2)v = 0$, which as earlier gives $v = \phi(y + m_1 x)$. Therefore, (27.5) takes the form $(D_1 - m_1 D_2)u = \phi(y + m_1 x)$, or

$$u_x - m_1 u_y = \phi(y + m_1 x).$$

This is again a Lagrange equation and its subsidiary equations are

$$\frac{dx}{1} = \frac{dy}{-m_1} = \frac{du}{\phi(y + m_1 x)},$$

giving $y + m_1 x = a$ and $du = \phi(a)dx$, i.e., $u = \phi(a)x + b$. Thus, the general solution of (27.1) in this case is

$$u = f(y + m_1 x) + x\phi(y + m_1 x).$$

Example 27.1. For the partial DE

$$2u_{xx} + 5u_{xy} + 2u_{yy} = 0$$

the auxiliary equation is $2m^2 + 5m + 2 = 0$, $m = D_1/D_2$, which gives $m_1 = -2$, $m_2 = -1/2$. Hence, its general solution can be written as $u = f_1(y - 2x) + f_2(2y - x)$.

Example 27.2. For the partial DE

$$u_{xx} + 6u_{xy} + 9u_{yy} = 0$$

the auxiliary equation is $m^2 + 6m + 9 = 0$, $m = D_1/D_2$, which gives $m_1 = -3$, $m_2 = -3$. Hence, its general solution can be written as $u = f_1(y - 3x) + xf_2(y - 3x)$.

Now we shall consider nonhomogeneous partial DEs of the form

$$\mathcal{L}(D_1, D_2)[u] = (D_1^2 + k_1 D_1 D_2 + k_2 D_2^2)u = F(x, y). \tag{27.6}$$

As in the case of ordinary DEs a particular solution $u_p(x, y)$ of (27.6) can be obtained by employing the operator method, i.e.,

$$u_p(x, y) = \frac{1}{\mathcal{L}(D_1, D_2)} F(x, y). \tag{27.7}$$

Case 1. $F(x, y) = e^{ax + by}$. Since

$$(D_1^2 + k_1 D_1 D_2 + k_2 D_2^2)e^{ax+by} = (a^2 + k_1 ab + k_2 b^2)e^{ax+by},$$

i.e.,
$$\mathcal{L}(D_1, D_2)[e^{ax+by}] = \mathcal{L}(a, b)[e^{ax+by}]$$

operating both sides by $1/\mathcal{L}(D_1, D_2)$, we get

$$u_p(x, y) = \frac{1}{\mathcal{L}(D_1, D_2)} e^{ax+by} = \frac{1}{\mathcal{L}(a, b)} e^{ax+by} \quad \text{if} \quad \mathcal{L}(a, b) \neq 0.$$

Case 2. $F(x, y) = \sin(mx + ny)$ or $\cos(mx + ny)$. Since

$$(D_1^2 + k_1 D_1 D_2 + k_2 D_2^2) \sin(mx + ny) = (-m^2 - k_1 mn - k_2 n^2) \sin(mx + ny)$$

operating both sides by $1/\mathcal{L}(D_1, D_2)$, we find

$$u_p(x, y) = \frac{1}{\mathcal{L}(D_1, D_2)} \sin(mx + ny) = \frac{1}{-m^2 - k_1 mn - k_2 n^2} \sin(mx + ny)$$

provided $m^2 + k_1 mn + k_2 n^2 \neq 0$. Similarly, we have

$$u_p(x, y) = \frac{1}{\mathcal{L}(D_1, D_2)} \cos(mx + ny) = \frac{1}{-m^2 - k_1 mn - k_2 n^2} \cos(mx + ny).$$

Case 3. $F(x, y) = x^m y^n$, where m and n are nonnegative integers. Since

$$u_p(x, y) = \frac{1}{\mathcal{L}(D_1, D_2)} x^m y^n = \mathcal{L}(D_1, D_2)^{-1} x^m y^n$$

we expand $\mathcal{L}(D_1, D_2)^{-1}$ in ascending powers of D_1 or D_2 by the binomial theorem and then operate on $x^m y^n$ term by term.

Case 4. $F(x, y)$ is any function of x and y. To evaluate (27.7) we resolve $1/\mathcal{L}(D_1, D_2)$ into partial fractions treating $\mathcal{L}(D_1, D_2)$ as a function of D_1 alone and operate each partial fraction on $F(x, y)$, remembering that

$$\frac{1}{D_1 - mD_2} F(x, y) = \int F(x, c - mx) dx,$$

where c is replaced by $y + mx$ after integration. To show this, we let

$$\frac{1}{D_1 - mD_2} F(x, y) = \phi(x, y)$$

so that $(D_1 - mD_2)\phi(x, y) = F(x, y)$ for which

$$\frac{dx}{1} = \frac{dy}{-m} = \frac{d\phi}{F(x, y)}$$

and hence $y + mx = c$, and

$$dx = \frac{d\phi}{F(x, c - mx)},$$

which gives

$$\phi(x,y) = \int F(x, c - mx)dx.$$

Example 27.3. For the partial DE

$$u_{xx} - u_{xy} = \cos x \cos 2y \qquad (27.8)$$

the auxiliary equation is $m^2 - m = 0$, $m = D_1/D_2$, which gives $m_1 = 0$, $m_2 = 1$. Hence, its complementary function is $u_h(x,y) = f_1(y) + f_2(y+x)$. Now for its particular solution, we have

$$
\begin{aligned}
u_p(x,y) &= \frac{1}{D_1^2 - D_1 D_2} \cos x \cos 2y \\
&= \frac{1}{2} \frac{1}{D_1^2 - D_1 D_2} [\cos(x + 2y) + \cos(x - 2y)] \\
&= \frac{1}{2} \left[\frac{1}{-1+2} \cos(x + 2y) + \frac{1}{-1-2} \cos(x - 2y) \right] \\
&= \frac{1}{2} \cos(x + 2y) - \frac{1}{6} \cos(x - 2y).
\end{aligned}
$$

Hence, the general solution of (27.8) is

$$u(x,y) = f_1(y) + f_2(y + x) + \frac{1}{2} \cos(x + 2y) - \frac{1}{6} \cos(x - 2y).$$

Example 27.4. For the partial DE

$$u_{xx} - 4u_{xy} + 4u_{yy} = e^{2x+y} \qquad (27.9)$$

the auxiliary equation is $(m - 2)^2 = 0$, $m = D_1/D_2$, which gives $m_1 = 2$, $m_2 = 2$. Hence, its complementary function is $u_h(x,y) = f_1(y + 2x) + xf_2(y + 2x)$. Now for its particular solution

$$u_p(x,y) = \frac{1}{(D_1 - 2D_2)^2} e^{2x+y}$$

clearly Case 1 fails. However, we can apply Case 4. For this we note that for the equation $(D_1 - 2D_2)v = e^{2x+y}$ the solution is

$$v(x,y) = \int F(x, c - mx)dx = \int e^{2x+(c-2x)}dx = xe^c = xe^{2x+y}$$

and since $(D_1 - 2D_2)u_p = v = xe^{2x+y}$, the particular solution is

$$u_p(x,y) = \int xe^{2x+(c-2x)}dx = \frac{1}{2}x^2 e^c = \frac{1}{2}x^2 e^{2x+y}.$$

Hence, the general solution of (27.9) is

$$u(x, y) = f_1(y + 2x) + x f_2(y + 2x) + \frac{1}{2} x^2 e^{2x+y}.$$

Example 27.5. For the partial DE

$$u_{xx} + u_{xy} - 6u_{yy} = y \cos x \qquad (27.10)$$

the auxiliary equation is $m^2 + m - 6 = 0$, $m = D_1/D_2$, which gives $m_1 = -3$, $m_2 = 2$. Hence, its complementary function is $u_h(x, y) = f_1(y - 3x) + f_2(y + 2x)$. Now to find its particular solution

$$u_p(x, y) = \frac{1}{(D_1 - 2D_2)(D_1 + 3D_2)} y \cos x$$

first we solve the equation $(D_1 + 3D_2)v = y \cos x$, to find

$$v(x, y) = \int (c + 3x) \cos x \, dx = (c + 3x) \sin x + 3 \cos x = y \sin x + 3 \cos x.$$

Now since $(D_1 - 2D_2)u_p = v = y \sin x + 3 \cos x$, the particular solution is

$$\begin{aligned} u_p(x, y) &= \int [(c - 2x) \sin x + 3 \cos x] dx \\ &= (c - 2x)(- \cos x) - (-2)(- \sin x) + 3 \sin x = \sin x - y \cos x. \end{aligned}$$

Hence, the general solution of (27.10) is

$$u(x, y) = f_1(y - 3x) + f_2(y + 2x) + \sin x - y \cos x.$$

From the above examples it is clear that the above procedure can be extended rather easily for the higher order partial DEs of the form

$$\frac{\partial^n u}{\partial x^n} + k_1 \frac{\partial^n u}{\partial x^{n-1} \partial y} + \cdots + k_n \frac{\partial^n u}{\partial y^n} = F(x, y) \qquad (27.11)$$

provided the function $F(x, y)$ is of a particular form. We illustrate the technique in the following example.

Example 27.6. For the partial DE

$$u_{xxx} - 2u_{xxy} = 2e^{2x} + 3x^2 y \qquad (27.12)$$

the auxiliary equation is $m^3 - 2m^2 = 0$, $m = D_1/D_2$, which gives $m_1 = 0$, $m_2 = 0$, $m_3 = 2$. Hence, its complementary function is $u_h(x, y) =$

$f_1(y) + xf_2(y) + f_3(y + 2x)$. Now for its particular solution, we have

$$
\begin{aligned}
u_p(x, y) &= \frac{1}{D_1^3 - 2D_1^2 D_2}(2e^{2x} + 3x^2 y) \\
&= 2\frac{1}{8 - 8\cdot 0}e^{2x} + \frac{3}{D_1^3}\left(1 + \frac{2D_2}{D_1} + \frac{4D_2^2}{D_1^2} + \cdots\right)x^2 y \\
&= \frac{1}{4}e^{2x} + \frac{3}{D_1^3}\left(x^2 y + \frac{2}{3}x^3\right), \\
&= \frac{1}{4}e^{2x} + \frac{1}{20}x^5 y + \frac{1}{60}x^6.
\end{aligned}
$$

Hence, the general solution of (27.12) is

$$u(x, y) = f_1(y) + xf_2(y) + f_3(y + 2x) + \frac{1}{4}e^{2x} + \frac{1}{20}x^5 y + \frac{1}{60}x^6.$$

Now we shall consider the general second-order partial DE

$$\mathcal{A}(D_1, D_2)[u] = (D_1^2 + k_1 D_1 D_2 + k_2 D_2^2 + k_3 D_1 + k_4 D_2 + k_5)[u] = F(x, y). \tag{27.13}$$

Having in mind that for homogeneous ordinary DEs with constant coefficients we can assume a solution of the form e^{mx}, for (27.13) with $F(x, y) = 0$ we assume that $e^{\lambda x + \nu y}$ is a solution. This is possible if and only if

$$\lambda^2 + k_1 \lambda\mu + k_2\mu^2 + k_3\lambda + k_4\mu + k_5 = 0. \tag{27.14}$$

Since (27.14) is a single algebraic equation in two unknowns (representing a conic section in $\lambda\mu$-plane), in general it will have an infinite number of solutions (λ, μ). In particular, if (λ_i, μ_i), $i = 1, 2, \cdots, n$ are n solutions of (27.14), then by the principle of superposition $\sum_{i=1}^{n} c_i e^{\lambda_i x + \mu_i y}$ is a solution of (27.13) with $F(x, y) = 0$.

Example 27.7. For the partial DE

$$u_{xx} + 2u_{xy} + u_{yy} + 3u_x + 3u_y + 2u = 0 \tag{27.15}$$

the equation corresponding to (27.14) is

$$\lambda^2 + 2\lambda\mu + \mu^2 + 3\lambda + 3\mu + 2 = 0.$$

For this equation it is easy to see that $\lambda = -(2 + \mu)$ is a solution. Hence, $e^{-(2+\mu)x + \mu y}$ is a particular solution of (27.15).

Finally, we note that we can always factorize $\mathcal{A}(D_1, D_2)$ into factors of the form $D_1 - mD_2 - c$, and to find the solution of $(D_1 - mD_2 - c)u = 0$ we write it as $u_x - mu_y = cu$. For this the subsidiary equations are

$$\frac{dx}{1} = \frac{dy}{-m} = \frac{du}{cu},$$

which can be integrated to obtain $u(x, y) = e^{cx}\phi(y + mx)$. The solutions corresponding to various factors added up give the solution of (27.13) with $F(x, y) = 0$. Finally, in some cases a particular solution of (27.13) can be obtained by using the operator method.

Example 27.8. For the partial DE

$$(D_1^2 + 2D_1D_2 + D_2^2 - 2D_1 - 2D_2)u = \sin(x + 2y) \tag{27.16}$$

we have $\mathcal{A}(D_1, D_2) = (D_1 + D_2)(D_1 + D_2 - 2)$. Thus, the solution of the homogeneous equation $\mathcal{A}(D_1, D_2)[u] = 0$ is $u_h(x, y) = \phi_1(y-x)+e^{2x}\phi_2(y-x)$. Now to find the particular solution, we have

$$
\begin{aligned}
u_p(x, y) &= \frac{1}{D_1^2 + 2D_1D_2 + D_2^2 - 2D_1 - 2D_2} \sin(x + 2y) \\
&= \frac{1}{-1 + 2(-2) + (-4) - 2D_1 - 2D_2} \sin(x + 2y) \\
&= -\frac{1}{2(D_1 + D_2) + 9} \sin(x + 2y) \\
&= -\frac{2(D_1 + D_2) - 9}{4(D_1^2 + 2D_1D_2 + D_2^2) - 81} \sin(x + 2y) \\
&= -\frac{2(D_1 + D_2) - 9}{4[-1 + 2(-2) - 4] - 81} \sin(x + 2y) \\
&= \frac{1}{117}[2\{\cos(x + 2y) + 2\cos(x + 2y)\} - 9\sin(x + 2y)] \\
&= \frac{1}{39}[2\cos(x + 2y) - 3\sin(x + 2y)].
\end{aligned}
$$

Hence, the general solution of (27.16) is

$$u(x, y) = \phi_1(y - x) + e^{2x}\phi_2(y - x) + \frac{1}{39}[2\cos(x + 2y) - 3\sin(x + 2y)].$$

Problems

27.1. Solve the following second-order linear partial DEs:

(i) $u_{xx} + u_{xy} - 2u_{yy} = 0$

(ii) $u_{xx} - 5u_{xy} + 6u_{yy} = e^{x+y}$

(iii) $u_{xx} - 2u_{xy} + u_{yy} = \sin x$

(iv) $u_{xx} + 4u_{xy} - 5u_{yy} = y^2$

(v) $u_{xx} - u_{xy} - 6u_{yy} = xy$

(vi) $u_{xx} - u_{yy} = \sin x \cos 2y$.

27.2. Solve the following third order linear partial DEs:

(i) $u_{xxx} + u_{xxy} - u_{xyy} - u_{yyy} = 0$

(ii) $u_{xxx} - 3u_{xxy} + 4u_{yyy} = e^{x+2y}$

(iii) $u_{xxx} - 2u_{xxy} = 2e^{2x} + 3x^2 y$.

27.3. Solve the following fourth order linear partial DEs:

(i) $u_{xxxx} - u_{yyyy} = 0$

(ii) $u_{xxxx} - 2u_{xxyy} + u_{yyyy} = 0$.

27.4. Solve the following nonhomogeneous partial DEs:

(i) $u_{xx} - u_{yy} + u_x - u_y = e^{2x+3y}$

(ii) $2u_{xy} + u_{yy} - 3u_y = 3\cos(3x - 2y)$

(iii) $u_{xx} - u_{xy} + u_y - u = \cos(x + 2y) + e^y$

(iv) $(D_1 + D_2 - 1)(D_1 + 2D_2 - 3)u = 4 + 3x + 6y$

(v) $D_1(D_1 + D_2 - 1)(D_1 + 3D_2 - 2)u = xy + e^{2x+3y}$.

27.5. In elasticity certain problems in plane stress can be solved with the aid of *Airy's stress function* ϕ, which satisfies the partial DE

$$\phi_{xxxx} + 2\phi_{xxyy} + \phi_{yyyy} = 0. \tag{27.17}$$

This equation is called *biharmonic equation* and also occurs in the study of hydrodynamics. Show that

(i) $\phi(x, y) = f_1(y - ix) + x f_2(y - ix) + f_3(y + ix) + x f_4(y + ix)$ is a solution of (27.17)

(ii) if $u(x, y)$ and $v(x, y)$ are any two harmonic functions, then $\phi(x, y) = u(x, y) + xv(x, y)$ is a solution of (27.17).

27.6. Solve the following partial DE:

$$x^2 u_{xx} - 2xy u_{xy} + y^2 u_{yy} = 0.$$

Answers or Hints

27.1. (i) $u = f_1(y+x) + f_2(y-2x)$ (ii) $u = f_1(y+2x) + f_2(y+3x) + \frac{1}{2}e^{x+y}$ (iii) $u = f_1(y+x) + x f_2(y+x) - \sin x$ (iv) $u = f_1(y+x) + f_2(y-5x) + \frac{1}{2}x^2 y^2 - \frac{4}{3}x^3 y + \frac{7}{4}x^4$ (v) $u = f_1(y-2x) + f_2(y+3x) + \frac{1}{6}x^3 y + \frac{1}{24}x^4$ (vi) $u = f_1(y) + f_2(y+x) + \frac{1}{3}(\sin x \cos 2y + 2\cos x \sin 2y)$.

27.2. (i) $u = f_1(y+x) + f_2(y-x) + x f_3(y-x)$ (ii) $u = f_1(y-x) + f_2(y+2x) + x f_3(y+2x) + \frac{1}{27}e^{x+2y}$ (iii) $u = f_1(y) + x f_2(y) + f_3(y+2x) + \frac{1}{4}e^{2x} + \frac{1}{20}x^5 y + \frac{1}{60}x^6$.

27.3. (i) $u = f_1(y + x) + f_2(y - x) + f_3(y + ix) + f_4(y - ix)$ (ii) $u = f_1(y + x) + xf_2(y + x) + f_3(y - x) + xf_4(y - x)$.

27.4. (i) $u = \phi_1(y+x)+e^{-x}\phi_2(y-x)-\frac{1}{6}e^{2x+3y}$ (ii) $u = \phi_1(x)+e^{3y}\phi_2(2y-x)+\frac{3}{50}\{4\cos(3x-2y)+3\sin(3x-2y)\}$ (iii) $u = e^x\phi_1(y)+e^{-x}\phi_2(x+y)+\frac{1}{2}\sin(x+2y)-xe^y$ (iv) $u = e^x\phi_1(y-x)+e^{3x}\phi_2(y-2x)+x+2y+6$ (v) $u = \phi_1(y)+e^x\phi_2(y-x)+e^{2x}\phi_3(y-3x)+\frac{1}{72}e^{2x+3y}+\frac{1}{8}(2x^2y+6xy+5x^2+22x)$.

27.5. (i) Verify directly (ii) Verify directly.

27.6. $\sum A_k x^{m_k} y^{n_k}$ where $2n_k = 2m_k + 1 \pm (8m_k + 1)^{1/2}$.

Lecture 28
The Canonical Forms

In this lecture we shall show that coordinate transformations can be employed successfully to reduce second-order linear partial DEs to some standard forms which are known as canonical forms. These transformed equations sometimes can be solved rather easily. Here the concept of characteristic of second-order partial DEs plays an important role.

To solve the one-dimensional wave equation (25.11) we introduce the new coordinates

$$\xi = x + ct, \quad \eta = x - ct. \tag{28.1}$$

Then, by the chain rule we have

$$
\begin{aligned}
u_{xx} &= u_{\xi\xi} + 2u_{\xi\eta} + u_{\eta\eta} \\
u_{tt} &= c^2[u_{\xi\xi} - 2u_{\xi\eta} + u_{\eta\eta}].
\end{aligned}
$$

Substituting these expression in (25.11), we obtain

$$-4c^2 u_{\xi\eta} = 0.$$

Now since $c \neq 0$, it follows that $u_{\xi\eta} = 0$ for which the solution can be written as

$$u(\xi, \eta) = f_1(\xi) + f_2(\eta),$$

where f_1 and f_2 are arbitrary functions.

Thus, the general solution of (25.11) appears as

$$u(x, t) = f_1(x + ct) + f_2(x - ct). \tag{28.2}$$

We shall now consider the general partial DE

$$\mathcal{L}[u] = A u_{tt} + B u_{tx} + C u_{xx} = 0, \tag{28.3}$$

where A, B, and C are given constants. We once again attempt to find a linear transformation of coordinates

$$\xi = \alpha x + \beta t, \quad \eta = \gamma x + \delta t, \quad \alpha\delta - \gamma\beta \neq 0 \tag{28.4}$$

so that the differential operator $\mathcal{L}[u]$ in (28.3) becomes a multiple of $u_{\xi\eta}$. For this, by the chain rule we have

$$
\begin{aligned}
\mathcal{L}[u] &= (A\beta^2 + B\alpha\beta + C\alpha^2)u_{\xi\xi} + (2A\beta\delta + B(\alpha\delta + \beta\gamma) + 2C\alpha\gamma)u_{\xi\eta} \\
&\quad + (A\delta^2 + B\gamma\delta + C\gamma^2)u_{\eta\eta}.
\end{aligned}
$$

R.P. Agarwal, D. O'Regan, *Ordinary and Partial Differential Equations*,
Universitext, DOI 10.1007/978-0-387-79146-3_28,
© Springer Science+Business Media, LLC 2009

Thus, to meet the desired form we need

$$A\beta^2 + B\alpha\beta + C\alpha^2 = 0$$
$$A\delta^2 + B\gamma\delta + C\gamma^2 = 0.$$

If $A = C = 0$, the trivial transformation $\xi = x$, $\eta = t$ gives $\mathcal{L}$ in the desired form. We now suppose that either A or C is not zero. In what follows we shall assume that $A \neq 0$. Then, $\alpha \neq 0$ and $\gamma \neq 0$ and we may divide the first equation by α^2 and the second equation by γ^2. In this way we obtain two identical quadratic equations for the ratios β/α and δ/γ. The solutions of these equations are

$$\frac{\beta}{\alpha} = \frac{1}{2A}[-B \pm \sqrt{B^2 - 4AC}]$$
$$\frac{\delta}{\gamma} = \frac{1}{2A}[-B \pm \sqrt{B^2 - 4AC}].$$

In order for the coordinate transformation (28.4) to be nonsingular, the ratios β/α and δ/γ must be different. Hence, we must take the plus sign in the solution in one case, and the minus sign in the other. Moreover, we must assume that the quantity $B^2 - 4AC$ is positive. For if it was zero, the two ratios would still coincide, while if it was negative, neither of them would be real.

Thus, we may transform $\mathcal{L}[u]$ to a multiple of $u_{\xi\eta}$ if and only if $B^2 - 4AC > 0$; i.e., the operator $\mathcal{L}$ must be hyperbolic. The transformation in this case is given by

$$\xi = 2Ax + [-B + \sqrt{B^2 - 4AC}]t$$
$$\eta = 2Ax + [-B - \sqrt{B^2 - 4AC}]t \tag{28.5}$$

and the operator becomes

$$\mathcal{L}[u] = -4A(B^2 - 4AC)u_{\xi\eta}. \tag{28.6}$$

The case $A = 0$ can be treated similarly, with $\xi = t$, $\eta = x - (C/B)t$.

Finally, from (28.6) it is clear that the general solution again can be written as $u = f_1(\xi) + f_2(\eta)$.

If $B^2 - 4AC = 0$, i.e., $\mathcal{L}$ is parabolic, then there is only one value of β/α which makes the coefficient of $u_{\xi\xi}$ vanish. This is $\beta/\alpha = -B/(2A)$. Since $B/(2A) = 2C/B$ this choice also makes the coefficient of $u_{\xi\eta}$ vanish. Indeed, we have

$$2A\beta\delta + B\alpha\delta + B\beta\gamma + 2C\alpha\gamma$$
$$= 2A\left(-\frac{B}{2A}\alpha\right)\delta + B\alpha\delta + B\left(-\frac{B}{2A}\alpha\right)\gamma + \left(\frac{B^2}{2A}\right)\alpha\gamma = 0.$$

Thus, the transformation

$$\xi = 2Ax - Bt, \quad \eta = t$$

transforms $\mathcal{L}[u]$ into

$$\mathcal{L}[u] = Au_{\eta\eta}.$$

The choice of η is quite arbitrary. We could choose any $\gamma x + \delta t$ as long as $\delta/\gamma \neq -B/(2A)$. If $A = 0$, we may choose $\eta = x$, $\xi = t$ to obtain $\mathcal{L}[u] = Cu_{\eta\eta}$. The general solution of $\mathcal{L}[u] = 0$ is now

$$u = f_1(\xi) + \eta f_2(\xi).$$

Finally, if $B^2 - 4AC < 0$, i.e., the operator $\mathcal{L}$ is elliptic, then no choice of β/α or δ/γ makes the coefficients of $u_{\xi\xi}$ or $u_{\eta\eta}$ vanish. However, the transformation

$$\xi = \frac{2Ax - Bt}{\sqrt{4AC - B^2}}, \quad \eta = t$$

makes

$$\mathcal{L}[u] = A[u_{\xi\xi} + u_{\eta\eta}].$$

Since $4AC > B^2$, $A \neq 0$. A standard form for an elliptic differential equation $\mathcal{L}[u] = 0$ is $u_{\xi\xi} + u_{\eta\eta} = 0$, i.e., Laplace's equation.

We shall now consider the general linear partial DE

$$\mathcal{L}[u] = Au_{tt} + Bu_{tx} + Cu_{xx} + Du_t + Eu_x + Fu = 0, \tag{28.7}$$

where $A, B, C, D, E,$ and F are functions of x and t. Our aim is to show that in (28.7) the second-order terms may be reduced to one of the standard forms obtained earlier in a whole domain where it is hyperbolic, parabolic or elliptic. For this we need a more general (nonlinear) coordinate transformation

$$\xi = \xi(x, t), \quad \eta = \eta(x, t)$$

where ξ and η are twice continuously differentiable. By the chain rule it follows that

$$
\begin{aligned}
\mathcal{L}[u] = {} & [A(\xi_t)^2 + B\xi_x\xi_t + C(\xi_x)^2]u_{\xi\xi} + [2A\xi_t\eta_t + B\xi_t\eta_x + B\xi_x\eta_t \\
& + 2C\xi_x\eta_x]u_{\xi\eta} + [A(\eta_t)^2 + B\eta_x\eta_t + C(\eta_x)^2]u_{\eta\eta} \\
& + [\mathcal{L}[\xi] - F\xi]u_\xi + [\mathcal{L}[\eta] - F\eta]u_\eta + Fu.
\end{aligned}
$$

If $\mathcal{L}$ is hyperbolic, the coefficients of $u_{\xi\xi}$ and $u_{\eta\eta}$ may be made equal to zero by putting (assuming $A \neq 0$)

$$\frac{\xi_t}{\xi_x} = \frac{-B + \sqrt{B^2 - 4AC}}{2A} \quad \text{and} \quad \frac{\eta_t}{\eta_x} = \frac{-B - \sqrt{B^2 - 4AC}}{2A}.$$

Then, the curve $\xi = $ constant $(d\xi = \xi_x dx + \xi_t dt = 0$, i.e., $\xi_t/\xi_x = -dx/dt)$ is a solution of

$$\frac{dx}{dt} = \frac{B - \sqrt{B^2 - 4AC}}{2A}, \tag{28.8}$$

while the curve $\eta = $ constant, satisfies

$$\frac{dx}{dt} = \frac{B + \sqrt{B^2 - 4AC}}{2A}. \tag{28.9}$$

In both cases dx/dt satisfies

$$A\left(\frac{dx}{dt}\right)^2 - B\frac{dx}{dt} + C = 0.$$

The ordinary DEs (28.8) and (28.9) give two families of curves, which are called the *characteristic* of $\mathcal{L}$. The values of ξ and η may be prescribed arbitrarily along the initial line $t = 0$.

If we obtain a one-parameter family of solutions $x = f(t, a)$ of (28.8) satisfying the initial condition $f(0, a) = a$, we can, in principle, solve for a in terms of x and t. Then, ξ can be chosen to be any monotone function of $a(x, t)$. Similarly, if $x = g(t, b)$ is the solution of (28.9) with $g(0, b) = b$, we can solve for $b(x, t)$ and choose η to be any monotone function of b.

If $\mathcal{L}$ is parabolic, the two characteristic equations (28.8) and (28.9) are the same, so that there is only one family of characteristic. If the coordinate ξ is chosen so that it is constant along the characteristic, i.e., the solution of $dx/dt = (B/2A)$, then the coefficients of $u_{\xi\xi}$ and $u_{\xi\eta}$ vanish, so that the only second derivative occurring in $\mathcal{L}$ is $u_{\eta\eta}$ (here η is arbitrary).

Finally, if $\mathcal{L}$ is elliptic, one can make the coefficient of $u_{\xi\eta}$ vanish by choosing η arbitrary and making ξ constant along solutions of

$$\frac{dx}{dt} = \frac{B\eta_t + 2C\eta_x}{2A\eta_t + B\eta_x}.$$

The other two second-order derivatives of u will then have coefficients with the same sign as A.

Example 28.1. Consider the partial DE

$$t^2 u_{tt} - x^2 u_{xx} = 0. \tag{28.10}$$

Since $B^2 - 4AC = 4t^2 x^2 > 0$, equation (28.10) is hyperbolic everywhere except on the t– and x–axes. We consider a region that does not include any part of either axis, e.g., a region in the first quadrant. Then, the characteristic equations (28.8) and (28.9) for (28.10) are

$$\frac{dx}{dt} = \pm\frac{x}{t}.$$

These DEs have solutions given by $x = c_1 t$ and $x = c_2/t$. Hence, we can make the coordinate transformation $\xi = x/t$, $\eta = tx$. This transformation reduces (28.10) to

$$u_{\xi\eta} - \frac{1}{2\eta} u_\xi = 0.$$

We can solve this partial DE by observing that it is a first-order linear DE in u_ξ. In fact, we get

$$u(\xi, \eta) = \eta^{1/2} f(\xi) + g(\eta),$$

and hence the general solution of (28.10) is

$$u(x, t) = \sqrt{tx} f\left(\frac{x}{t}\right) + g(tx).$$

Example 28.2. Consider the partial DE

$$t^2 u_{tt} + 2tx u_{tx} + x^2 u_{xx} = 4t^2. \tag{28.11}$$

Since $B^2 - 4AC = 0$, equation (28.11) is parabolic everywhere. Thus, the characteristic equations (28.8) and (28.9) for (28.11) are

$$\frac{dx}{dt} = \frac{B}{2A} = \frac{x}{t}.$$

This DE can be solved to obtain the solution $x/t = c$. Hence, we can choose the coordinate transformation $\xi = x/t$, $\eta = x$. This transformation reduces (28.11) to

$$x^2 u_{\eta\eta} = 4t^2,$$

which is the same as

$$u_{\eta\eta} = 4\frac{t^2}{x^2} = \frac{4}{\xi^2}.$$

This partial DE can be solved rather easily, to obtain

$$u(\xi, \eta) = \frac{2\eta^2}{\xi^2} + \eta f(\xi) + g(\xi).$$

Hence, in terms of the original variables the general solution of (28.11) is

$$u(x, t) = 2t^2 + x f\left(\frac{x}{t}\right) + g\left(\frac{x}{t}\right).$$

Example 28.3. Consider the partial DE

$$t u_{tt} + u_{xx} = t^2. \tag{28.12}$$

Since $B^2 - 4AC = -4t$, equation (28.12) is elliptic in the half-plane $t > 0$. In this region the characteristic equations are

$$\frac{dx}{dt} = \pm it^{-1/2}$$

with solutions $x = 2it^{1/2} + c_1$ and $x = -2it^{1/2} + c_2$. We first make the coordinate transformation $\sigma = x + 2it^{1/2}$, $\tau = x - 2it^{1/2}$, but since these are complex valued, we make a second coordinate transformation defined by $\sigma = \xi + i\eta$, $\tau = \xi - i\eta$. As a result of these two transformations, we have $\xi = x$, $\eta = 2t^{1/2}$. This transformation reduces (28.12) to the canonical form

$$u_{\xi\xi} + u_{\eta\eta} = \frac{1}{\eta}u_\eta + \frac{\eta^4}{16}.$$

Problems

28.1. Find the characteristic of the following partial DEs through the point $(0, 1)$:

(i) $u_{tt} - tu_{xx}$

(ii) $u_{tt} + 2e^x u_{tx} + e^{2x} u_{xx} + \cos x \, u_t + \sin x \, u_x + x^2 u$

(iii) $(\cos^2 x - \sin^2 x)u_{tt} + 2\cos x \, u_{tx} + u_{xx} + u.$

28.2. Find characteristic coordinates ξ, η for

$$u_{tt} + (e^x + t^2)u_{tx} + t^2 e^x u_{xx}$$

such that $\xi(x, 0) = \eta(x, 0) = x$.

28.3. Transform the following partial DEs to canonical form:

(i) $u_{tt} - 5u_{tx} + 6u_{xx} = 0$

(ii) $u_{tt} + 2u_{tx} + u_{xx} + 3u_t + 9u = 0$

(iii) $2u_{tt} - 2u_{tx} + 5u_{xx} + u = 0$

(iv) $u_{tt} - t^2 u_{xx} = 0$

(v) $u_{tt} + t^2 u_{xx} = 0$

(vi) $u_{tt} + 2tu_{tx} + t^2 u_{xx} = 0$

(vii) $u_{tt} + (2t + 3)u_{tx} + 6tu_{xx} = 0$

(viii) $u_{tt} + (5 + 2x^2)u_{tx} + (1 + x^2)(4 + x^2)u_{xx} = 0.$

28.4. Transform the following partial DEs to canonical forms and then solve

(i) $u_{tt} + 2u_{tx} + u_{xx} = 0$

(ii) $u_{tt} - 2\sin t\, u_{tx} - \cos^2 t\, u_{xx} - \cos t\, u_x = 0$

(iii) $x^2 u_{tt} - 2txu_{tx} + t^2 u_{xx} = (x^2/t)u_t + (t^2/x)u_x.$

28.5. (i) Use the substitution $u(x,y) = w(x,y)e^{-(bx+ay)}$ to show that the hyperbolic equation

$$\frac{\partial^2 u}{\partial x \partial y} + a\frac{\partial u}{\partial x} + b\frac{\partial u}{\partial y} + cu = 0$$

can be written as

$$\frac{\partial^2 w}{\partial x \partial y} + (c - ab)w = 0. \tag{28.13}$$

(ii) Show that (28.13) reduces to a Bessel equation if we assume a solution of the form $w(x,y) = f(xy)$.

28.6. Use the substitution $u(x,y) = w(x,y)e^{-(ax+by)/2}$ to show that the elliptic equation

$$\frac{\partial^2 u}{\partial x^2} + \frac{\partial^2 u}{\partial y^2} + a\frac{\partial u}{\partial x} + b\frac{\partial u}{\partial y} + cu = 0$$

can be written as

$$\frac{\partial^2 w}{\partial x^2} + \frac{\partial^2 w}{\partial y^2} + \left(c - \frac{a^2}{4} - \frac{b^2}{4}\right)w = 0.$$

28.7. A partial DE which describes the advective transport of a chemical u subject to first-order reaction is

$$R\frac{\partial u}{\partial t} = -V\frac{\partial u}{\partial x} - Ku, \quad x > 0, \quad t > 0,$$

where R is a retardation coefficient, V the velocity of the solution carrying the chemical, and K the first-order reaction coefficient.

(i) Write this equation as

$$u_t + cu_x + \lambda u = 0, \quad c = V/R, \quad \lambda = K/R \tag{28.14}$$

and make the change of variables $\xi = x - ct$, $\eta = t$ to find its general solution $u(x,t) = f(x - ct)e^{-\lambda t}$ where f is an arbitrary function.

(ii) Show that the choice of f as

$$f(\xi) = u_1(-\xi/c)e^{-\lambda\xi/c} + [u_0(\xi) - u_1(-\xi/c)e^{-\lambda\xi/c}]H(\xi),$$

where $H(\xi)$ is the unit step function, i.e., $H(\xi) = 1$ if $\xi > 0$ and 0 otherwise, gives a solution of (28.14) that satisfies the initial condition $u(x,0) = u_0(x)$ and the boundary condition $u(0,t) = u_1(t)$.

Answers or Hints

28.1. (i) $x + \frac{2}{3}t^{3/2} = \frac{2}{3}$, $x - \frac{2}{3}t^{3/2} = -\frac{2}{3}$ (ii) $t = 2 - e^{-x}$ (one characteristic only) (iii) $t = \sin x + \cos x$, $t = 2 + \sin x - \cos x$.

28.2. $\xi = -\ln(e^{-x} + t)$, $\eta = x - \frac{1}{3}t^3$.

28.3. (i) Hyperbolic, $\xi = 2t + x$, $\eta = 3t + x$, $u_{\xi\eta} = 0$ (ii) Parabolic, $\xi = -t + x$; $\eta = x$, $u_{\eta\eta} = 3u_\xi - 9u$ (iii) Elliptic, $\xi = \frac{1}{2}t + x$, $\eta = \frac{3}{2}t$, $u_{\xi\xi} + u_{\eta\eta} = -\frac{2}{9}u$ (iv) Hyperbolic for $t > 0$, $\xi = x + \frac{1}{2}t^2$, $\eta = x - \frac{1}{2}t^2$, $u_{\xi\eta} = \frac{1}{4(\xi - \eta)}(u_\xi - u_\eta)$ (v) Elliptic for $t > 0$, $\xi = x$, $\eta = \frac{1}{2}t^2$, $u_{\eta\eta} + u_{\xi\xi} = -\frac{1}{2\eta}u_\eta$ (vi) Parabolic, $\xi = x - \frac{1}{2}t^2$, $\eta = t$, $u_{\eta\eta} = u_\xi$ (vii) Parabolic for $t = \frac{3}{2}$, $\xi = -3t + x$, $\eta = t$, $u_{\eta\eta} = 0$; Hyperbolic for $t \neq \frac{3}{2}$, $\xi = x - 3t$, $\eta = x - t^2$, $u_{\xi\eta} = \frac{2}{4(\eta - \xi) - 9}u_\eta$ (viii) Hyperbolic, $\xi = \tan^{-1} x - t$, $\eta = \frac{1}{2}\tan^{-1}\frac{1}{2}x - t$, $u_{\xi\eta} + \frac{2x}{9}(4 + x^2)^2 u_\xi + \frac{2x}{9}(1 + x^2)^2 u_\eta = 0$, where $x = \tan^{1/3}\left(\frac{1}{4}\pi + \xi - \eta\right) - \cot^{1/3}\left(\frac{1}{4}\pi + \xi - \eta\right)$.

28.4. (i) Parabolic, $\xi = t - x$, $\eta = t + x$, $u_{\eta\eta} = 0$, $u = (x + y)f_1(x - y) + f_2(x - y)$ (ii) Hyperbolic, $\xi = \cos t + t - x$, $\eta = \cos t - t - x$, the equation reduces to $u_{\xi\eta} = 0$, the solution is $u = f(\cos t - t - x) + g(\cos t + t - x)$ (iii) Parabolic, $\xi = t^2 + x^2$, $\eta = x$, the equation reduces to $u_{\eta\eta} = (1/\eta)u_\eta$, the solution is $u = x^2 f(t^2 + x^2) + g(t^2 + x^2)$.

28.5. Verify directly.

28.6. Verify directly.

Lecture 29
The Method of
Separation of Variables

The method of separation of variables involves a solution which breaks up into a product of functions each of which contains only one of the variables. This widely used method for finding solutions of linear homogeneous partial DEs we shall explain through several examples.

Example 29.1. For the one–dimensional wave equation (25.11) with $c = 1$, we assume that the trial solution can be written as $u = u(x,t) = X(x)T(t)$, where the functions X, T are to be determined. For this, we have $u_{tt} = X(x)T''(t)$ $('= d/dt)$, $u_{xx} = X''(x)T(t)$ $('= d/dx)$. Substitution of these in (25.11) leads to the relation

$$X(x)T''(t) = X''(x)T(t).$$

Dividing this equation by $u = X(x)T(t)$ (assuming that $u \neq 0$), we obtain

$$\frac{T''(t)}{T(t)} = \frac{X''(x)}{X(x)}. \tag{29.1}$$

Since $T''(t)/T(t)$ does not contain the variable x, we note that changes in x will not have any effect on the expression $T''(t)/T(t)$. Thus, if (29.1) is to be an equality, it must happen that changes in the variable x do not affect the expression $X''(x)/X(x)$ either. Similarly, changes in t should not affect the expression $T''(t)/T(t)$. Thus, we can conclude that in order for (29.1) to be an equality, the expressions $T''(t)/T(t)$ and $X''(x)/X(x)$ must be constants. In fact, they must be the same constant. If the constant is denoted by λ, we can write

$$\frac{T''(t)}{T(t)} = \lambda \quad \text{and} \quad \frac{X''(x)}{X(x)} = \lambda.$$

Thus, we obtain two second-order ordinary DEs:

$$X''(x) - \lambda X(x) = 0$$

and

$$T''(t) - \lambda T(t) = 0.$$

R.P. Agarwal, D. O'Regan, *Ordinary and Partial Differential Equations*,
Universitext, DOI 10.1007/978-0-387-79146-3_29,
© Springer Science+Business Media, LLC 2009

These equations can be solved to obtain the solutions

$$X(x) = \begin{cases} c_1 e^{\sqrt{\lambda}x} + c_2 e^{-\sqrt{\lambda}x}, & \lambda > 0 \\ c_1 + c_2 x, & \lambda = 0 \\ c_1 \cos \sqrt{-\lambda}x + c_2 \sin \sqrt{-\lambda}x, & \lambda < 0 \end{cases}$$

and

$$T(t) = \begin{cases} c_3 e^{\sqrt{\lambda}t} + c_4 e^{-\sqrt{\lambda}t}, & \lambda > 0 \\ c_3 + c_4 t, & \lambda = 0 \\ c_3 \cos \sqrt{-\lambda}t + c_4 \sin \sqrt{-\lambda}t, & \lambda < 0. \end{cases}$$

Thus, the solution of (25.11) can be written as

$$\begin{aligned} u &= u(x,t) = X(x)T(t) \\ &= \begin{cases} \left(c_1 e^{\sqrt{\lambda}x} + c_2 e^{-\sqrt{\lambda}x}\right)\left(c_3 e^{\sqrt{\lambda}t} + c_4 e^{-\sqrt{\lambda}t}\right), & \lambda > 0 \\ (c_1 + c_2 x)(c_3 + c_4 t), & \lambda = 0 \\ \left(c_1 \cos \sqrt{-\lambda}x + c_2 \sin \sqrt{-\lambda}x\right)\left(c_3 \cos \sqrt{-\lambda}t + c_4 \sin \sqrt{-\lambda}t\right), \\ \hspace{8cm} \lambda < 0. \end{cases} \end{aligned}$$

Without further information we have no way of knowing the value of λ; hence we cannot specify the form of the solution. In many practical problems there are other conditions that the solution must satisfy. These conditions usually dictate the value of λ and the form of the solution.

Example 29.2. For the partial DE

$$u_{xx} - 2u_x + u_y = 0, \quad u = u(x,y) \tag{29.2}$$

we assume that the trial solution can be written as $u = u(x,y) = X(x)Y(y)$. Then, it is necessary that

$$X''Y - 2X'Y + XY' = 0.$$

Separating the variables, we get

$$\frac{X'' - 2X'}{X} = -\frac{Y'}{Y}. \tag{29.3}$$

Since x and y are independent variables, (29.3) can be true if each side is equal to the same constant, λ (say). Therefore, it follows that

$$\frac{X'' - 2X'}{X} = -\frac{Y'}{Y} = \lambda$$

and hence

$$Y' + \lambda Y = 0 \tag{29.4}$$

and

$$X'' - 2X' - \lambda X = 0. \tag{29.5}$$

The solution of the first-order ordinary differential equation (29.4) is

$$Y(y) = c_1 e^{-\lambda y}.$$

For the second-order ordinary differential equation (29.5) the auxiliary equation is $m^2 - 2m - \lambda = 0$, and its roots are $m = 1 \pm \sqrt{(1 + \lambda)}$. Thus, the solution of (29.5) can be written as

$$X(x) = c_2 e^{(1+\sqrt{1+\lambda})x} + c_3 e^{(1-\sqrt{1+\lambda})x}.$$

Therefore, the solution of the partial DE (29.2) is

$$
\begin{aligned}
u = u(x,y) = X(x)Y(y) &= \left(c_2 e^{(1+\sqrt{1+\lambda})x} + c_3 e^{(1-\sqrt{1+\lambda})x} \right) c_1 e^{-\lambda y} \\
&= \left(A e^{(1+\sqrt{1+\lambda})x} + B e^{(1-\sqrt{1+\lambda})x} \right) e^{-\lambda y}.
\end{aligned}
$$

Example 29.3. Consider a thin, tapered rod of constant density ρ lying along the x-axis with cross-sectional area αx^2 and moment of inertia βx for some constants α and β. Suppose that the rod is undergoing oscillatory motion. The displacement $u(x,t)$ of a point x at time t satisfies the partial DE

$$\rho \alpha x^2 \frac{\partial^2 u}{\partial t^2} = -\frac{\partial^2}{\partial x^2} \left\{ E\beta x^4 \frac{\partial^2 u}{\partial x^2} \right\}, \tag{29.6}$$

where E is Young's modulus. We assume that the trial solution of (29.6) can be written as $u = u(x,t) = X(x)T(t)$. Then, it is necessary that

$$\frac{T''}{T} = -\frac{E\beta}{\rho \alpha} \frac{\frac{d^2}{dx^2}(x^4 X'')}{x^2 X} = \lambda,$$

which leads to the DEs $T'' - \lambda T = 0$, and

$$x^2 X'''' + 8x X''' + 12 X'' - k^4 X = 0, \quad k^4 = -\frac{\lambda \rho \alpha}{E\beta}. \tag{29.7}$$

Now for simplicity we assume that $\lambda < 0$, so that $T(t) = c_1 \cos \omega t + c_2 \sin \omega t$, $\omega = \sqrt{-\lambda}$ and to find the solution of (29.7) we rewrite it as

$$(xD^2 + 3D + k^2)(xD^2 + 3D - k^2)X = 0, \quad D = \frac{d}{dx}$$

or

$$(xD^2 + 3D - k^2)(xD^2 + 3D + k^2)X = 0.$$

Thus, if X_1 and X_2 are solutions of

$$(xD^2 + 3D - k^2)X = 0$$

and

$$(xD^2 + 3D + k^2)X = 0$$

respectively, then X_1 and X_2 are solutions of the DE (29.7). Now from the considerations of Lecture 9 it follows that

$$X(x) = x^{-1}\left[AJ_2(2kx^{1/2}) + BJ_{-2}(2kx^{1/2}) + CI_2(2kx^{1/2}) + DK_2(2kx^{1/2})\right].$$

Example 29.4. Suppose that the potential energy of a particle at x is given by the function $V(x)$. Then, the partial DE satisfied by the wave function is the Schrödinger equation

$$-\frac{h^2}{8\pi^2 m}\frac{\partial^2 u}{\partial x^2} + V(x)u = \frac{ih}{2\pi}\frac{\partial u}{\partial t}. \tag{29.8}$$

Now following the notational tradition in quantum mechanics we assume that the trial solution of (29.8) can be written as $u = u(x,t) = \psi(x)F(t)$. Then, it is necessary that

$$-\frac{h^2}{8\pi^2 m}\frac{\psi''(x)}{\psi(x)} + V(x) = \frac{ih}{2\pi}\frac{F'(t)}{F(t)} = E,$$

where from the physical reasons E is a real constant. This leads to two DEs. The first equation,

$$F' + \frac{2\pi i E}{h}F = 0,$$

can be easily solved and yields $F(t) = Ce^{-2\pi iEt/h}$. The second equation, which is known as time-independent Schrödinger equation, appears as

$$\psi''(x) + \frac{8\pi^2 m}{h^2}[E - V(x)]\psi(x) = 0. \tag{29.9}$$

Now recall that if the particle moves under the influence of a force $F(x)$, the potential energy is given by $V(x) = -\int_0^x F(s)ds$. In particular, for a free particle we have $F(x) = 0$, then $V(x) = 0$. Thus, the equation (29.9) simply reduces to

$$\psi''(x) + \frac{8\pi^2 m}{h^2}E\psi(x) = 0,$$

which can be solved rather easily.

For a particle on a spring $F(x) = -kx$, then $V(x) = kx^2/2$, and the equation (29.9) becomes

$$\psi''(x) + \frac{8\pi^2 m}{h^2}\left[E - \frac{1}{2}kx^2\right]\psi(x) = 0,$$

which is the same as (3.20) in Problem 3.10, and hence can be transformed to the Hermite equation (3.16).

We remark that the method of separation of variables should be used with caution. In fact, often it fails to work even for simple partial DEs. For example, consider the equation

$$u_{xy} + u_{xx} + 4u = 0.$$

We try a solution of the form $u = X(x)Y(y)$. Then,

$$X'Y' + X''Y + 4XY = 0.$$

Clearly, it is not possible to manipulate this equation algebraically to write it in a form $P(x) = Q(y)$. Similarly, the equation $u_{xx} + (x+y)u_{yy} = 0$ is not separable.

Problems

29.1. Use the method of separation of variables to solve the following partial DEs:

(i) $u_y = yu_x$

(ii) $xu_x = u + yu_y$

(iii) $xu_x = yu_y$

(iv) $x^2u_{xx} + 2xu_x + u_{yy} = 0$

(v) $u_{xx} + 4xu_x + u_{yy} = 0$

(vi) $u_{xx} - (1 + y^2)u_{xy} = 0.$

29.2. Use the method of separation of variables to solve the following partial DEs:

(i) $u_x = 2u_y + u$ where $u(x, 0) = 6e^{-3x}$

(ii) $4u_x + u_y = 3u$ where $u(0, y) = 3e^{-y} - e^{-5y}$

(iii) $u_{xx} = u_y + 2u$ where $u(0, y) = 0$, $u_x(0, y) = 1 + e^{-3y}$

29.3. Find separated solutions of the equation $u_{xx} - u_y = 0$ in the form

(i) $u(x, y) = e^{i\mu x}e^{\beta y}$ where μ, β are real

(ii) $u(x, y) = e^{\alpha x}e^{i\omega y}$ where ω real and positive.

29.4. Show that for the partial DE (27.1) with $k_1 \neq 0$ the method of separation of variables cannot be applied. However, a solution of the form $u = u(x, y) = e^{ry}X(x)$ can always be obtained and appears as $u =$

$e^{ry}(AX_1(x)+BX_2(x))$, where X_1 and X_2 are linearly independent solutions of the ordinary DE $X''+k_1rX'+k_2r^2X = 0$. In particular, find the solution for the case $k_1 = 4$, $k_2 = 8$.

29.5. Show that

$$u_{rr} + \frac{2}{r}u_r = u_{tt}$$

has solutions of the form

$$u(r,t) = \frac{R(r)}{r}\cos nt, \quad n = 0, 1, 2, \cdots.$$

Find a DE that $R(r)$ must satisfy and find its general solution.

29.6. The nonlinear partial DE

$$(u_x)^{n+1}u_{tt} = c^2 u_{xx}, \quad u = u(x,t)$$

occurs in the study of the propagation of sound in a medium. Here the constant c represents the velocity of sound in the medium. Set $u = X(x)T(t)$ to determine the ordinary DEs for X and T.

29.7. In the study of the supersonic flow of an ideal compressible fluid past an obstacle, the velocity potential satisfies the linear partial DE

$$(M^2 - 1)u_{xx} - u_{yy} = 0, \quad u = u(x,y)$$

where the constant $M > 1$ is known as the *Mach number* of the flow. Set $u = X(x)Y(y)$ to determine the ordinary DEs for X and Y.

Answers or Hints

29.1. (i) $Ae^{k(x+y^2/2)}$ and in general $f(2x+y^2)$ (ii) Ax^ky^{k-1} (iii) Ax^ky^k and in general $f(x^ky^k)$ (iv) $|x|^{-1/2}[A_1\cos(c_\lambda \ln|x|) + A_2\sin(c_\lambda \ln|x|)]$ $\times(A_3e^{\sqrt{-\lambda}y} + A_4e^{-\sqrt{-\lambda}y})$, $\lambda < -1/4$, $c_\lambda = \sqrt{-\lambda - 1/4}$; $(A_1|x|^{-1/2} + A_2|x|^{-1/2}\ln|x|)(A_3e^{y/2}+A_4e^{-y/2})$, $\lambda = -1/4$; $(A_1|x|^{\gamma_1}+A_2|x|^{\gamma_2})(A_3e^{\sqrt{-\lambda}y} +A_4e^{-\sqrt{-\lambda}y})$, $-1/4 < \lambda < 0$; $(A_1 + A_2x^{-1})(A_3 + A_4y)$, $\lambda = 0$; $(A_1|x|^{\gamma_1} + A_2|x|^{\gamma_2})(A_3\cos\sqrt{\lambda}y + A_4\sin\sqrt{\lambda}y)$, $\lambda > 0$ where $\gamma_i = \frac{1\pm\sqrt{1+4\lambda}}{2}$, $i = 1,2$ (v) $(Ae^{2ky} + Be^{-2ky})(CX_1(x) + DX_2(x))$ where X_1 and X_2 are linearly independent solutions of the DE $X'' + 4xX' + 4k^2X = 0$ (vi) $(c_1 + c_2e^{\lambda x})\exp(\lambda\tan^{-1}y)$.

29.2. (i) $6e^{-3x-2y}$ (ii) $3e^{x-y} - e^{2x-5y}$ (iii) $\frac{1}{\sqrt{2}}\sinh\sqrt{2}x + e^{-3y}\sin x$

29.3. (i) $\cos\mu x\, e^{-\mu^2 y} + i(\sin\mu x\, e^{-\mu^2 y})$ (ii) $e^x\sqrt{\omega/2}\cos(\omega y + x\sqrt{\omega/2})$,

$e^x\sqrt{\omega/2}\sin(\omega y + x\sqrt{\omega/2})$, $e^{-x}\sqrt{\omega/2}\cos(\omega y - x\sqrt{\omega/2})$, $e^{-x}\sqrt{\omega/2}\sin(\omega y - x\sqrt{\omega/2})$.

29.4. $e^{r(y-2x)}(A\cos 2rx + B\sin 2rx)$.

29.5. $R'' + n^2 R = 0$, $R(r) = c_1\cos nr + c_2\sin nr$.

29.6. $X'' + \frac{\lambda}{c^2}(X')^{n+1}X = 0$, $T'' + \lambda T^{-n} = 0$.

29.7. $X'' + \frac{\lambda}{(M^2-1)}X = 0$, $Y'' + \lambda Y = 0$.

Lecture 30
The One-Dimensional Heat Equation

We begin this lecture with the derivation of the one-dimensional heat equation. Then we shall formulate initial-boundary value problems, which involve the heat equation, the initial condition, and the homogeneous and nonhomogeneous boundary conditions. We shall use the method of separation of variables also known as the *Fourier method* to solve these problems.

We recall that the fundamental principles involved in the problems of heat conduction are—

(1). Heat flows from a higher temperature to the lower temperature.

(2). The quantity of heat in a body is proportional to its mass and temperature.

(3). The rate of heat flow across an area is proportional to the area and to the rate of change of temperature with respect to its distance normal to the area.

Consider a homogeneous bar of uniform cross section S (cm^2). Suppose that the sides are covered with a material impervious to heat, so that the stream lines of heat flow are all parallel and perpendicular to the area S. Take one end of the bar as the origin and the direction of flow as the positive x-axis (see Figure 30.1). Let ρ be the density (g/cm^3), s the specific heat (cal/g deg), and k the thermal conductivity (cal/cm deg sec).

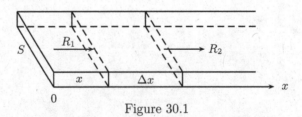

Figure 30.1

Let $u(x,t)$ be the temperature at a distance x from O. If Δu is the temperature change in a slab of thickness Δx of the bar, then by principle (2) the quantity of heat in this slab $= s\rho S \Delta x \Delta u$. Hence, the rate of increase of heat in this slab, i.e., $s\rho S \Delta x u_t = R_1 - R_2$, where R_1 and R_2 are, respectively, the rates (cal/sec) of inflow and outflow of heat. Now since

R.P. Agarwal, D. O'Regan, *Ordinary and Partial Differential Equations*,
Universitext, DOI 10.1007/978-0-387-79146-3_30,
© Springer Science+Business Media, LLC 2009

the rate of propagation of heat (i.e., the quantity of heat passing through a cross-sectional area S with abscissa x in unit time) in view of (3) is given by *Fourier's Law* $q = -ku_xS$, where k is a constant depending upon the material of the body and called as the *thermal conductivity*, it follows that

$$R_1 = -kS\left(\frac{\partial u}{\partial x}\right)_x \quad \text{and} \quad R_2 = -kS\left(\frac{\partial u}{\partial x}\right)_{x+\Delta x};$$

here the negative sign appears as a result of (1). Hence, we have

$$s\rho S\Delta x\frac{\partial u}{\partial t} = -kS\left(\frac{\partial u}{\partial x}\right)_x + kS\left(\frac{\partial u}{\partial x}\right)_{x+\Delta x},$$

which is the same as

$$\frac{\partial u}{\partial t} = \frac{k}{s\rho}\left\{\frac{\left(\frac{\partial u}{\partial x}\right)_{x+\Delta x} - \left(\frac{\partial u}{\partial x}\right)_x}{\Delta x}\right\}.$$

Denoting the constant $k/s\rho = c^2$, known as the *diffusivity* of the substance (cm^2/sec), and taking the limit as $\Delta x \to 0$, we obtain the *equation of heat conduction* in a homogeneous rod

$$\frac{\partial u}{\partial t} = c^2\frac{\partial^2 u}{\partial x^2}, \quad 0 < x < a, \quad t > 0, \quad c > 0. \tag{30.1}$$

For the solution of (30.1) to be definite, the function $u(x,t)$ must satisfy some initial and boundary conditions corresponding to the physical conditions of the problem. Let initially, i.e., when $t = 0$ a temperature be given in various cross sections of the rod equal to $f(x)$, which gives the initial condition

$$u(x,0) = f(x), \quad 0 < x < a \tag{30.2}$$

and let for simplicity the ends of the rod, i.e., $x = 0$ and $x = a$, be held at zero temperature all the time, which gives the boundary conditions

$$u(0,t) = 0, \quad t > 0 \tag{30.3}$$

$$u(a,t) = 0, \quad t > 0. \tag{30.4}$$

These boundary conditions are of the *first kind* and are known as *Dirichlet conditions*.

Now to solve the initial-boundary value problem (30.1)–(30.4) we shall use the method of separation of variables. For this, we assume a solution of (30.1) of the form $u(x,t) = X(x)T(t) \neq 0$, so that

$$X(x)T'(t) - c^2X''(x)T(t) = 0,$$

or

$$\frac{T'(t)}{T(t)} = c^2 \frac{X''(x)}{X(x)} = \lambda,$$

which leads to the differential equations

$$X'' - \frac{\lambda}{c^2} X = 0 \tag{30.5}$$

and

$$T' - \lambda T = 0. \tag{30.6}$$

The boundary condition (30.3) demands that $X(0)T(t) = 0$ for all $t \geq 0$, thus $X(0) = 0$. Similarly, the boundary condition (30.4) requires that $X(a)T(t) = 0$ and hence $X(a) = 0$. Thus, the function X has to be a solution of the eigenvalue problem (30.5),

$$X(0) = 0, \quad X(a) = 0. \tag{30.7}$$

The eigenvalues and eigenfunctions of (30.5), (30.7) are

$$\lambda_n = -\frac{n^2 \pi^2 c^2}{a^2}, \quad n = 1, 2, \cdots \tag{30.8}$$

$$X_n(x) = \sin \frac{n \pi x}{a}, \quad n = 1, 2, \cdots. \tag{30.9}$$

With λ given by (30.8), equation (30.6) takes the form

$$T' + \frac{n^2 \pi^2 c^2}{a^2} T = 0$$

whose general solution appears as

$$T_n(t) = c_n e^{-(n^2 \pi^2 c^2 / a^2) t},$$

where c_n is an arbitrary constant.

We conclude that for each specific value of n ($n = 1, 2, \cdots$) the function $X_n(x) T_n(t)$ is a solution of (30.1) that satisfies conditions (30.3) and (30.4). Now the condition (30.2), when $u(x, t) = X_n(x) T_n(t)$ with n not specified, but otherwise considered fixed, is satisfied provided $X_n(x) T_n(0) = f(x)$, i.e.,

$$\left(\sin \frac{n \pi x}{a} \right) c_n = f(x).$$

But the only way this can happen is for $f(x)$ to be restricted to the form $A \sin(n \pi x / a)$, where A is a constant. This places too great a restriction on the permissible forms of f; therefore we consider an alternative approach. Since $X_n(x) T_n(t)$ is a solution of (30.1) for each value of n ($n = 1, 2, \cdots$) and since (30.1) is a linear partial differential equation, it seems reasonable to expect that $\sum_{n=1}^{\infty} X_n(x)T_n(t)$ is a solution of (30.1). Naturally, the question of whether this infinite series converges is always there. We will not investigate this question here, but rather emphasize the method of solution. Thus, we consider

$$u(x,t) = \sum_{n=1}^{\infty} X_n(x)T_n(t) = \sum_{n=1}^{\infty} c_n e^{-(n^2\pi^2 c^2/a^2)t} \sin \frac{n\pi x}{a} \qquad (30.10)$$

as a solution of (30.1) that satisfies conditions (30.3) and (30.4). Now condition (30.2) is satisfied if and only if

$$\sum_{n=1}^{\infty} c_n \sin \frac{n\pi x}{a} = f(x),$$

i.e., the Fourier sine series for $f(x)$ in the interval $0 \le x \le a$ be $\sum_{n=1}^{\infty} c_n \sin(n\pi x/a)$. Consequently, c_n is given by

$$c_n = \frac{2}{a} \int_0^a f(x) \sin \frac{n\pi x}{a} dx, \quad n = 1, 2, \cdots. \qquad (30.11)$$

Hence, the solution of (30.1)–(30.2) can be written as (30.10) where c_n is given by (30.11).

In particular, we consider the initial-boundary value problem (30.1)–(30.2) with $c = 1$, $a = 1$, $f(x) = x$. Clearly,

$$c_n = 2 \int_0^1 x \sin n\pi x\, dx = \frac{2(-1)^{n+1}}{n\pi}.$$

Thus, in this case the solution is

$$u(x,t) = \sum_{n=1}^{\infty} \frac{2(-1)^{n+1}}{n\pi} e^{-n^2\pi^2 t} \sin n\pi x.$$

Next we shall assume that the ends of the rod, i.e., $x = 0$ and $x = a$, are insulated, which gives the boundary conditions

$$u_x(0,t) = 0, \quad t > 0 \qquad (30.12)$$

$$u_x(a,t) = 0, \quad t > 0. \qquad (30.13)$$

These boundary conditions are of the *second kind* and known as *Neumann conditions*.

Clearly in this case also we have the same DEs (30.5) and (30.6); however, instead of (30.7) the new boundary conditions are

$$X'(0) = 0, \quad X'(a) = 0 \qquad (30.14)$$

For (30.5), (30.14) the eigenvalues and eigenfunctions are

$$\lambda_n = -\frac{n^2\pi^2c^2}{a^2}, \quad n = 0,1,\cdots \tag{30.15}$$

$$X_n(x) = \cos\frac{n\pi x}{a}, \quad n = 0,1,\cdots \tag{30.16}$$

and correspondingly

$$T_0(t) = \frac{a_0}{2} \quad \text{and} \quad T_n(t) = a_n e^{-(n^2\pi^2c^2/a^2)t}. \tag{30.17}$$

Therefore,

$$
\begin{aligned}
u(x,t) &\doteq X_0(x)T_0(t) + \sum_{n=1}^{\infty} X_n(x)T_n(t) \\
&= \frac{a_0}{2} + \sum_{n=1}^{\infty} a_n e^{-(n^2\pi^2c^2/a^2)t}\cos\frac{n\pi x}{a}.
\end{aligned}
\tag{30.18}
$$

Finally, condition (30.2) implies

$$f(x) = \frac{a_0}{2} + \sum_{n=1}^{\infty} a_n \cos\frac{n\pi x}{a},$$

which is the Fourier cosine series for $f(x)$, and hence

$$a_n = \frac{2}{a}\int_0^a f(x)\cos\frac{n\pi x}{a}dx, \quad n \geq 0. \tag{30.19}$$

Hence, the solution of (30.1), (30.2), (30.12), (30.13) can be written as (30.18) where a_n is given by (30.19).

In particular, we consider the initial-boundary value problem (30.1), (30.2), (30.12), (30.13) with $c = 2$, $a = 1$, $f(x) = x$. Clearly,

$$
\begin{aligned}
a_0 &= 2\int_0^1 x\,dx = 1 \\
a_n &= 2\int_0^1 x\cos n\pi x\,dx = \frac{2}{n^2\pi^2}\left((-1)^n - 1\right) \\
&= \begin{cases} -\dfrac{4}{\pi^2(2n-1)^2} & n \text{ odd} \\ 0, & n \text{ even.} \end{cases}
\end{aligned}
$$

Hence,

$$u(x,t) = \frac{1}{2} - \frac{4}{\pi^2}\sum_{n=1}^{\infty}\frac{1}{(2n-1)^2}e^{-4(2n-1)^2\pi^2 t}\cos(2n-1)\pi x.$$

Finally, we assume that the ends of the rod, i.e., $x = 0$ and $x = a$, are kept at the fixed temperatures A and B, respectively. This means that we have the boundary conditions

$$u(0,t) = A, \quad t > 0 \tag{30.20}$$

$$u(a,t) = B, \quad t > 0. \tag{30.21}$$

Now let $u(x,t)$ be a solution of (30.1), (30.2), (30.20), (30.21). We claim that the function

$$v(x,t) = u(x,t) + \left(\frac{x-a}{a}\right) A - \frac{x}{a} B \tag{30.22}$$

is a solution of the initial-boundary value problem

$$
\begin{aligned}
&v_t - c^2 v_{xx} = 0, \quad 0 < x < a, \quad t > 0, \quad c > 0 \\
&v(0,t) = 0, \quad t > 0 \\
&v(a,t) = 0, \quad t > 0 \\
&v(x,0) = f(x) + \left(\frac{x-a}{a}\right) A - \frac{x}{a} B, \quad 0 < x < a.
\end{aligned}
\tag{30.23}
$$

For this, it suffices to note that

$$v_x = u_x + \frac{A}{a} - \frac{B}{a}, \quad v_{xx} = u_{xx}, \quad v_t = u_t;$$

and hence $v_t - c^2 v_{xx} = u_t - c^2 u_{xx} = 0$, i.e., v satisfies the same differential equation as u. Further, we have

$$v(0,t) = u(0,t) + \left(\frac{0-a}{a}\right) A - \frac{0}{a} B = A - A = 0$$

$$v(a,t) = u(a,t) + \left(\frac{a-a}{a}\right) A - \frac{a}{a} B = B - B = 0$$

$$v(x,0) = u(x,0) + \left(\frac{x-a}{a}\right) A - \frac{x}{a} B = f(x) + \left(\frac{x-a}{a}\right) A - \frac{x}{a} B.$$

Since the problem (30.23) is of the type (30.1)–(30.4), we can find its solution $v(x,t)$. The solution $u(x,t)$ of (30.1), (30.2), (30.20), (30.21) is then obtained by the relation (30.22).

In particular, we consider the initial-boundary value problem (30.1), (30.2), (30.20), (30.21) with $c^2 = 5$, $a = \pi$, $f(x) = x$, $A = 10$, $B = 0$. For this problem, (30.23) becomes

$$v_t - 5v_{xx} = 0, \quad 0 < x < \pi, \quad t > 0$$
$$v(0, t) = 0, \quad t > 0$$
$$v(\pi, t) = 0, \quad t > 0 \tag{30.24}$$
$$v(x, 0) = x + \left(\frac{x - \pi}{\pi}\right) 10, \quad 0 < x < \pi$$

and

$$v(x, t) = u(x, t) + \left(\frac{x - \pi}{\pi}\right) 10. \tag{30.25}$$

Now from the consideration of (30.1)–(30.4) the solution of (30.24) can be written as (see (30.10))

$$v(x, t) = \sum_{n=1}^{\infty} c_n e^{-5n^2 t} \sin nx, \tag{30.26}$$

where (see (30.11))

$$c_n = \frac{2}{\pi} \int_0^{\pi} \left[x + \left(\frac{x - \pi}{\pi}\right) 10 \right] \sin nx\, dx = \frac{2}{n\pi} \left[-10 + \pi(-1)^{n+1} \right]. \tag{30.27}$$

Hence, in view of (30.25)–(30.27) the solution $u(x, t)$ of the given problem appears as

$$u(x, t) = \frac{10(\pi - x)}{\pi} + \sum_{n=1}^{\infty} \frac{2}{n\pi} \left[-10 + (-1)^{n+1} \pi \right] e^{-5n^2 t} \sin nx.$$

Lecture 31
The One-Dimensional
Heat Equation (Cont'd.)

In this lecture we shall use the method of separation of variables to solve the general one-dimensional heat equation with the boundary conditions of the third kind.

The partial differential equation that governs the temperature $u(x,t)$ in the rod whose material properties vary with position can be written as

$$\frac{\partial}{\partial x}\left(k(x)\frac{\partial u}{\partial x}\right) = \rho(x)c(x)\frac{\partial u}{\partial t}, \quad \alpha < x < \beta, \quad t > 0. \tag{31.1}$$

We shall consider (31.1) with the initial condition

$$u(x,0) = f(x), \quad \alpha < x < \beta \tag{31.2}$$

and the boundary conditions

$$a_0 u(\alpha,t) - a_1 \frac{\partial u}{\partial x}(\alpha,t) = c_1, \quad t > 0, \quad a_0^2 + a_1^2 > 0 \tag{31.3}$$

$$d_0 u(\beta,t) + d_1 \frac{\partial u}{\partial x}(\beta,t) = c_2, \quad t > 0, \quad d_0^2 + d_1^2 > 0 \tag{31.4}$$

Equations (31.1)–(31.4) make up an initial-boundary value problem.

Boundary conditions (31.3) and (31.4) are of the *third kind* and are known as *Robin's conditions*. These boundary conditions appear when each face loses heat to a surrounding medium according to Newton's law of cooling, which states that a body radiates heat from its surface at a rate proportional to the difference between the skin temperature of the body and the temperature of the surrounding medium.

Experience indicates that after a long time "under the same conditions" the variation of temperature with time dies away. In terms of the function $u(x,t)$ that represents temperature, we expect that the limit of $u(x,t)$, as t tends to infinity, exists and depends only on x: $\lim_{t\to\infty} u(x,t) = v(x)$ and also that $\lim_{t\to\infty} u_t = 0$. The function $v(x)$, called the *steady-state temperature distribution*, must still satisfy the boundary conditions and the

R.P. Agarwal, D. O'Regan, *Ordinary and Partial Differential Equations*,
Universitext, DOI 10.1007/978-0-387-79146-3_31,
© Springer Science+Business Media, LLC 2009

heat equation, which are valid for all $t > 0$. Therefore, $v(x)$ *(steady–state solution)* should be the solution to the problem

$$\frac{d}{dx}\left(k(x)\frac{dv}{dx}\right) = 0, \quad \alpha < x < \beta \tag{31.5}$$

$$\begin{aligned} a_0 v(\alpha) - a_1 v'(\alpha) &= c_1 \\ d_0 v(\beta) + d_1 v'(\beta) &= c_2. \end{aligned} \tag{31.6}$$

Equation (31.5) can be solved to obtain

$$v(x) = A \int_\alpha^x \frac{d\xi}{k(\xi)} + B, \tag{31.7}$$

which satisfies the boundary conditions (31.6) if and only if

$$a_0 B - a_1 \frac{A}{k(\alpha)} = c_1$$

$$d_0 \left(A \int_\alpha^\beta \frac{d\xi}{k(\xi)} + B \right) + d_1 \frac{A}{k(\beta)} = c_2. \tag{31.8}$$

Clearly, we can solve (31.8) if and only if

$$a_0 \left(d_0 \int_\alpha^\beta \frac{d\xi}{k(\xi)} + \frac{d_1}{k(\beta)} \right) + \frac{a_1 d_0}{k(\alpha)} \neq 0,$$

or

$$a_0 d_0 \int_\alpha^\beta \frac{d\xi}{k(\xi)} + \frac{a_0 d_1}{k(\beta)} + \frac{a_1 d_0}{k(\alpha)} \neq 0. \tag{31.9}$$

Thus, the problem (31.5), (31.6) has a unique solution if and only if condition (31.9) is satisfied.

Example 31.1. To find the steady-state solution of the problem (30.1), (30.2), (30.20), (30.21) we need to solve the problem

$$\frac{d^2 v}{dx^2} = 0, \quad v(0) = A, \quad v(a) = B,$$

whose solution is $v(x) = A + (B - A)x/a$.

Now we define the function

$$w(x, t) = u(x, t) - v(x), \tag{31.10}$$

where $u(x, t)$ and $v(x)$ are the solutions of (31.1)–(31.4) and (31.5), (31.6) respectively. Clearly,

$$\frac{\partial w(x, t)}{\partial x} = \frac{\partial u(x, t)}{\partial x} - \frac{dv(x)}{dx},$$

and hence

$$k(x)\frac{\partial w(x,t)}{\partial x} = k(x)\frac{\partial u(x,t)}{\partial x} - k(x)\frac{dv(x)}{dx},$$

which gives

$$\frac{\partial}{\partial x}\left(k(x)\frac{\partial w(x,t)}{\partial x}\right) = \frac{\partial}{\partial x}\left(k(x)\frac{\partial u(x,t)}{\partial x}\right) - \frac{d}{dx}\left(k(x)\frac{dv(x)}{dx}\right). \quad (31.11)$$

We also have

$$\rho(x)c(x)\frac{\partial w(x,t)}{\partial t} = \rho(x)c(x)\frac{\partial u(x,t)}{\partial t}. \quad (31.12)$$

Subtraction of (31.12) from (31.11) gives

$$\frac{\partial}{\partial x}\left(k(x)\frac{\partial w}{\partial x}\right) - \rho(x)c(x)\frac{\partial w}{\partial t}$$

$$= \left[\frac{\partial}{\partial x}\left(k(x)\frac{\partial u}{\partial x}\right) - \rho(x)c(x)\frac{\partial u}{\partial t}\right] - \frac{d}{dx}\left(k(x)\frac{dv}{dx}\right) = 0 - 0 = 0.$$

Therefore,

$$\frac{\partial}{\partial x}\left(k(x)\frac{\partial w}{\partial x}\right) = \rho(x)c(x)\frac{\partial w}{\partial t}, \quad \alpha < x < \beta, \quad t > 0. \quad (31.13)$$

We also have

$$a_0 w(\alpha,t) - a_1\frac{\partial w}{\partial x}(\alpha,t) = a_0(u(\alpha,t) - v(\alpha)) - a_1\left(\frac{\partial u}{\partial x}(\alpha,t) - v'(\alpha)\right)$$

$$= \left[a_0 u(\alpha,t) - a_1\frac{\partial u}{\partial x}(\alpha,t)\right] - [a_0 v(\alpha) - a_1 v'(\alpha)]$$

$$= c_1 - c_1 = 0;$$

i.e.,

$$a_0 w(\alpha,t) - a_1\frac{\partial w}{\partial x}(\alpha,t) = 0, \quad t > 0, \quad (31.14)$$

and similarly

$$d_0 w(\beta,t) + d_1\frac{\partial w}{\partial x}(\beta,t) = 0, \quad t > 0. \quad (31.15)$$

Now we shall solve (31.13) – (31.15) by using the method of separation of variables. We assume $w(x,t) = X(x)T(t) \neq 0$, to obtain

$$\frac{d}{dx}\left(k(x)\frac{dX(x)}{dx}\right)T(t) = \rho(x)c(x)X(x)T'(t),$$

or

$$\frac{\frac{d}{dx}\left(k(x)\frac{dX}{dx}\right)}{\rho(x)c(x)X(x)} = \frac{T'(t)}{T(t)} = -\lambda,$$

which in view of (31.14) and (31.15) gives

$$(k(x)X')' + \lambda \rho(x)c(x)X = 0$$
$$a_0 X(\alpha) - a_1 X'(\alpha) = 0 \qquad (31.16)$$
$$d_0 X(\beta) + d_1 X'(\beta) = 0$$

and

$$T' + \lambda T = 0. \qquad (31.17)$$

Clearly, (31.16) is a Sturm–Liouville problem, for which we already know that

1. there are infinite number of eigenvalues $0 \le \lambda_1 < \lambda_2 < \cdots$,
2. for each eigenvalue λ_n there exists a unique eigenfunction $X_n(x)$,
3. the set of eigenfunctions $\{X_n(x)\}$ is orthogonal with respect to the weight function $\rho(x)c(x)$, i.e.,

$$\int_\alpha^\beta \rho(x)c(x)X_n(x)X_m(x)dx = 0, \quad n \ne m.$$

For $\lambda = \lambda_n$ equation (31.17) becomes $T_n' + \lambda_n T_n = 0$ and gives

$$T_n(t) = ce^{-\lambda_n t}, \quad n \ge 1.$$

Hence, the solution of (31.13) – (31.15) can be written as

$$w(x,t) = \sum_{n=1}^\infty a_n X_n(x)e^{-\lambda_n t}. \qquad (31.18)$$

Finally, we note that condition (31.2) gives

$$w(x,0) = u(x,0) - v(x) = f(x) - v(x) = F(x), \quad \text{say.} \qquad (31.19)$$

The solution (31.18) satisfies (31.19) if and only if

$$w(x,0) = F(x) = \sum_{n=1}^\infty a_n X_n(x),$$

which gives

$$a_n = \frac{\int_\alpha^\beta \rho(x)c(x)X_n(x)F(x)dx}{\int_\alpha^\beta \rho(x)c(x)X_n^2(x)dx}, \quad n \ge 1. \qquad (31.20)$$

Therefore, in view of (31.10) and (31.18) the solution of (31.1)–(31.4) appears as

$$u(x,t) = v(x) + \sum_{n=1}^\infty a_n X_n(x)e^{-\lambda_n t}, \qquad (31.21)$$

where a_n, $n \geq 1$ are given by (31.20).

From the representation (31.21) the following properties are immediate:

(i)   Since each $\lambda_n > 0$, $u(x,t) \to v(x)$ as $t \to \infty$.

(ii)   For any $t_1 > 0$, the series for $u(x,t_1)$ converges uniformly in $\alpha \leq x \leq \beta$ because of the exponential factors; therefore $u(x,t_1)$ is a continuous function of x.

(iii) For large t we can approximate $u(x,t)$ by

$$v(x) + a_1 X_1(x)e^{-\lambda_1 t}.$$

Example 31.2. For the problem (30.1)–(30.3) and

$$hu(a,t) + u_x(a,t) = 0, \quad t > 0, \quad h > 0, \tag{31.22}$$

which is a particular case of (31.1)–(31.4), we have $v(x) \equiv 0$. The eigenvalues are $c^2 \lambda_n^2$, where λ_n is the root of the equation $h \tan \lambda a + \lambda = 0$, and the eigenfunctions are $X_n(x) = \sin \lambda_n x$, (see Problem 18.1 (iv)). Thus, the solution can be written as

$$u(x,t) = \sum_{n=1}^{\infty} a_n e^{-c^2 \lambda_n^2 t} \sin \lambda_n x, \tag{31.23}$$

where

$$a_n = \frac{\int_0^a f(x) \sin \lambda_n x \, dx}{\int_0^a \sin^2 \lambda_n x \, dx} = \frac{2h \int_0^a f(x) \sin \lambda_n x \, dx}{ah + \cos^2 \lambda_n a}, \quad n \geq 1. \tag{31.24}$$

For $a = 1$, $h = 2$, $f(x) = 1$ this solution becomes

$$u(x,t) = 4 \sum_{n=1}^{\infty} \frac{1}{\lambda_n} \left(\frac{1 - \cos \lambda_n}{2 + \cos^2 \lambda_n} \right) e^{-c^2 \lambda_n^2 t} \sin \lambda_n x.$$

Problems

31.1. Solve the initial-boundary value problem (30.1)–(30.4) when

(i) $a = 1$, $c = 4$, $f(x) = 1 + x$

(ii) $a = \pi$, $c = 2$, $f(x) = x^2$

(iii) $a = 1$, $c^2 = 5$, $f(x) = e^x$.

31.2. Solve the initial-boundary value problem (30.1), (30.2), (30.12), (30.13) when

(i) $f(x) = T_0 \sin^2 \dfrac{\pi x}{a}, \; 0 < x < a$

(ii) $f(x) = T_0 x^2, \; 0 < x < a$

(iii) $f(x) = T_0 e^x, \; 0 < x < a.$

31.3.   Solve the initial-boundary value problem (30.1), (30.2), (30.20), (30.21) when

(i) $a = 10, \; c = 10, \; f(x) = 0, \; A = 10, \; B = 30$

(ii) $a = 1, \; c = 1, \; f(x) = 3(1 - x), \; A = 3, \; B = 1$

(iii) $a = \pi, \; c = \sqrt{3}, \; f(x) = -\cos 7x, \; A = -1, \; B = 1.$

31.4.   Find the solution of the initial-boundary value problem (30.1)–(30.3), (30.13), and in particular solve when $a = \pi, \; c = 1, \; f(x) = x(\pi - x)$.

31.5.   Find the solution of the initial-boundary value problem (30.1), (30.2), (30.4), (30.12), and in particular solve when $a = \pi, \; c = 1, \; f(x) = x(\pi - x)$.

31.6.   Heat conduction in a thin circular ring (consider it as a rod, bent into the shape of a circular ring by tightly joining the two ends) of length $2a$, labeled from $-a$ to a leads to the equation $u_t = c^2 u_{xx}, \; -a < x < a, \; t > 0, \; c > 0$ with the initial condition $u(x, 0) = f(x), \; -a < x < a$ and the *periodic boundary conditions*

$$
\begin{aligned}
u(-a, t) &= u(a, t) \\
u_x(-a, t) &= u_x(a, t), \quad t > 0.
\end{aligned}
\tag{31.25}
$$

Find the solution of this initial-boundary value problem, and in particular solve when $a = \pi, \; f(x) = |x|$.

31.7.   Find the steady-state solution of the problem

$$
\frac{\partial}{\partial x}\left(k(x)\frac{\partial u}{\partial x} \right) = c\rho \frac{\partial u}{\partial t}, \quad 0 < x < a, \quad t > 0
$$

$$
u(0, t) = T_0, \quad u(a, t) = T_1, \quad t > 0,
$$

where $k(x) = k_0 + \beta x$ and k_0 and β are constants.

31.8.   Find the steady-state solution of the problem

$$
\frac{\partial^2 u}{\partial x^2} + \gamma^2(U(x) - u) = \frac{1}{k}\frac{\partial u}{\partial t}, \quad 0 < x < a, \quad t > 0
$$

$$
u(0, t) = U_0, \quad \frac{\partial u}{\partial x}(a, t) = 0, \quad t > 0,
$$

where $U(x) = U_0 + Sx$ and U_0 and S are constants.

31.9. Find the solution of the initial-boundary value problem (30.1), (30.2), (31.22) and $hu(0,t) - u_x(0,t) = 0$, $t > 0$.

31.10. If the lateral surface of the rod is not insulated, there is a heat exchange by convection into the surrounding medium. If the surrounding medium has constant temperature T_0, the rate at which heat is lost from the rod is proportional to the difference $u - T_0$. The governing partial DE in this situation is

$$c^2 u_{xx} = u_t + b(u - T_0), \quad 0 < x < a, \quad b > 0. \tag{31.26}$$

Show that the change of variable $u(x,t) = T_0 + v(x,t)e^{-bt}$ leads to the heat equation (30.1) in v. In particular, find the solution of (31.26) when $c = 1$, $b = 4$, $T_0 = 5$, $a = 1$, satisfying the initial and boundary conditions $u(x,0) = 5 + x$, $u(0,t) = u(1,t) = 5$.

31.11. Find the solution of the partial DE

$$u_t = u_{xx} + 2ku_x, \quad 0 < x < a, \quad t > 0$$

subject to the initial-boundary conditions (30.2)–(30.4); here k is a constant.

Answers or Hints

31.1. (i) $\sum_{n=1}^{\infty} \frac{2}{n\pi}(2(-1)^{n+1}+1)e^{-16n^2\pi^2 t}\sin n\pi x$

(ii) $\sum_{n=1}^{\infty} -\frac{\pi}{n}e^{-16n^2 t}\sin 2nx + \sum_{n=1}^{\infty}\left[\frac{2\pi}{2n-1} - \frac{8}{\pi(2n-1)^3}\right]e^{-4(2n-1)^2 t}$

$\times \sin(2n-1)x$ (iii) $\sum_{n=1}^{\infty} \frac{2n\pi}{n^2\pi^2+1}\left[1 + e(-1)^{n+1}\right]e^{-5n^2\pi^2 t}\sin n\pi x$.

31.2. (i) $\frac{T_0}{2}(1-\cos\frac{2\pi x}{a}e^{-(4\pi^2 c^2/a^2)t})$ (ii) $\frac{1}{3}a^2 T_0 + \sum_{n=1}^{\infty}\frac{4T_0 a^2(-1)^n}{n^2\pi^2}\cos\frac{n\pi x}{a}$

$\times e^{-(n^2\pi^2 c^2/a^2)t}$ (iii) $\frac{T_0}{a}(e^a-1) + \sum_{n=1}^{\infty}\frac{2T_0 a(e^a(-1)^n-1)}{a^2+n^2\pi^2}\cos\frac{n\pi x}{a}e^{-(n^2\pi^2 c^2/a^2)t}$.

31.3. (i) $10 + 2x + \frac{20}{\pi}\sum_{n=1}^{\infty}\frac{3(-1)^n-1}{n}\sin\left(\frac{n\pi x}{10}\right)e^{-n^2\pi^2 t}$ (ii) $3 - 2x +$ $\sum_{n=1}^{\infty}\frac{2(-1)^n}{n\pi}e^{-n^2\pi^2 t}\sin n\pi x$ (iii) $\frac{2}{\pi}x-1+\sum_{n=1,n\neq 7}^{\infty}\frac{98(n-(-1)^n)}{\pi n(n^2-49)}e^{-3n^2 t}\sin nx$.

31.4. $u(x,t) = \sum_{n=1}^{\infty}c_n\sin\frac{(2n-1)\pi x}{2a}e^{-(2n-1)^2\pi^2 c^2 t/4a^2}$, where $c_n =$ $\frac{2}{a}\int_0^a f(x)\sin\frac{(2n-1)\pi x}{2a}dx$, $\sum_{n=1}^{\infty}\left[\frac{32}{\pi(2n-1)^3} + \frac{8(-1)^n}{(2n-1)^2}\right]\sin\frac{(2n-1)x}{2}$

$\times e^{-(2n-1)^2 t/4}$.

31.5. $u(x,t) = \sum_{n=1}^{\infty}c_n\cos\frac{(2n-1)\pi x}{2a}e^{-(2n-1)^2\pi^2 c^2 t/4a^2}$, where $c_n =$ $\frac{2}{a}\int_0^a f(x)\cos\frac{(2n-1)\pi x}{2a}dx$, $\sum_{n=1}^{\infty}\left[\frac{32(-1)^{n-1}}{\pi(2n-1)^3} - \frac{8}{(2n-1)^2}\right]\cos\frac{(2n-1)x}{2}$

$\times e^{-(2n-1)^2 t/4}$. The same solution can be obtained by replacing x in Problem 31.4 by $a - x$.

31.6. $\frac{\pi}{2} - \frac{4}{\pi} \sum_{n=0}^{\infty} \frac{1}{(2n+1)^2} \cos(2n+1)x e^{-(2n+1)^2 c^2 t}$.

31.7. $v(x) = A \ln(k_0 + \beta x) + B$, where $A = (T_1 - T_0)/\ln\left(1 + \frac{\beta a}{k_0}\right)$, $B = T_0 - A \ln k_0$.

31.8. $v(x) = A \cosh \gamma x + B \sinh \gamma x + U_0 + Sx$ where $A = 0$, $B = -S/(\gamma \cosh \gamma a)$.

31.9. $\sum_{n=1}^{\infty} \frac{\lambda_n \cos \lambda_n x + h \sin \lambda_n x}{(\lambda_n^2 + h^2)a + 2h} e^{-c^2 \lambda_n^2 t} \int_0^a f(x)(\lambda_n \cos \lambda_n x + h \sin \lambda_n x) dx$ where λ_n are the zeros of the equation $\tan \lambda a = 2h\lambda/(\lambda^2 - h^2)$.

31.10. $5 + \frac{2e^{-4t}}{\pi} \sum_{n=1}^{\infty} \frac{(-1)^{n+1}}{n} e^{-n^2 \pi^2 t} \sin n\pi x$.

31.11. $u(x,t) = \sum_{n=1}^{\infty} c_n \exp\left\{-\left[\left(\frac{n\pi}{a}\right)^2 + k^2\right]t\right\} e^{-kx} \sin \frac{n\pi x}{a}$, where $c_n = \frac{2}{a} \int_0^a f(x)e^{kx} \sin \frac{n\pi x}{a} dx$.

Lecture 32
The One-Dimensional Wave Equation

In this lecture we shall provide two different derivations of the one-dimensional wave equation. The first derivation comes from the oscillation of a elastic string, whereas the second one is from the electric oscillations in wires. Then, we shall formulate an initial-boundary value problem, which involves the wave equation, the initial conditions, and the boundary conditions. Finally, we shall use the method of separation of variables to solve the initial-boundary value problem.

Consider a tightly stretched elastic string of length a, initially directed along a segment of the x-axis from O to a. We assume that the ends of the string are fixed at the points $x = 0$ and $x = a$. If the string is deflected from its original position and then let loose, or if we give to its points a certain velocity at the initial time, or if we deflect the string and give a velocity to its points, then the points of the string will perform certain motions. In such a stage we say that the string is set into oscillation, or allowed to vibrate. The problem of interest is then to find the shape of the string at any instant of time.

We assume that the string is subjected to a constant tension T, which is directed along the tangent to its profile. We also assume that T is large compared to the weight of the string so that the effects of gravity are negligible. We further assume that no external forces are acting on the string, and each point of the string makes only small vibrations at right angles to the equilibrium position so that the motion takes place entirely in the xu-plane. Figure 32.1 shows the string in the position $OPQa$ at time t.

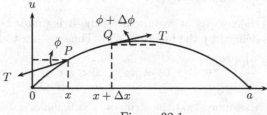

Figure 32.1

Consider the motion of the element PQ of the string between its points

R.P. Agarwal, D. O'Regan, *Ordinary and Partial Differential Equations*,
Universitext, DOI 10.1007/978-0-387-79146-3_32,
© Springer Science+Business Media, LLC 2009

$P(x, u)$ and $Q(x+\Delta x, u+\Delta u)$ where the tangents make angles ϕ and $\phi+\Delta\phi$ with the x-axis. Clearly, this element is moving upwards with acceleration $\partial^2 u/\partial t^2$. Also the vertical component of the force acting on this element is

$$
\begin{aligned}
&= T\sin(\phi + \Delta\phi) - T\sin\phi \\
&\simeq T[\tan(\phi + \Delta\phi) - \tan\phi], \quad \text{since } \phi \text{ is small} \\
&= T\left[\left(\frac{\partial u}{\partial x}\right)_{x+\Delta x} - \left(\frac{\partial u}{\partial x}\right)_x\right].
\end{aligned}
$$

If m is the mass per unit length of the string, then by Newton's second law of motion, we have

$$
m\Delta x\frac{\partial^2 u}{\partial t^2} = T\left[\left(\frac{\partial u}{\partial x}\right)_{x+\Delta x} - \left(\frac{\partial u}{\partial x}\right)_x\right],
$$

which is the same as

$$
\frac{\partial^2 u}{\partial t^2} = \frac{T}{m}\left[\frac{\left(\frac{\partial u}{\partial x}\right)_{x+\Delta x} - \left(\frac{\partial u}{\partial x}\right)_x}{\Delta x}\right].
$$

Finally, taking the limit as $Q \to P$, i.e., $\Delta x \to 0$, we obtain

$$
\frac{\partial^2 u}{\partial t^2} = c^2\frac{\partial^2 u}{\partial x^2}, \quad c^2 = \frac{T}{m}. \tag{32.1}
$$

This partial DE gives the transverse vibrations of the string. It is called the one-dimensional *wave equation*.

Equation (32.1) by itself does not describe the motion of the string. The required function $u(x,t)$ must also satisfy the *initial conditions* which describe the state of the string at the initial time $t = 0$ and the *boundary conditions* which indicate to what occurs at the ends of the string, i.e., $x = 0$ and $x = a$. At $t = 0$ the string has a definite shape, that which we gave it. We assume that this shape is defined by the function $f(x)$. This leads to the condition

$$
u(x,0) = f(x), \quad 0 < x < a. \tag{32.2}
$$

Further, at $t = 0$ the velocity at each point of the string must be given, we assume that it is defined by the function $g(x)$. Thus, we must also have

$$
\frac{\partial u}{\partial t}\bigg|_{t=0} = u_t(x,0) = g(x), \quad 0 < x < a. \tag{32.3}
$$

Now since we have assumed that the string at $x = 0$ and $x = a$ is fixed, for any t the following conditions must be satisfied

$$
u(0,t) = 0, \quad t > 0 \tag{32.4}
$$

$$u(a,t) = 0, \quad t > 0. \tag{32.5}$$

The partial DE (32.1) together with the initial conditions (32.2), (32.3) and the boundary conditions (32.4), (32.5) constitutes a typical *initial-boundary value problem*.

Now we shall show that the problem of electric oscillations in wires also leads to equation (32.1). The electric current in a wire is characterized by the current flow $i(x,t)$ and the voltage $v(x,t)$, which are dependent on the coordinate x of the point of the wire and on the time t. On an element Δx of the wire the drop in voltage is equal to $v(x,t) - v(x+\Delta x, t) \simeq -(\partial v/\partial x)\Delta x$. This voltage drop consists of the ohmic drop which is equal to $iR\Delta x$, and the inductive drop which is the same as $(\partial i/\partial t)L\Delta x$. Thus, we have

$$-\frac{\partial v}{\partial x}\Delta x = iR\Delta x + \frac{\partial i}{\partial t}L\Delta x, \tag{32.6}$$

where R and L are the resistance and the coefficient of self-induction per unit length of wire. In (32.6) the minus sign indicates that the current flow is in a direction opposite to the build-up of v. From (32.6) it follows that

$$\frac{\partial v}{\partial x} + iR + L\frac{\partial i}{\partial t} = 0. \tag{32.7}$$

Further, the difference between the current leaving the element Δx and entering it during the time Δt is

$$i(x,t) - i(x+\Delta x, t) \simeq -\frac{\partial i}{\partial x}\Delta x\Delta t.$$

In charging the element Δx it requires $C\Delta x(\partial v/\partial t)\Delta t$, and in leakage through the lateral surface of the wire due to imperfect insulation we have $Av\Delta x\Delta t$, where A is the leak coefficient and C is the capacitance. Equating these expressions and canceling out $\Delta x\Delta t$, we get the equation

$$\frac{\partial i}{\partial x} + C\frac{\partial v}{\partial t} + Av = 0. \tag{32.8}$$

Equations (32.7) and (32.8) are called *telegraph equations*.

Differentiating equation (32.8) with respect to x, (32.7) with respect to t and multiplying it by C, and subtracting, we obtain

$$\frac{\partial^2 i}{\partial x^2} + A\frac{\partial v}{\partial x} - CR\frac{\partial i}{\partial t} - CL\frac{\partial^2 i}{\partial t^2} = 0.$$

Substituting in this equation the expression $\partial v/\partial x$ from (32.7), we get an equation only in $i(x,t)$,

$$\frac{\partial^2 i}{\partial x^2} = CL\frac{\partial^2 i}{\partial t^2} + (CR + AL)\frac{\partial i}{\partial t} + ARi. \tag{32.9}$$

Similarly, we obtain an equation for determining $v(x, t)$,

$$\frac{\partial^2 v}{\partial x^2} = CL\frac{\partial^2 v}{\partial t^2} + (CR + AL)\frac{\partial v}{\partial t} + ARv. \tag{32.10}$$

If we neglect the leakage through the insulation $(A = 0)$ and the resistance $(R = 0)$, then equations (32.9) and (32.10) reduce to wave equations

$$c^2\frac{\partial^2 i}{\partial x^2} = \frac{\partial^2 i}{\partial t^2}, \quad c^2\frac{\partial^2 v}{\partial x^2} = \frac{\partial^2 v}{\partial t^2}, \tag{32.11}$$

where $c^2 = 1/(CL)$. Again the physical conditions dictate the formulation of the initial and boundary conditions of the problem.

Now to solve the initial-boundary value problem (32.1)–(32.5) we shall use the method of separation of variables also known as the *Fourier method*. For this, we assume a solution of (32.1) to be of the form $u(x, t) = X(x)T(t) \neq 0$ where X, T are unknown functions to be determined. Substitution of this into (32.1) yields

$$XT'' - c^2 X''T = 0.$$

Thus, we obtain

$$\frac{T''}{T} = \frac{c^2 X''}{X} = \lambda,$$

where λ is a constant. Consequently, we have two separate equations:

$$T'' = \lambda T \tag{32.12}$$

and

$$c^2 X'' = \lambda X. \tag{32.13}$$

The boundary condition (32.4) demands that $X(0)T(t) = 0$ for all $t \geq 0$, thus $X(0) = 0$. Similarly, the boundary condition (32.5) leads to $X(a)T(t) = 0$ and hence $X(a) = 0$. Thus, in view of (32.13) the function X has to be a solution of the eigenvalue problem

$$X'' - \frac{\lambda}{c^2}X = 0, \quad X(0) = 0, \quad X(a) = 0. \tag{32.14}$$

The eigenvalues and eigenfunctions of (32.14) are

$$\lambda_n = -\frac{n^2\pi^2 c^2}{a^2}, \quad X_n(x) = \sin\frac{n\pi x}{a}, \quad n = 1, 2, \cdots. \tag{32.15}$$

With λ given by (32.15), equation (32.12) takes the form

$$T'' + \frac{n^2\pi^2 c^2}{a^2}T = 0$$

whose solution appears as

$$T_n(t) = a_n \cos \frac{n\pi ct}{a} + b_n \sin \frac{n\pi ct}{a}, \quad n = 1, 2, \cdots \qquad (32.16)$$

where a_n and b_n are the integration constants in the general solution.

Therefore, it follows that

$$u(x,t) = \sum_{n=1}^{\infty} X_n(x)T_n(t) = \sum_{n=1}^{\infty} \left(a_n \cos \frac{n\pi ct}{a} + b_n \sin \frac{n\pi ct}{a} \right) \sin \frac{n\pi x}{a}$$

$$(32.17)$$

is a solution of (32.1). Clearly, $u(x,t)$ satisfies conditions (32.4) and (32.5), and it will satisfy (32.2) provided

$$\sum_{n=1}^{\infty} a_n \sin \frac{n\pi x}{a} = f(x), \qquad (32.18)$$

which is the Fourier sine series for $f(x)$. Consequently, a_n is given by

$$a_n = \frac{2}{a} \int_0^a f(x) \sin \frac{n\pi x}{a} dx, \quad n = 1, 2, \cdots. \qquad (32.19)$$

Likewise condition (32.3) will be satisfied provided that

$$\sum_{n=1}^{\infty} \sin \frac{n\pi x}{a} \left(\frac{n\pi c}{a} b_n \right) = g(x) \qquad (32.20)$$

and hence

$$\frac{n\pi c}{a} b_n = \frac{2}{a} \int_0^a g(x) \sin \frac{n\pi x}{a} dx,$$

which gives

$$b_n = \frac{2}{n\pi c} \int_0^a g(x) \sin \frac{n\pi x}{a} dx, \quad n = 1, 2, \cdots. \qquad (32.21)$$

We conclude that the solution of the initial-boundary value problem (32.1)–(32.5) is given by (32.17) where a_n and b_n are as in (32.19) and (32.21) respectively. This solution is due to Daniel Bernoulli.

Example 32.1.   We shall find the solution of (32.1)–(32.5) with $c = 2$, $a = \pi$, $f(x) = x(\pi - x)$, $g(x) = 0$. From (32.21) it is clear that $b_n = 0$, $n \geq 1$. Now from (32.19) we have

$$
\begin{aligned}
a_n &= \frac{2}{\pi} \int_0^\pi x(\pi - x) \sin nx\, dx \\
&= \frac{2}{\pi} \left[\pi \int_0^\pi x \sin nx\, dx - \int_0^\pi x^2 \sin nx\, dx \right] \\
&= \frac{2}{\pi} \left[\pi \left\{ \frac{-x \cos nx}{n} \Big|_0^\pi + \frac{1}{n} \int_0^\pi \cos nx\, dx \right\} \right. \\
&\quad \left. - \left\{ -\frac{x^2 \cos nx}{n} \Big|_0^\pi + \frac{2}{n} \int_0^\pi x \cos nx\, dx \right\} \right] \\
&= \frac{2}{\pi} \left[\pi \frac{-\pi \cos n\pi}{n} + \frac{\pi^2 \cos n\pi}{n} + \frac{2}{n} \left\{ x \frac{\sin nx}{n} \Big|_0^\pi - \int_0^\pi \frac{\sin nx}{n} \right\} \right] \\
&= \frac{2}{\pi} \left[\frac{\pi^2 (-1)^{n+1}}{n} - \frac{(-1)^{n+1} \pi^2}{n} + \frac{2}{n^2} \times -\frac{\cos nx}{n} \Big|_0^\pi \right] \\
&= \frac{2}{\pi} \frac{2}{n^2} \left[\frac{1 - (-1)^n}{n} \right] = \frac{4}{\pi n^3} [1 - (-1)^n].
\end{aligned}
$$

Thus, the solution of (32.1)–(32.5) in this particular case is

$$
\begin{aligned}
u(x,t) &= \sum_{n=1}^\infty \frac{4}{\pi n^3} [1 - (-1)^n] \cos 2nt \sin nx \\
&= \sum_{n=0}^\infty \frac{8}{\pi (2n+1)^3} \cos 2(2n+1)t \sin(2n+1)x.
\end{aligned}
$$

Now for simplicity we assume that $g(x) \equiv 0$, i.e., the string is initially at rest. We further define $f(x)$ for all x by its Fourier series (32.18). Then, $f(x)$ is an odd function of period $2a$, i.e., $f(-x) = -f(x)$ and $f(x + 2a) = f(x)$. With these assumptions $b_n = 0$, $n \geq 1$ and thus the solution (32.17) by the trigonometric identity

$$
\sin \frac{n\pi x}{a} \cos \frac{n\pi ct}{a} = \frac{1}{2} \left(\sin \frac{n\pi}{a}(x + ct) + \sin \frac{n\pi}{a}(x - ct) \right)
$$

can be written as

$$
u(x,t) = \frac{1}{2} \sum_{n=1}^\infty a_n \left(\sin \frac{n\pi}{a}(x + ct) + \sin \frac{n\pi}{a}(x - ct) \right),
$$

which in view of (32.18) is the same as

$$
u(x,t) = \frac{1}{2}[f(x + ct) + f(x - ct)]. \tag{32.22}
$$

This is d'Alembert's solution. It is easy to verify that this indeed satisfies (32.1)–(32.5) with $g(x) \equiv 0$ provided $f(x)$ is twice differentiable. To realize the significance of this solution, consider the term $f(x-ct)$ and evaluate it at two pairs of values (x_1, t_1) and (x_2, t_2), where $t_2 = t_1 + \tau$, and $x_2 = x_1 + c\tau$. Then $x_1 - ct_1 = x_2 - ct_2$ and $f(x_1 - ct_1) = f(x_2 - ct_2)$, which means that this displacement travels along the string with velocity c. Thus, $f(x - ct)$ represents a wave traveling to the right with velocity c, and similarly, $f(x + ct)$ represents a wave traveling to the left with velocity c. It is for this reason that (32.1) is called the one-dimensional wave equation.

Lecture 33
The One-Dimensional Wave Equation (Cont'd.)

In this lecture we continue using the method of separation of variables to solve other initial-boundary value problems related to the one-dimensional wave equation.

Suppose that the vibrating string is subject to a damping force that is proportional at each instance to the velocity at each point. This results in a partial DE of the form

$$\frac{\partial^2 u}{\partial x^2} = \frac{1}{c^2}\left(\frac{\partial^2 u}{\partial t^2} + 2k\frac{\partial u}{\partial t}\right), \quad 0 < x < a, \quad t > 0, \quad c > 0. \tag{33.1}$$

We shall consider this equation together with the initial-boundary conditions (32.2)–(32.5). In (33.1) the constant k is small and positive. Clearly, if $k = 0$ the equation (33.1) reduces to (32.1).

Again we assume that the solution of (33.1) can be written as $u(x, t) = X(x)T(t) \neq 0$, so that

$$X''(x)T(t) = \frac{1}{c^2}(X(x)T''(t) + 2kX(x)T'(t))$$

and hence

$$\frac{X''}{X} = \frac{T'' + 2kT'}{c^2T} = \lambda,$$

which leads to

$$X'' - \lambda X = 0, \quad X(0) = X(a) = 0 \tag{33.2}$$

$$T'' + 2kT' - \lambda c^2 T = 0. \tag{33.3}$$

For (33.2), we have

$$\lambda_n = -\frac{n^2\pi^2}{a^2}, \quad X_n(x) = \sin\frac{n\pi x}{a}.$$

With $\lambda = \lambda_n = -n^2\pi^2/a^2$ equation (33.3) takes the form

$$T_n'' + 2kT_n' + \frac{n^2\pi^2c^2}{a^2}T_n = 0. \tag{33.4}$$

R.P. Agarwal, D. O'Regan, *Ordinary and Partial Differential Equations*,
Universitext, DOI 10.1007/978-0-387-79146-3_33,
© Springer Science+Business Media, LLC 2009

The auxiliary equation for (33.4) is

$$m^2 + 2km + \frac{n^2\pi^2c^2}{a^2} = 0,$$

or

$$(m+k)^2 = -\left(\frac{n^2\pi^2c^2}{a^2} - k^2\right)$$

and hence

$$m = -k \pm i\mu_n \quad \text{where} \quad \mu_n = \left(\frac{n^2\pi^2c^2}{a^2} - k^2\right)^{1/2}.$$

Recall $k > 0$ and small, so $\mu_n > 0$, $n \geq 1$.

Thus, the solution of (33.4) appears as

$$T_n(t) = e^{-kt}\left(a_n \cos \mu_n t + b_n \sin \mu_n t\right).$$

Therefore, the solution of (33.1) which satisfies (32.4) and (32.5) can be written as

$$u(x,t) = \sum_{n=1}^{\infty} e^{-kt}\left(a_n \cos \mu_n t + b_n \sin \mu_n t\right) \sin \frac{n\pi x}{a}. \tag{33.5}$$

This solution satisfies (32.2) if

$$f(x) = \sum_{n=1}^{\infty} a_n \sin \frac{n\pi x}{a},$$

which gives

$$a_n = \frac{2}{a}\int_0^a f(x)\sin \frac{n\pi x}{a}dx, \quad n = 1, 2, \cdots. \tag{33.6}$$

Finally, condition (32.3) is satisfied if

$$g(x) = \sum_{n=1}^{\infty}(-ka_n + b_n\mu_n)\sin \frac{n\pi x}{a}$$

and hence

$$-ka_n + b_n\mu_n = \frac{2}{a}\int_0^a g(x)\sin \frac{n\pi x}{a}dx,$$

which gives

$$b_n = k\frac{a_n}{\mu_n} + \frac{2}{\mu_n a}\int_0^a g(x)\sin \frac{n\pi x}{a}dx, \quad n = 1, 2, \cdots. \tag{33.7}$$

In particular, we shall find a_n and b_n when $g(x) = 0$, and

$$f(x) = \begin{cases} h\dfrac{2x}{a}, & 0 < x < \dfrac{a}{2} \\[2mm] h\left(2 - \dfrac{2x}{a}\right), & \dfrac{a}{2} < x < a, \quad h > 0. \end{cases}$$

From (33.6), we have

$$
\begin{aligned}
a_n &= \frac{2}{a}\left[\int_0^{a/2} h\frac{2x}{a}\sin\frac{n\pi x}{a}\,dx + \int_{a/2}^a h\left(2 - \frac{2x}{a}\right)\sin\frac{n\pi x}{a}\,dx\right] \\[2mm]
&= \frac{2h}{a}\left[\left\{\left(\frac{2x}{a}\right)\cos\frac{n\pi x}{a}\left(-\frac{a}{n\pi}\right) - \frac{2}{a}\sin\frac{n\pi x}{a}\left(-\frac{a^2}{n^2\pi^2}\right)\right\}\Big|_{x=0}^{x=a/2}\right. \\[2mm]
&\quad \left. + \left\{\left(2 - \frac{2x}{a}\right)\cos\frac{n\pi x}{a}\left(-\frac{a}{n\pi}\right) - \left(-\frac{2}{a}\right)\sin\frac{n\pi x}{a}\left(-\frac{a^2}{n^2\pi^2}\right)\right\}\Big|_{x=a/2}^{a}\right] \\[2mm]
&= \frac{2h}{a}\left[\cos\frac{n\pi}{2}\left(-\frac{a}{n\pi}\right) - \frac{2}{a}\sin\frac{n\pi}{2}\left(-\frac{a^2}{n^2\pi^2}\right)\right. \\[2mm]
&\quad \left. - \cos\frac{n\pi}{2}\left(-\frac{a}{n\pi}\right) + \left(-\frac{2}{a}\right)\sin\frac{n\pi}{2}\left(-\frac{a^2}{n^2\pi^2}\right)\right] \\[2mm]
&= \frac{2h}{a}\cdot\frac{2}{a}\cdot\frac{a^2}{n^2\pi^2}\cdot 2\cdot\sin\frac{n\pi}{2} \\[2mm]
&= \frac{8h}{\pi^2}\cdot\frac{\sin n\pi/2}{n^2}.
\end{aligned}
$$

Finally, from (33.7) in view of $g(x) = 0$, we find $b_n = ka_n/\mu_n$.

Now we shall assume that for the vibrating string the ends are free—they are allowed to slide without friction along the vertical lines $x = 0$ and $x = a$. This may seem impossible, but it is a standard mathematically modeled case. This leads to the initial-boundary value problem (32.1)–(32.3), and the Neumann boundary conditions

$$u_x(0, t) = 0, \quad t > 0 \tag{33.8}$$

$$u_x(a, t) = 0, \quad t > 0. \tag{33.9}$$

In this problem conditions (33.8), (33.9) are different from (32.4), (32.5); therefore, if we assume a solution in the form $u(x, t) = X(x)T(t) \neq 0$, X must satisfy the eigenvalue problem

$$X'' - \frac{\lambda}{c^2}X = 0$$

$$X'(0) = X'(a) = 0 \quad \text{(instead of } X(0) = X(a) = 0\text{)}.$$

For this problem the eigenvalues are

$$\lambda_0 = 0, \quad \lambda_n = -\frac{n^2\pi^2c^2}{a^2}, \quad n = 1, 2, \cdots$$

and the corresponding eigenfunctions are

$$X_0(x) = 1, \quad X_n(x) = \cos\frac{n\pi x}{a}, \quad n = 1, 2, \cdots.$$

For $\lambda_0 = 0$ the equation $T'' - \lambda T = 0$ reduces to $T_0'' = 0$, and hence

$$T_0(t) = b_0 t + a_0.$$

For $\lambda_n = -n^2\pi^2c^2/a^2$ the equation $T'' - \lambda T = 0$ is $T_n'' + (n^2\pi^2c^2/a^2)T_n = 0$ and hence the solution is the same as (32.16). Thus, the solution of (32.1) satisfying (33.8), (33.9) can be written as

$$u(x,t) = \sum_{n=0}^{\infty} X_n(x)T_n(t) = X_0(x)T_0(t) + \sum_{n=1}^{\infty} X_n(x)T_n(t),$$

or

$$u(x,t) = (b_0 t + a_0) + \sum_{n=1}^{\infty} \left(a_n \cos\frac{n\pi c}{a}t + b_n \sin\frac{n\pi c}{a}t\right) \cos\frac{n\pi x}{a}. \quad (33.10)$$

The solution (33.10) satisfies (32.2) if and only if

$$f(x) = a_0 + \sum_{n=1}^{\infty} a_n \cos\frac{n\pi x}{a}$$

and hence

$$\begin{aligned} a_0 &= \frac{1}{a}\int_0^a f(x)dx \\ a_n &= \frac{2}{a}\int_0^a f(x)\cos\frac{n\pi x}{a}dx, \quad n = 1, 2, \cdots. \end{aligned} \quad (33.11)$$

Finally, the solution (33.10) satisfies (32.3) if and only if

$$g(x) = b_0 + \sum_{n=1}^{\infty} b_n \frac{n\pi c}{a}\cos\frac{n\pi x}{a}$$

and hence

$$\begin{aligned} b_0 &= \frac{1}{a}\int_0^a g(x)dx \\ b_n &= \frac{2}{n\pi c}\int_0^a g(x)\cos\frac{n\pi x}{a}dx, \quad n = 1, 2, \cdots. \end{aligned} \quad (33.12)$$

Thus, the solution of (32.1)–(32.3), (33.8), (33.9) can be written as (33.10), where the constants a_n, b_n, $n = 0, 1, \cdots$ are given by (33.11) and (33.12), respectively.

Next we shall consider the equation of the general vibrating string

$$\frac{\partial}{\partial x}\left(k(x)\frac{\partial u}{\partial x}\right) = \frac{\rho(x)}{c^2}\frac{\partial^2 u}{\partial t^2}, \quad \alpha < x < \beta, \quad t > 0, \quad c > 0 \tag{33.13}$$

subject to the initial conditions (31.2),

$$u_t(x, 0) = g(x), \quad \alpha < x < \beta \tag{33.14}$$

and Robin's boundary conditions (31.3), (31.4). These boundary conditions describe some type of an elastic or spring attachment at both ends of the string. Now following as in Lecture 31, although there is no steady state for the wave equation (33.13), we let $v(x)$ be the solution of the problem (31.5), (31.6). Again, we define the function $w(x, t)$ as in (31.10), which satisfies the wave equation

$$\frac{\partial}{\partial x}\left(k(x)\frac{\partial w}{\partial x}\right) = \frac{\rho(x)}{c^2}\frac{\partial^2 w}{\partial t^2} \quad \alpha < x < \beta, \quad t > 0, \tag{33.15}$$

the initial conditions (31.19),

$$w_t(x, 0) = g(x), \quad \alpha < x < \beta, \tag{33.16}$$

and the boundary conditions (31.14), (31.15). We use the substitution $w(x, t) = X(t)T(t) \neq 0$, which leads to solving

$$\begin{aligned} (k(x)X')' + \frac{\lambda}{c^2}\rho(x)X &= 0 \\ a_0 X(\alpha) - a_1 X'(\alpha) &= 0 \\ d_0 X(\beta) + d_1 X'(\beta) &= 0 \end{aligned} \tag{33.17}$$

and

$$T'' + \lambda T = 0. \tag{33.18}$$

Thus, the solution of (33.15), (31.14), (31.15) in terms of the eigenvalues $0 \leq \lambda_1 < \lambda_2 < \cdots$ and eigenfunctions $X_n(x)$ of (33.17) appears as

$$w(x, t) = \sum_{n=1}^{\infty}(a_n \cos\sqrt{\lambda_n}t + b_n \sin\sqrt{\lambda_n}t)X_n(x). \tag{33.19}$$

This solution satisfies the initial conditions (31.19), (33.16) if and only if

$$a_n = \frac{\int_\alpha^\beta \rho(x)X_n(x)F(x)dx}{\int_\alpha^\beta \rho(x)X_n^2(x)dx}, \quad b_n = \frac{\int_\alpha^\beta \rho(x)X_n(x)g(x)dx}{\sqrt{\lambda_n}\int_\alpha^\beta \rho(x)X_n^2(x)dx}, \quad n \geq 1.$$

Finally, the solution of (33.13), (31.2), (33.14), (31.3), (31.4) is obtained from the relation $u(x,t) = w(x,t) + v(x)$.

Problems

33.1. Solve the initial-boundary value problem (32.1)–(32.5) when

(i) $a = \pi$, $c = 5$, $f(x) = \sin 3x$, $g(x) = 4$

(ii) $a = \pi$, $c = 1$, $f(x) = x(\pi - x)$, $g(x) = 3$

(iii) $a = 3$, $c = 2$, $f(x) = \begin{cases} x, & 0 < x < 1 \\ 1, & 1 < x < 2 \\ 3 - x, & 2 < x < 3 \end{cases}$, $g(x) = 0$

(iv) $a = \pi$, $c = 2/3$, $f(x) = \sin^2 x$, $g(x) = \sin x$

(v) $a = \pi$, $c = 1$, $f(x) = x^2(\pi - x)$, $g(x) = 0$.

33.2. A tightly stretched string with fixed end points $x = 0$ and $x = a$ is initially in a position given by $u = u_0 \sin^3 \pi x/a$. If it is released from rest from this position, find the displacement $u(x, t)$.

33.3. The points of trisection of a string of length a are pulled aside through the same distance h on opposite sides of the position of equilibrium and the string is released from rest. Derive an expression for the displacement of the string of subsequent time and show that the midpoint of the string remains at rest.

33.4. Solve the initial-boundary value problem (32.1)–(32.4), $u_x(a, t) = 0$, $t > 0$, i.e., the string is fixed at the end $x = 0$ and free at the end $x = a$.

33.5. Solve the initial–boundary value problem (32.1)–(32.3), (32.5), $u_x(0, t) = 0$, $t > 0$ i.e., the string is free at the end $x = 0$ and fixed at the end $x = a$.

33.6. Suppose that u is a solution of the initial–boundary value problem (32.1)–(32.3),

$$\begin{aligned} u(0,t) &= A, \quad t > 0 \\ u(a,t) &= B, \quad t > 0 \end{aligned} \tag{33.20}$$

where A and B are constants. Show that if

$$v(x,t) = u(x,t) + \left(\frac{x-a}{a}\right) A - \frac{x}{a} B, \tag{33.21}$$

then v is a solution of the initial-boundary value problem (32.1),

$$v(x,0) = f(x) + \left(\frac{x-a}{a}\right)A - \frac{x}{a}B, \quad 0 < x < a \qquad (33.22)$$

(32.3)–(32.5). In particular, solve the problem (32.1)–(32.3), (33.20) when

(i) $a = \pi$, $c = 1$, $f(x) = x(\pi - x)$, $g(x) = 3$, $A = 0$, $B = 5$
(ii) $a = \pi$, $c = 1$, $f(x) = x^2(\pi - x)$, $g(x) = 0$, $A = B = 2$.

33.7. Consider the particular case of the telegraph equation (32.10),

$$v_{tt} + 2av_t + a^2v = c^2v_{xx},$$

with the initial conditions $v(x,0) = \phi(x)$, $v_t(x,0) = 0$ and the boundary conditions $v(0,t) = v(a,0) = 0$. Show that the change of variable $v(x,t) = e^{-at}u(x,t)$ transforms this initial–boundary value problem to (32.1)–(32.5) with $f(x) = \phi(x)$ and $g(x) = a\phi(x)$.

33.8. Solve the initial-boundary value problem (33.1), (32.2)–(32.5) when

(i) $a = 1$, $c = 1$, $k = 1$, $f(x) = A\sin\pi x$, $g(x) = 0$
(ii) $a = \pi$, $c = 1$, $k = 1$, $f(x) = x$, $g(x) = 0$.

33.9. Solve the initial–boundary value problem (32.1)–(32.3), (33.8), (33.9) when a, c, $f(x)$ and $g(x)$ are the same as in Problem 33.1 (i)–(v).

33.10. Show that the solution (32.17) of (32.1)–(32.5) can be written as

$$u(x,t) = \frac{1}{2}[f(x+ct) + f(x-ct)] + \frac{1}{2c}\int_{x-ct}^{x+ct} g(z)dz.$$

This is *d'Alembert's solution*. Thus, to find the solution $u(x,t)$, we need to know only the initial displacement $f(x)$ and the initial velocity $g(x)$. This makes d'Alembert's solution easy to apply as compared to the infinite series (32.24). In particular, find the solution on $-\infty < x < \infty$, $t > 0$ when

(i) $f(x) = 1/(1 + 2x^2)$, $g(x) = 0$
(ii) $f(x) = e^{-|x|}$, $g(x) = xe^{-x^2}$
(iii) $f(x) = \operatorname{sech} x$, $g(x) = x/(1 + x^2)$.

33.11. The partial DE which describes the small displacement $w = w(x,t)$ of a heavy flexible chain of length a from equilibrium is

$$\frac{\partial^2 w}{\partial t^2} = -g\frac{\partial w}{\partial x} + g(a-x)\frac{\partial^2 w}{\partial x^2},$$

where g is the gravitational constant. This equation was studied extensively by Daniel Bernoulli around 1732 and later by Leonhard Euler in 1781.

(i)   Set $y = a - x$, $u(y,t) = w(a - x, t)$ to transform the above partial DE to

$$\frac{\partial^2 u}{\partial t^2} = g\frac{\partial u}{\partial y} + gy\frac{\partial^2 u}{\partial y^2}. \tag{33.23}$$

(ii)   Use separation of variables to show that the solution of (33.23) which is bounded for $0 \le y \le a$ and satisfies $u(a,t) = 0$ is

$$\sum_{n=1}^{\infty} J_0\left(2\lambda_n\sqrt{\frac{y}{g}}\right)(a_n \cos\lambda_n t + b_n \sin\lambda_n t),$$

where $\lambda_n = (1/2)b_{0,n}\sqrt{g/a}$ and $b_{0,n}$ is a positive root of $J_0(x)$.

33.12.   A bar has length a, density δ, cross-sectional area A, Young's modulus E, and total mass $M = \delta Aa$. Its end $x = 0$ is fixed and a mass m is attached to its free end. The bar initially is stretched linearly by moving m a distance $d = ba$ to the right, and at time $t = 0$ the system is released from rest. Find the subsequent vibrations of the bar by solving the initial-boundary value problem (32.1), $u(x,0) = bx$, $u_t(x,0) = 0$, (32.4) and $mu_{tt}(a,t) = -AEu_x(a,t)$.

33.13.   Small transverse vibrations of a beam are governed by the partial DE

$$\frac{\partial^2 u}{\partial t^2} + c^2\frac{\partial^4 u}{\partial x^4} = 0, \quad 0 < x < a, \quad t > 0,$$

where $c^2 = EI/A\mu$, and E is the modulus of elasticity, I is the moment of inertia of any cross section about the x-axis, A is the area of cross section, and μ is the mass per unit length. Boundary conditions at the ends of the beam are usually of the following type:

(1)   A *fixed end* also known as *built-in* or a *clamped end* has its displacement and slope equal to zero (see Figure 33.1a):

$$u(a,t) = \frac{\partial u}{\partial x}(a,t) = 0.$$

(2)   A *simply supported end* has displacement and moment equal to zero (see Figure 33.1b):

$$u(a,t) = \frac{\partial^2 u}{\partial x^2}(a,t) = 0.$$

(3)   A *free end* has zero moment and zero shear (see Figure 33.1c):

$$\frac{\partial^2 u}{\partial x^2}(a,t) = \frac{\partial^3 u}{\partial x^3}(a,t) = 0.$$

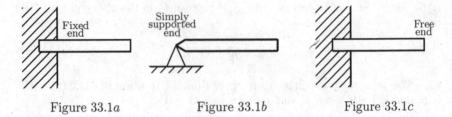

Figure 33.1*a*   Figure 33.1*b*   Figure 33.1*c*

Find the solution for the vibration of a beam that has simply supported ends at $x = 0$ and $x = a$ with the initial conditions $u(x,0) = f(x)$, $u_t(x,0) = g(x)$, $0 < x < a$. In particular, compute the solution when $f(x) = Ax(a-x)$ and $g(x) = 0$.

Answers or Hints

33.1. (i) $\cos 15t \sin 3x + \sum_{n=0}^{\infty} \frac{16}{5(2n+1)^2 \pi} \sin 5(2n+1)t \sin(2n+1)x$

(ii) $\sum_{n=1}^{\infty} \left[\frac{8}{(2n-1)^3 \pi} \cos(2n-1)t + \frac{12}{(2n-1)^2 \pi} \sin(2n-1)t \right] \sin(2n-1)x$

(iii) $\sum_{n=1}^{\infty} \frac{12}{n^2 \pi^2} \sin \frac{n\pi}{2} \cos \frac{n\pi}{6} \sin \frac{n\pi x}{3} \cos \frac{2n\pi t}{3}$

(iv) $\frac{3}{2} \sin x \sin \frac{2t}{3} - \frac{8}{\pi} \sum_{n=1}^{\infty} \frac{1}{(2n-1)[(2n-1)^2-4]} \sin(2n-1)x \cos \frac{2(2n-1)t}{3}$

(v) $\frac{\pi^3}{12} + \sum_{n=1}^{\infty} -\frac{\pi}{2n^2} \cos 2nt \cos 2nx + \sum_{n=1}^{\infty} \left[\frac{2\pi}{(2n-1)^2} - \frac{24}{(2n-1)^4 \pi} \right]$
$\times \cos(2n-1)t \cos(2n-1)x.$

33.2. Use $\sin^3 \theta = \frac{1}{4}(3 \sin \theta - \sin 3\theta)$, $u(x,t) = \frac{u_0}{2} \left(3 \sin \frac{\pi}{a}x \cos \frac{c\pi}{a}t \right.$
$\left. - \sin \frac{3\pi}{a}x \cos \frac{3c\pi}{a}t \right).$

33.3. In (32.1)–(32.5), $f(x) = \frac{3h}{a} \begin{cases} x, & 0 \le x \le a/3 \\ (a - 2x), & a/3 \le x \le 2a/3, \\ (x - a), & 2a/3 \le x \le a \end{cases}$ $g(x) = 0,$

$u(x,t) = \frac{9h}{\pi^2} \sum_{n=1}^{\infty} \frac{1}{n^2} \sin \frac{2n\pi}{3} \sin \frac{2n\pi}{a}x \cos \frac{2n\pi c}{a}t.$

33.4 $\sum_{n=1}^{\infty} \left[a_n \cos \frac{(2n-1)\pi ct}{2a} + b_n \sin \frac{(2n-1)\pi ct}{2a} \right] \sin \frac{(2n-1)\pi x}{2a},$

$a_n = \frac{2}{a} \int_0^a f(x) \sin \frac{(2n-1)\pi x}{2a} dx, \ n \ge 1,$

$b_n = \frac{4}{(2n-1)\pi c} \int_0^a g(x) \sin \frac{(2n-1)\pi x}{2a} dx, \ n \ge 1.$

33.5 $\sum_{n=1}^{\infty} \left[a_n \cos \frac{(2n-1)\pi ct}{2a} + b_n \sin \frac{(2n-1)\pi ct}{2a} \right] \cos \frac{(2n-1)\pi x}{2a},$

$a_n = \frac{2}{a} \int_0^a f(x) \cos \frac{(2n-1)\pi x}{2a} dx, \ n \ge 1$

$b_n = \frac{4}{(2n-1)\pi c} \int_0^a g(x) \cos \frac{(2n-1)\pi x}{2a} dx, \ n \ge 1.$

33.6. (i) $\frac{5x}{\pi} + \sum_{n=1}^{\infty} \frac{5}{n\pi} \cos 2nt \sin 2nx + \sum_{n=1}^{\infty} \left\{ \left[\frac{8}{(2n-1)^3\pi} - \frac{10}{(2n-1)\pi} \right] \right.$
$\left. \times \cos(2n-1)t + \frac{12}{(2n-1)^2\pi} \sin(2n-1)t \right\} \sin(2n-1)x$

(ii) $2 + \sum_{n=1}^{\infty} -\frac{3}{2n^3} \cos 2nt \sin 2nx + \sum_{n=1}^{\infty} \left[\frac{4}{(2n-1)^3} - \frac{8}{(2n-1)\pi} \right]$
$\times \cos(2n-1)t \sin(2n-1)x.$

33.7. Verify directly.

33.8. (i) $Ae^{-t} \left(\cos \sqrt{\pi^2-1}t + \frac{1}{\sqrt{\pi^2-1}} \sin \sqrt{\pi^2-1}t \right) \sin \pi x$ (ii) $2e^{-t} \times$
$\left[(1+t) \sin x + \sum_{n=2}^{\infty} \frac{(-1)^{n+1}}{n} \left(\cos \sqrt{n^2-1}t + \frac{1}{\sqrt{n^2-1}} \sin \sqrt{n^2-1}t \right) \sin nx \right].$

33.9. Compute a_n and b_n using (33.11) and (33.12) and substitute in (33.10).

33.10. Use $\sin A \sin B = \frac{1}{2}[\cos(A-B) - \cos(A+B)]$ and (32.22).
(i) $\frac{1}{2} \left[\frac{1}{1+2(x+ct)^2} + \frac{1}{1+2(x-ct)^2} \right]$ (ii) $\frac{1}{2} \left[e^{-|x+ct|} + e^{-|x-ct|} \right]$
$-\frac{1}{4c} \left(e^{-(x+ct)^2} - e^{-(x-ct)^2} \right)$ (iii) $\frac{1}{2} \left(\mathrm{sech}\,(x+ct) + \mathrm{sech}\,(x-ct) \right)$
$+ \frac{1}{4c} \left[\ln(1+(x+ct)^2) - \ln(1+(x-ct)^2) \right].$

33.11. Compare the ordinary DE with (9.19).

33.12. $u(x,t) = \sum_{n=1}^{\infty} b_n \cos \frac{\alpha_n ct}{a} \sin \frac{\alpha_n x}{a}$ where $b_n = \frac{4ab \sin \alpha_n}{\alpha_n(2\alpha_n + \sin 2\alpha_n)}$ and α_n is a root of the equation $\tan \alpha = \frac{AEa}{mc^2}\frac{1}{\alpha}$. Use the fact that the set $\{\sin \frac{\alpha_n x}{a}\}$ is not orthogonal on $[0,a]$, however, in view of Problem 18.4 the set $\{\cos \frac{\alpha_n x}{a}\}$ is orthogonal on $[0,a]$.

33.13 See Problem 18.9(i), $\sum_{n=1}^{\infty} \left(a_n \cos \frac{cn^2\pi^2 t}{a^2} + b_n \sin \frac{cn^2\pi^2 t}{a^2} \right) \sin \frac{n\pi x}{a}$,
where $a_n = \frac{2}{a} \int_0^a f(x) \sin \frac{n\pi x}{a} dx$, $b_n = \frac{2a}{cn^2\pi^2} \int_0^a g(x) \sin \frac{n\pi x}{a} dx$.
$\frac{8Aa^2}{\pi^3} \sum_{n=0}^{\infty} \frac{1}{(2n+1)^3} \cos \frac{c(2n+1)^2\pi^2 t}{a^2} \sin \frac{(2n+1)\pi x}{a}.$

Lecture 34
Laplace Equation in Two Dimensions

In this lecture we give a derivation of the two-dimensional Laplace equation and formulate the Dirichlet problem on a rectangle. Then we use the method of separation of variables to solve this problem.

Consider the flow of heat in a metal plate of uniform thickness α (cm), density ρ (g/cm^3), specific heat s (cal/g deg) and thermal conductivity k (cal/cm sec deg). Let the XY-plane be taken in one face of the plate. If the temperature at any point is independent of the z-coordinate and depends only on x, y, and time t (for instance, its two parallel faces are insulated), then the flow is said to be two-dimensional. In this case, the heat flow is in the XY-plane only and is zero along the normal to the XY-plane.

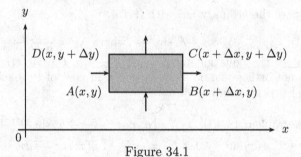

Figure 34.1

Consider a rectangular element $ABCD$ of the plate with sides Δx and Δy as shown in Figure 34.1. By Fourier's law, the amount of heat entering the element in 1 sec from the side AB is

$$= -k\alpha\Delta x \left(\frac{\partial u}{\partial y}\right)_y ;$$

and the amount of heat entering the element in 1 sec from the side AD is

$$= -k\alpha\Delta y \left(\frac{\partial u}{\partial x}\right)_x .$$

The quantity of heat flowing out through the side CD in 1 sec is

$$= -k\alpha\Delta x \left(\frac{\partial u}{\partial y}\right)_{y+\Delta y} ;$$

R.P. Agarwal, D. O'Regan, *Ordinary and Partial Differential Equations*,
Universitext, DOI 10.1007/978-0-387-79146-3_34,
© Springer Science+Business Media, LLC 2009

and the quantity of heat flowing out through the side BC in 1 sec is

$$= -k\alpha\Delta y \left(\frac{\partial u}{\partial x}\right)_{x+\Delta x}.$$

Hence, the total gain of heat by the rectangular element $ABCD$ in 1 sec is

$$= -k\alpha\Delta x \left(\frac{\partial u}{\partial y}\right)_{y} - k\alpha\Delta y \left(\frac{\partial u}{\partial x}\right)_{x} + k\alpha\Delta x \left(\frac{\partial u}{\partial y}\right)_{y+\Delta y} + k\alpha\Delta y \left(\frac{\partial u}{\partial x}\right)_{x+\Delta x}$$

$$= k\alpha\Delta x\Delta y \left[\frac{\left(\frac{\partial u}{\partial x}\right)_{x+\Delta x} - \left(\frac{\partial u}{\partial x}\right)_{x}}{\Delta x} + \frac{\left(\frac{\partial u}{\partial y}\right)_{y+\Delta y} - \left(\frac{\partial u}{\partial y}\right)_{y}}{\Delta y}\right].$$

$$(34.1)$$

Also the rate of gain of heat by the element is

$$= \rho\Delta x\Delta y\alpha s \frac{\partial u}{\partial t}. \qquad (34.2)$$

Thus, equating (34.1) and (34.2), dividing both sides by $\alpha\Delta x\Delta y$, and taking limits as $\Delta x \to 0$, $\Delta y \to 0$, we get

$$k\left(\frac{\partial^2 u}{\partial x^2} + \frac{\partial^2 u}{\partial y^2}\right) = \rho s \frac{\partial u}{\partial t},$$

which is the same as

$$\frac{\partial u}{\partial t} = c^2 \left(\frac{\partial^2 u}{\partial x^2} + \frac{\partial^2 u}{\partial y^2}\right), \qquad (34.3)$$

where $c^2 = k/(\rho s)$ is the *diffusivity coefficient*.

Equation (34.3) gives the temperature distribution of the plate in the *transient state*. In the *steady state*, u is independent of t, so that $u_t = 0$ and the equation (34.3) reduces to

$$\Delta_2 u = u_{xx} + u_{yy} = 0, \qquad (34.4)$$

which is the well-known *Laplace equation* in two dimensions. Since there is no time dependence in (34.4), no initial conditions are required to be satisfied by its solution $u(x, y)$. However, certain boundary conditions on the boundary of the region must be satisfied. Thus, a typical problem associated with Laplace's equation is a boundary value problem. A common way is to specify $u(x, y)$ at each point (x, y) on the boundary, which is known as a *Dirichlet problem*.

Now we shall use the method of separation of variables to solve the Dirichlet problem on the rectangle $R = 0 < x < a$, $0 < y < b$, i.e., find the solution $u(x, y)$ of (34.4) on R satisfying the boundary conditions

$$u(x, 0) = f(x), \quad 0 < x < a \tag{34.5}$$

$$u(x, b) = g(x), \quad 0 < x < a \tag{34.6}$$

$$u(0, y) = 0, \quad 0 < y < b \tag{34.7}$$

$$u(a, y) = 0, \quad 0 < y < b. \tag{34.8}$$

This problem is illustrated in Figure 34.2.

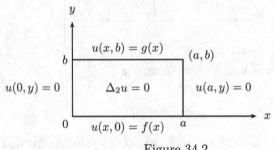

Figure 34.2

We seek a solution of (34.4) in the form $u(x, y) = X(x)Y(y) \neq 0$. Thus, it follows that

$$X''(x)Y(y) + X(x)Y''(y) = 0,$$

or

$$\frac{X''(x)}{X(x)} + \frac{Y''(y)}{Y(y)} = 0,$$

which is the same as

$$-\frac{X''(x)}{X(x)} = \frac{Y''(y)}{Y(y)} = \lambda \quad \text{(constant)}.$$

Hence, we have

$$X'' + \lambda X = 0, \tag{34.9}$$

and the conditions (34.7) and (34.8) imply

$$X(0) = 0, \quad X(a) = 0. \tag{34.10}$$

Also Y satisfies the differential equation

$$Y'' - \lambda Y = 0. \tag{34.11}$$

The eigenvalues and eigenfunctions of the problem (34.9), (34.10) are respectively given by

$$\lambda_n = \frac{n^2 \pi^2}{a^2}, \quad n = 1, 2, \cdots \tag{34.12}$$

and
$$X_n(x) = \sin \frac{n\pi x}{a}, \quad n = 1, 2, \cdots. \tag{34.13}$$

For λ as given in (34.12) the general solution of the differential equation (34.11) is
$$Y_n(y) = a_n \cosh \frac{n\pi y}{a} + b_n \sinh \frac{n\pi y}{a}. \tag{34.14}$$

Thus, the solution of (34.4) satisfying (34.7) and (34.8) can be written as

$$u(x,y) = \sum_{n=1}^{\infty} \left(a_n \cosh \frac{n\pi y}{a} + b_n \sinh \frac{n\pi y}{a} \right) \sin \frac{n\pi x}{a}. \tag{34.15}$$

Now (34.15) satisfies (34.5) if and only if

$$f(x) = \sum_{n=1}^{\infty} a_n \sin \frac{n\pi x}{a},$$

which gives
$$a_n = \frac{2}{a} \int_0^a f(x) \sin \frac{n\pi x}{a} dx, \quad n = 1, 2, \cdots. \tag{34.16}$$

Finally, (34.15) satisfies (34.6) provided

$$g(x) = \sum_{n=1}^{\infty} \left(a_n \cosh \frac{n\pi b}{a} + b_n \sinh \frac{n\pi b}{a} \right) \sin \frac{n\pi x}{a},$$

which gives

$$a_n \cosh \frac{n\pi b}{a} + b_n \sinh \frac{n\pi b}{a} = \frac{2}{a} \int_0^a g(x) \sin \frac{n\pi x}{a} dx$$

and therefore

$$b_n \sinh \frac{n\pi b}{a} = \frac{2}{a} \int_0^a g(x) \sin \frac{n\pi x}{a} dx - a_n \cosh \frac{n\pi b}{a},$$

which in view of (34.16) gives

$$b_n = \frac{1}{\sinh \frac{n\pi b}{a}} \left[\frac{2}{a} \int_0^a g(x) \sin \frac{n\pi x}{a} dx \right.$$
$$\left. - \left(\cosh \frac{n\pi b}{a} \right) \frac{2}{a} \int_0^a f(x) \sin \frac{n\pi x}{a} dx \right], \quad n \geq 1. \tag{34.17}$$

Hence, the solution of the boundary value problem (34.4)–(34.8) is given by (34.15) where a_n and b_n are as in (34.16) and (34.17), respectively.

In particular, we shall solve the boundary value problem (34.4)–(34.8) with $f(x) = 0$, $g(x) = x$, $a = 1$, $b = 1$. Clearly, from (34.16) and (34.17), we have

$$a_n = 0, \quad n \geq 1$$

$$b_n = \frac{1}{\sinh n\pi} \left[2 \int_0^1 x \sin n\pi x\, dx \right]$$

$$= \frac{2}{\sinh n\pi} \left[-\frac{x \cos n\pi x}{n\pi} \right]\Big|_{x=0}^1 = \frac{2(-1)^{n+1}}{n\pi \sinh n\pi}.$$

Thus, the solution in this particular case is

$$u(x,y) = \sum_{n=1}^{\infty} \frac{2(-1)^{n+1}}{n\pi \sinh n\pi} \sinh n\pi y \sin n\pi x.$$

Next we note that as for the problem (34.4)–(34.8) the solution $u(x, y)$ of the Dirichlet problem (34.4) on the rectangle R satisfying the boundary conditions

$$u(x, 0) = 0, \quad 0 < x < a \tag{34.18}$$

$$u(x, b) = 0, \quad 0 < x < a \tag{34.19}$$

$$u(0, y) = h(y), \quad 0 < y < b \tag{34.20}$$

$$u(a, y) = k(y), \quad 0 < y < b \tag{34.21}$$

(see Figure 34.3) can be written as

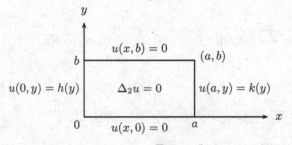

Figure 34.3

$$u(x, y) = \sum_{n=1}^{\infty} \left(\alpha_n \cosh \frac{n\pi x}{b} + \beta_n \sinh \frac{n\pi x}{b} \right) \sin \frac{n\pi y}{b}, \tag{34.22}$$

where

$$\alpha_n = \frac{2}{b} \int_0^b h(y) \sin \frac{n\pi y}{b} dy, \quad n = 1, 2, \cdots \tag{34.23}$$

and

$$\beta_n = \frac{1}{\sinh \frac{n\pi a}{b}} \left[\frac{2}{b} \int_0^b k(y) \sin \frac{n\pi y}{b} dy \right.$$

$$\left. - \left(\cosh \frac{n\pi a}{b} \right) \frac{2}{b} \int_0^b h(y) \sin \frac{n\pi y}{b} dy \right], \quad n = 1, 2, \cdots.$$

$$(34.24)$$

In the particular case $h(y) = Ay(b-y)$, $k(y) = 0$ this solution simplifies to

$$u(x,y) = \frac{8Ab^2}{\pi^3} \sum_{n=0}^{\infty} \frac{1}{(2n+1)^3} \frac{\sinh \frac{(2n+1)\pi}{b}(a-x)}{\sinh \frac{(2n+1)\pi}{b}a} \sin \frac{(2n+1)\pi}{b} y.$$

Finally, from the linearity of the problem as well as by direct substitution it is clear that if $u_1(x, y)$ is the solution of the problem (34.4)–(34.8) and $u_2(x, y)$ is the solution of the problem (34.4), (34.18)–(34.21) then

$$u(x, y) = u_1(x, y) + u_2(x, y) \qquad (34.25)$$

is the solution of the Dirichlet problem (34.4) on the rectangle R satisfying the boundary conditions (34.5), (34.6), (34.20), (34.21) (see Figure 34.4).

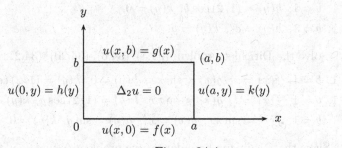

Figure 34.4

In particular we shall solve the boundary value problem (34.4), (34.5), (34.6), (34.20), (34.21) with $f(x) = x$, $g(x) = 0$, $h(y) = \sin y$, $k(y) = 0$, $a = b = 1$. From (34.16), (34.17), (34.23) and (34.24) it follows that

$$a_n = \frac{2(-1)^{n+1}}{n\pi}, \quad b_n = \frac{2(-1)^n}{n\pi} \coth n\pi$$

$$\alpha_n = \frac{(-1)^{n+1} 2n\pi \sin 1}{2n^2\pi^2 - 1}, \quad \beta_n = -\coth n\pi \frac{(-1)^{n+1} 2n\pi \sin 1}{2n^2\pi^2 - 1}.$$

Thus, the solution in this case appears as

$$u(x,y) = u_1(x,y) + u_2(x,y)$$

$$= \sum_{n=1}^{\infty} \left[\frac{2(-1)^{n+1}}{n\pi} \cosh n\pi y + \frac{2(-1)^n}{n\pi} \coth n\pi \sinh n\pi y \right] \sin n\pi x$$

$$+ \sum_{n=1}^{\infty} \left[\frac{(-1)^{n+1}2n\pi \sin 1}{2n^2\pi^2 - 1} \cosh n\pi x \right.$$

$$\left. - \coth n\pi \frac{(-1)^{n+1}2n\pi \sin 1}{2n^2\pi^2 - 1} \sinh n\pi x \right] \sin n\pi y.$$

Problems

34.1. Solve the Dirichlet problem (34.4)–(34.8) with

(i) $a = 2,\ b = 1,\ f(x) = 0,\ g(x) = e^x$

(ii) $a = 1,\ b = 1,\ f(x) = x,\ g(x) = \sin \pi x$

(iii) $a = 2,\ b = 2,\ f(x) = e^x,\ g(x) = 4\cos x.$

34.2. Solve the Dirichlet problem (34.4), (34.18)–(34.21) with

(i) $a = 1,\ b = 1,\ h(y) = 0,\ k(y) = (1/2)\cos y$

(ii) $a = 1,\ b = 1,\ h(y) = (1/2)\cos y,\ k(y) = 0$

(iii) $a = 1,\ b = 1,\ h(y) = e^y,\ k(y) = y.$

34.3. Solve the Dirichlet problem (34.4)–(34.6), (34.20), (34.21) with

(i) $a = 1,\ b = 1,\ f(x) = x,\ g(x) = \sin \pi x,\ h(y) = 0,\ k(y) = (1/2)\cos y$

(ii) $a = 1,\ b = 1,\ f(x) = x,\ g(x) = \sin \pi x,\ h(y) = (1/2)\cos y,\ k(y) = 0$

(iii) $a = 1,\ b = 1,\ f(x) = \sin \pi x,\ g(x) = x^2,\ h(y) = \sin y,\ k(y) = 0.$

34.4. Show that Neumann boundary value problem $\Delta_2 u = 0,\ u_y(x,0) = f(x),\ u_y(x,b) = g(x),\ u_x(0,y) = 0 = u_x(a,y)$ has an infinite number of solutions.

34.5. Solve the Laplace equation (34.4) in the rectangle $R = 0 < x < \pi,\ 0 < y < 1$ subject to the mixed boundary conditions $u(x,0) = T_0 \cos x,\ u(x,1) = T_0 \cos^2 x,\ u_x(0,y) = 0,\ u_x(\pi,y) = 0.$

34.6. Solve the Laplace equation (34.4) in the rectangle $R = 0 < x < 1,\ 0 < y < 1$ subject to the mixed boundary conditions $u(x,0) = x^2,\ u_y(x,1) = 0,\ u_x(0,y) = 0,\ u_x(1,y) = 0.$

34.7. Solve the Laplace equation (34.4) in the rectangle $R = 0 < x < 1$, $0 < y < 1$ subject to the mixed boundary conditions $u(x,0) = 0$, $u_y(1/2) - x$, $u_y(x,1) = (1/2) - x$, $u_x(0,y) = 0$, $u_x(1,y) = 0$.

34.8. A rectangular plate with insulated surface is a cm wide and so long compared to its width that it may be considered infinite in length without introducing an appreciable error. If the temperature of the short edge $y = 0$ is given by $f(x)$, $0 < x < a$ and the two edges $x = 0$, $x = a$ are kept at $0°C$, determine the temperature at any point of the plate in the steady state. In particular, solve this problem for

(i) $f(x) = T_0$

(ii) $f(x) = cx$

(iii) $a = 10$, $f(x) = \begin{cases} 20x, & 0 < x < 5 \\ 20(10 - x), & 5 < x < 10. \end{cases}$

Answers or Hints

34.1. (i) $\sum_{n=1}^{\infty} \frac{2n\pi}{(n^2\pi^2+4)\sinh(n\pi/2)}\left[1 - (-1)^n e^2\right]\sinh\frac{n\pi y}{2}\sin\frac{n\pi x}{2}$

(ii) $\left[\frac{2}{\pi}\cosh\pi y + \left(1 - \frac{2\cosh\pi}{\pi}\right)\frac{\sinh\pi y}{\sinh\pi}\right]\sin\pi x$

$+ \sum_{n=2}^{\infty}\frac{2(-1)^{n+1}}{n\pi}[\cosh n\pi y - \coth n\pi\sinh n\pi y]\sin n\pi x$

(iii) $\sum_{n=1}^{\infty}\left\{\frac{2n\pi}{n^2\pi^2+4}(1 - (-1)^n e^2)\cosh\frac{n\pi y}{2} + \left[\frac{2n\pi(1-(-1)^n\cos 2)}{(n^2\pi^2-1)\sinh n\pi}\right.\right.$

$\left.\left. - \frac{2n\pi(1-(-1)^n e^2)\coth n\pi}{n^2\pi^2+4}\right]\sinh\frac{n\pi y}{2}\right\}\sin\frac{n\pi x}{2}$.

34.2. (i) $\sum_{n=1}^{\infty}\frac{1}{\sinh n\pi}\frac{n\pi(1-(-1)^n\cos 1)}{n^2\pi^2-1}\sinh n\pi x\sin n\pi y$

(ii) $\sum_{n=1}^{\infty}\frac{n\pi(1-(-1)^n\cos 1)}{n^2\pi^2-1}(\cosh n\pi x - \coth n\pi\sinh n\pi x)\sin n\pi y$

(iii) $\sum_{n=1}^{\infty}\left\{\frac{2n\pi}{n^2\pi^2+1}(1 - (-1)^n e)\cosh n\pi x - \left[\frac{2(-1)^n}{n\pi\sinh n\pi}\right.\right.$

$\left.\left. + \frac{2n\pi\coth n\pi}{n^2\pi^2+1}(1 - (-1)^n e)\right]\sinh n\pi x\right\}\sin n\pi y$.

34.3. (i) $u_1(x,y) + u_2(x,y)$ where u_1 is the solution of Problem 34.1(ii) and u_2 is the solution of Problem 34.2(i) (ii) $u_1(x,y) + u_2(x,y)$ where u_1 is the solution of Problem 34.1(ii) and u_2 is the solution of Problem 34.2(ii)

(iii) $\left[\cosh\pi y + \left(\frac{2}{\pi} - \frac{8}{\pi^3} - \cosh\pi\right)\frac{\sinh\pi y}{\sinh\pi}\right]\sin\pi x$

$+ \sum_{n=2}^{\infty}\left[\frac{2-n^2\pi^2}{n^3\pi^3}(-1)^n - \frac{2}{n^3\pi^3}\right]\frac{2\sinh n\pi y}{\sinh n\pi}\sin n\pi x$

$+ \sum_{n=1}^{\infty}\left\{\frac{(-1)^{n+1}2n\pi\sin 1}{n^2\pi^2-1}[\cosh n\pi x - \coth n\pi\sinh n\pi x]\right\}\sin n\pi y$.

34.4. If u is a solution, then $u + K$ is also a solution.

34.5. $T_0 \left(\frac{1}{2} y + \frac{\sinh(1-y)}{\sinh 1} \cos x + \frac{\sinh 2y}{2 \sinh 2} \cos 2x \right).$

34.6. $\frac{1}{3} + \frac{4}{\pi^2} \sum_{n=1}^{\infty} \frac{(-1)^n}{n^2 \cosh n\pi} \cosh n\pi(1-y) \cos n\pi x.$

34.7. $u(x,y) = \frac{4}{\pi^3} \sum_{n=0}^{\infty} \frac{1}{(2n+1)^3} \cdot \frac{\sinh((2n+1)\pi y)}{\cosh((2n+1)\pi)} \cos((2n+1)\pi x).$

34.8. $u(x,t) = \sum_{n=1}^{\infty} e^{-n\pi y/a} \sin \frac{n\pi x}{a}$, $a_n = \frac{2}{a} \int_0^a f(x) \sin \frac{n\pi x}{a} dx,$
(i) $\frac{4T_0}{\pi} \sum_{n=0}^{\infty} \frac{1}{2n+1} e^{-(2n+1)\pi y/a} \sin \frac{(2n+1)\pi x}{a}$ (ii) $\frac{2ca}{\pi} \sum_{n=1}^{\infty} \frac{(-1)^{n+1}}{n} \times$
$e^{-n\pi y/a} \sin \frac{n\pi x}{a}$ (iii) $\frac{800}{\pi^2} \sum_{n=1}^{\infty} \frac{(-1)^{n+1}}{(2n-1)^2} e^{-(2n-1)\pi y/10} \sin \frac{(2n-1)\pi x}{10}.$

Lecture 35
Laplace Equation
in Polar Coordinates

In this lecture we shall discuss the steady-state heat flow problem in a disk. For this, it is convenient to consider the Laplace equation in polar coordinates instead of rectangular coordinates.

Consider the steady-state heat conduction problem for a flat plate in the shape of a circular disk with the boundary curve $x^2 + y^2 = a^2$. In what follows we assume that the plate is isotropic; i.e., the flat surfaces are insulated, and that the temperature is known everywhere on the circular boundary. The temperature inside the disk is then a solution of the *Dirichlet problem* (see Figure 35.1) consisting of Laplace's equation in polar coordinates (see Problem 35.1)

$$\frac{\partial^2 u}{\partial r^2} + \frac{1}{r}\frac{\partial u}{\partial r} + \frac{1}{r^2}\frac{\partial^2 u}{\partial \theta^2} = 0, \quad 0 < r < a, \quad -\pi < \theta \leq \pi \tag{35.1}$$

and the boundary condition

$$u(a, \theta) = f(\theta), \quad -\pi < \theta \leq \pi. \tag{35.2}$$

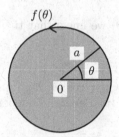

Figure 35.1

In problem (35.1), (35.2) we notice that $r = 0$ is not a physical boundary; rather we recognize it as a "mathematical boundary," and for a solution $u(r, \theta)$ to be physically meaningful we need to impose at $r = 0$ the *implicit boundary condition*

$$|u(0, \theta)| < \infty; \tag{35.3}$$

i.e., the solution remains bounded at the origin. We also wish to allow θ to assume any value rather than restrict it to the interval $-\pi < \theta \leq \pi$, and

R.P. Agarwal, D. O'Regan, *Ordinary and Partial Differential Equations*,
Universitext, DOI 10.1007/978-0-387-79146-3_35,
© Springer Science+Business Media, LLC 2009

hence we assume that $f(\theta)$, and consequently $u(r, \theta)$ to be periodic with period 2π. Thus, we also need the conditions

$$u(r, \pi) = u(r, -\pi), \quad 0 < r < a \tag{35.4}$$

$$\frac{\partial u}{\partial \theta}(r, \pi) = \frac{\partial u}{\partial \theta}(r, -\pi), \quad 0 < r < a, \tag{35.5}$$

which are actually continuity requirements along the slit $\theta = \pi$. The problem (35.1)–(35.5) is often called as an *interior problem*.

To solve (35.1)–(35.5) we assume $u(r, \theta) = R(r)\Theta(\theta) \neq 0$. Clearly, equation (35.1) becomes

$$R''\Theta + \frac{1}{r}R'\Theta + \frac{1}{r^2}R\Theta'' = 0,$$

which gives

$$\frac{r^2 R'' + r R'}{R} = -\frac{\Theta''}{\Theta} = \lambda$$

and hence

$$\Theta'' + \lambda\Theta = 0, \quad -\pi < \theta \leq \pi \tag{35.6}$$

and

$$r^2 R'' + r R' - \lambda R = 0, \quad 0 < r < a. \tag{35.7}$$

Now (35.4) implies

$$\Theta(-\pi) = \Theta(\pi), \tag{35.8}$$

whereas (35.5) gives

$$\Theta'(-\pi) = \Theta'(\pi). \tag{35.9}$$

For (35.6), (35.8), (35.9) we know that the eigenvalues and eigenfunctions are

$$\begin{cases} \lambda_0 = 0, \quad \Theta_0 = 1 \\ \lambda_n = n^2 \ (n \geq 1), \quad \Theta_n = \cos n\theta \quad \text{and} \quad \sin n\theta \\ \qquad\qquad\qquad\qquad \text{(two linearly independent eigenfunctions).} \end{cases} \tag{35.10}$$

Next for $\lambda = 0$, equation (35.7) is

$$r^2 R_0'' + r R_0' = 0 \tag{35.11}$$

for which the auxiliary equation is $m(m-1) + m = 0$, or $m^2 = 0$ and hence $m = 0, 0$. Thus, two linearly independent solutions of (35.11) are 1 and $\ln r$. However, in view of (35.3) the solution $\ln r$ is discarded because of its behavior at $r = 0$. Thus, we have

$$R_0(r) = 1. \tag{35.12}$$

For $\lambda = \lambda_n = n^2$, equation (35.7) is

$$r^2 R_n'' + r R_n' - n^2 R_n = 0 \tag{35.13}$$

for which the auxiliary equation is $m(m-1)+m-n^2 = 0$, or $m^2 - n^2 = 0$ and hence $m = n, -n$. Thus, two linearly independent solutions of (35.13) are r^n and r^{-n}. However, since the solution r^{-n} is unbounded as r approaches zero, to fulfill condition (35.3) we need to discard it. Thus, we obtain

$$R_n(r) = r^n. \tag{35.14}$$

Therefore, the solution $u(r, \theta)$ can be written as

$$u(r, \theta) = \frac{a_0}{2} + \sum_{n=1}^{\infty} r^n (a_n \cos n\theta + b_n \sin n\theta). \tag{35.15}$$

This solution satisfies (35.2) if

$$u(a, \theta) = f(\theta) = \frac{a_0}{2} + \sum_{n=1}^{\infty} a^n (a_n \cos n\theta + b_n \sin n\theta). \tag{35.16}$$

Clearly, (35.16) is a Fourier trigonometric series, and hence

$$
\begin{aligned}
a_n &= \frac{1}{\pi a^n} \int_{-\pi}^{\pi} f(\phi) \cos n\phi \, d\phi, \quad n \geq 0 \\
b_n &= \frac{1}{\pi a^n} \int_{-\pi}^{\pi} f(\phi) \sin n\phi \, d\phi, \quad n \geq 1.
\end{aligned}
\tag{35.17}
$$

In conclusion the solution of (35.1)–(35.5) can be written as (35.15), where a_n and b_n are given in (35.17).

As an example we shall solve (35.1)–(35.5) with

$$f(\theta) = \begin{cases} 0, & -\pi < \theta < -\pi/2 \\ 1, & -\pi/2 < \theta < \pi/2 \\ 0, & \pi/2 < \theta < \pi. \end{cases}$$

From (35.17), we have

$$a_0 = \frac{1}{\pi} \int_{-\pi/2}^{\pi/2} 1 \cdot d\phi = \frac{1}{\pi} \cdot \pi = 1$$

$$
\begin{aligned}
a_n &= \frac{1}{\pi a^n} \int_{-\pi/2}^{\pi/2} 1 \cdot \cos n\phi \, d\phi = \frac{1}{\pi a^n} \left. \frac{\sin n\phi}{n} \right|_{-\pi/2}^{\pi/2} \\
&= \frac{1}{n\pi a^n} \left[\sin \frac{n\pi}{2} - \sin \left(-\frac{n\pi}{2} \right) \right] = \frac{2 \sin(n\pi/2)}{n\pi a^n}
\end{aligned}
$$

$$b_n = \frac{1}{\pi a^n} \int_{-\pi/2}^{\pi/2} 1 \cdot \sin n\phi \, d\phi = \frac{1}{n\pi a^n} \cos n\phi \Big|_{-\pi/2}^{\pi/2} = 0.$$

Hence, the solution of (35.1)–(35.5) in this particular case is

$$u(r,\theta) = \frac{1}{2} + \sum_{n=1}^{\infty} \frac{2\sin(n\pi/2)}{n\pi} \frac{r^n}{a^n} \cos n\theta.$$

Now in (35.15) we substitute the coefficients a_n, b_n from (35.17), interchange the order of summation and integration, and use some elementary identities, to get

$$
\begin{aligned}
u(r,\theta) &= \frac{1}{2\pi} \int_{-\pi}^{\pi} f(\phi) d\phi + \frac{1}{\pi} \sum_{n=1}^{\infty} \frac{r^n}{a^n} \left[\cos n\theta \left(\int_{-\pi}^{\pi} f(\phi) \cos n\phi \, d\phi \right) \right. \\
&\quad \left. + \sin n\theta \left(\int_{-\pi}^{\pi} f(\phi) \sin n\phi \, d\phi \right) \right] \\
&= \frac{1}{\pi} \int_{-\pi}^{\pi} f(\phi) \left[\frac{1}{2} + \sum_{n=1}^{\infty} \frac{r^n}{a^n} (\cos n\theta \cos n\phi + \sin n\theta \sin n\phi) \right] d\phi \\
&= \frac{1}{\pi} \int_{-\pi}^{\pi} f(\phi) \left[\frac{1}{2} + \sum_{n=1}^{\infty} \frac{r^n}{a^n} \cos n(\theta - \phi) \right] d\phi \\
&= \frac{1}{\pi} \int_{-\pi}^{\pi} f(\phi) \left[\frac{1}{2} + \sum_{n=1}^{\infty} \frac{r^n}{a^n} \frac{1}{2} \left(e^{n(\theta-\phi)i} + e^{-n(\theta-\phi)i} \right) \right] d\phi \\
&= \frac{1}{2\pi} \int_{-\pi}^{\pi} f(\phi) \left[1 + \sum_{n=1}^{\infty} \left\{ \left(\frac{r}{a} e^{(\theta-\phi)i} \right)^n + \left(\frac{r}{a} e^{-(\theta-\phi)i} \right)^n \right\} \right] d\phi.
\end{aligned}
$$

Now since $|e^{i\psi}| = 1$, for $r < a$ we can sum the geometric series, to obtain

$$u(r,\theta) = \frac{1}{2\pi} \int_{-\pi}^{\pi} f(\phi) \left[1 + \frac{\frac{r}{a} e^{(\theta-\phi)i}}{1 - \frac{r}{a} e^{(\theta-\phi)i}} + \frac{\frac{r}{a} e^{-(\theta-\phi)i}}{1 - \frac{r}{a} e^{-(\theta-\phi)i}} \right] d\phi,$$

which is the same as

$$u(r,\theta) = \frac{(a^2 - r^2)}{2\pi} \int_{-\pi}^{\pi} \frac{f(\phi)}{a^2 + r^2 - 2ra\cos(\theta - \phi)} d\phi, \quad r < a. \qquad (35.18)$$

This formula is called the *Poisson integral formula*. It shows that the temperature at any interior point (r, θ) of the disk of radius a may be obtained by integrating the boundary temperatures according to the formula (35.18). In particular, if $r = 0$, then the temperature at the center of the disk is

$$u(0,\theta) = \frac{1}{2\pi} \int_{-\pi}^{\pi} f(\phi) d\phi, \qquad (35.19)$$

i.e., the temperature at the center is the integral average of the boundary temperatures. This fact is called the *mean value theorem* and holds for all functions that satisfy Laplace's equation on the disk.

Now we shall find the solution of the Laplace equation (35.1) outside the disk $r = a$ (see Figure 35.2). For this, again we assume that the conditions (34.2), (35.4), (35.5) are satisfied, but the condition (35.3) has to be replaced by

$$\lim_{r\to\infty} |u(r,\theta)| < \infty. \qquad (35.20)$$

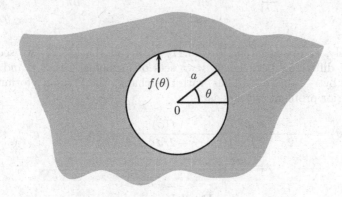

Figure 35.2

Clearly, for this *exterior problem* also all the steps remain the same as for the case $r < a$, except that the solution of (35.13) which satisfies the condition (35.20) is now r^{-n}. This change leads to the solution

$$u(r,\theta) = \frac{\alpha_0}{2} + \sum_{n=1}^{\infty} r^{-n}(\alpha_n \cos n\theta + \beta_n \sin n\theta), \qquad (35.21)$$

where

$$\alpha_n = \frac{a^n}{\pi} \int_{-\pi}^{\pi} f(\phi) \cos n\phi d\phi, \quad n \geq 0$$

$$\beta_n = \frac{a^n}{\pi} \int_{-\pi}^{\pi} f(\phi) \sin n\phi d\phi, \quad n \geq 1. \qquad (35.22)$$

As an example, we let $a = 1$, and

$$f(\theta) = \begin{cases} 1, & -\pi < \theta < 0 \\ \theta, & 0 < \theta < \pi. \end{cases}$$

Then, from Example 19.3, we have

$$a_0 = \alpha_0 = 1 + \frac{\pi}{2}, \quad a_n = \alpha_n = \frac{(-1)^n - 1}{\pi n^2}, \quad b_n = \beta_n = \frac{-1 + (1-\pi)(-1)^n}{n\pi}.$$

Thus, the solution of the interior problem can be written as

$$u(r, \theta) = \frac{1 + \frac{\pi}{2}}{2} + \sum_{n=1}^{\infty} r^n \left(\frac{(-1)^n - 1}{\pi n^2} \cos n\theta + \frac{-1 + (1-\pi)(-1)^n}{n\pi} \sin n\theta \right),$$
$$-\pi < \theta < \pi$$

whereas the solution of the exterior problem is

$$u(r, \theta) = \frac{1 + \frac{\pi}{2}}{2} + \sum_{n=1}^{\infty} r^{-n} \left(\frac{(-1)^n - 1}{\pi n^2} \cos n\theta + \frac{-1 + (1-\pi)(-1)^n}{n\pi} \sin n\theta \right),$$
$$-\pi < \theta < \pi.$$

Finally, comparing (35.15), (35.17) with (35.21), (35.22) we see that the only difference between the two sets of formulas is that r and a are replaced by r^{-1} and a^{-1}. Thus, with this change the Poisson's formula for the exterior problem appears as

$$u(r, \theta) = \frac{(r^2 - a^2)}{2\pi} \int_{-\pi}^{\pi} \frac{f(\phi)}{a^2 + r^2 - 2ra\cos(\theta - \phi)} d\phi, \quad r > a. \qquad (35.23)$$

Problems

35.1. Make the change of variables $x = r\cos\theta$, $y = r\sin\theta$ to show that Laplace's equation (34.4) in rectangular coordinates becomes

$$\frac{\partial^2 u}{\partial r^2} + \frac{1}{r}\frac{\partial u}{\partial r} + \frac{1}{r^2}\frac{\partial^2 u}{\partial \theta^2} = 0$$

in polar coordinates.

35.2. A circular plate of unit radius, whose faces are insulated, has upper half of its boundary kept at constant temperature T_1 and the lower half at constant temperature T_2. Find the steady-state temperature of the plate.

35.3. Solve the Dirichlet problem (35.1)–(35.5) when

(i) $f(\theta) = \frac{1}{2}(1 + \cos\theta), \quad -\pi < \theta < \pi$

(ii) $f(\theta) = \frac{1}{2}(1 + \cos^3\theta), \quad -\pi < \theta < \pi$

(iii) $f(\theta) = |\theta|, \quad -\pi < \theta < \pi$

(iv) $f(\theta) = \begin{cases} \cos\theta, & -\pi/2 < \theta < \pi/2 \\ 0, & \text{otherwise.} \end{cases}$

35.4. Show that a necessary condition for the existence of a solution to the Neumann problem (35.1),

$$\frac{\partial u}{\partial r}(a, \theta) = f(\theta), \quad -\pi < \theta \leq \pi \tag{35.25}$$

is that

$$\int_{-\pi}^{\pi} f(\phi)d\phi = 0,$$

i.e., the mean value of the normal derivative on the boundary is zero.

35.5. Solve the Laplace equation (35.1) in the wedge with three sides $\theta = 0$, $\theta = \beta$, and $r = a$ (see Figure 35.3) and the boundary conditions $u(r, 0) = 0 = u(r, \beta)$, $0 < r < a$, and (35.2) for $0 < \theta < \beta$.

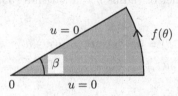

Figure 35.3

35.6. Solve the same problem as in Problem 35.5 with condition (35.2) replaced by the Neumann condition (35.25) for $0 < \theta < \beta$.

35.7. The diameter of a semi-circular plate of radius a is kept at $0°C$ and the temperature at the semi-circular boundary at $T°C$. Show that the steady–state temperature in the plate is given by

$$u(r, \theta) = \frac{4T}{\pi} \sum_{n=1}^{\infty} \frac{1}{2n-1} \left(\frac{r}{a}\right)^{2n-1} \sin(2n-1)\theta.$$

35.8. A semi-circular plate of radius a has its circumference kept at temperature $k\theta(\pi - \theta)$, while the boundary diameter is kept at zero temperature. Find the steady-state temperature distribution $u(r, \theta)$ of the plate, assuming the lateral surfaces of the plate to be insulated.

35.9. Solve the Laplace equation (35.1) in the annulus $0 < a^2 < x^2 + y^2 < b^2$ (see Figure 35.4) with the Dirichlet conditions $u(a, \theta) = f(\theta)$, $u(b, \theta) = g(\theta)$, $-\pi < \theta < \pi$.

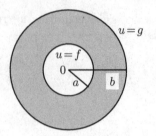

Figure 35.4

Further, show that in the particular case $f(\theta) = T_0$, $g(\theta) = T_1$ the solution reduces to

$$u(r, \theta) = T_0 + \frac{\ln r/a}{\ln b/a}(T_1 - T_0).$$

35.10.   The velocity potential function $u(r, \theta)$ for steady flow of an ideal fluid around a cylinder of radius $r = a$ satisfies (35.1) for $r > a$ with the boundary conditions

$$u_r(a, \theta) = 0, \quad u(r, \theta) = u(r, -\theta)$$
$$\lim_{r \to \infty} [u(r, \theta) - U_0 r \cos \theta] = 0.$$

Find its solution and the components of the velocity.

35.11.   From the real part of the solution of Laplace's equation in two independent variables

$$f(x + iy) = \frac{ae^{i\phi} + x + iy}{ae^{i\phi} - (x + iy)}$$

show that *Poisson's integral*

$$\frac{a^2 - r^2}{2\pi} \int_0^{2\pi} \frac{V(\phi)}{a^2 + r^2 - 2ar \cos(\theta - \phi)} d\phi,$$

where $x = r \cos \theta$, $y = r \sin \theta$ and V is an arbitrary function, is a solution.

Answers or Hints

35.1. Verify directly.

35.2. $\frac{T_1+T_2}{2} + \frac{2}{\pi}(T_1 - T_2) \sum_{n=1}^{\infty} \frac{1}{2n-1} r^{2n-1} \sin(2n - 1)\theta$
$= \frac{T_1+T_2}{2} + \frac{T_1-T_2}{\pi} \tan^{-1}\left(\frac{2r \sin \theta}{1-r^2}\right)$.

35.3. (i) $\frac{1}{2}\left(1 + \frac{r}{a}\cos\theta\right)$ (ii) $\frac{1}{2}\left(1 + \frac{3}{4}\frac{r}{a}\cos\theta + \frac{1}{4}\frac{r^3}{a^3}\cos 3\theta\right)$ (iii) $\frac{\pi}{2} + \sum_{n=1}^{\infty}\frac{2((-1)^n - 1)}{\pi n^2}\frac{r^n}{a^n}\cos n\theta$ (iv) $\frac{1}{\pi} + \frac{1}{2a}\cos\theta + \sum_{n=2}^{\infty}\frac{2\sin(n-1)\pi/2}{\pi(n^2-1)}\frac{r^n}{a^n}\cos n\theta$.

35.4. Use Green's theorem $\iint_S (f\Delta_2 g - g\Delta_2 f)dS = \int_\Gamma \left(f\frac{\partial g}{\partial n} - g\frac{\partial f}{\partial n}\right)ds$.

35.5. $u(r,\theta) = \sum_{n=1}^{\infty} A_n r^{n\pi/\beta}\sin\frac{n\pi}{\beta}\theta$, $A_n = \frac{2}{\beta}a^{-n\pi/\beta}\int_0^\beta f(\phi)\sin\frac{n\pi}{\beta}\phi d\phi$.

35.6. $u(r,\theta) = \sum_{n=1}^{\infty} A_n r^{n\pi/\beta}\sin\frac{n\pi}{\beta}\theta$, $A_n = \frac{2}{n\pi}a^{1-n\pi/\beta}\int_0^\beta f(\phi)\sin\frac{n\pi}{\beta}\phi d\phi$.

35.7. Use Problem 35.5.

35.8. $u(r,\theta) = \frac{8k}{\pi}\sum_{n=1}^{\infty}\frac{1}{(2n-1)^3}\left(\frac{r}{a}\right)^{2n-1}\sin(2n-1)\theta$.

35.9. $u(r,\theta) = \frac{1}{2}(c_0 + d_0\ln r) + \sum_{n=1}^{\infty}(a_n r^n + d_n r^{-n})(A_n\cos n\theta + B_n\sin n\theta)$, where the unknowns are determined by using the boundary conditions.

35.10. $u(r,\theta) = \frac{U_0}{r}(r^2 + a^2)\cos\theta$, $u_x = \frac{U_0}{r^2}(r^2 - a^2\cos 2\theta)$, $u_y = -\frac{U_0}{r^2}a^2\sin 2\theta$.

35.11. Verify directly.

Lecture 36
Two-Dimensional Heat Equation

In this lecture we shall use the method of separation of variables to find the temperature distribution of rectangular and circular plates in the transient state.

Suppose that for a thin rectangular plate which occupies the plane region $0 \leq x \leq a$, $0 \leq y \leq b$, the top and bottom faces are insulated, and that its four edges are held at zero temperature. If the plate has the initial temperature function $u(x, y, 0) = f(x, y)$, then in the transient state its temperature function $u(x, y, t)$ is the solution of the following initial-boundary value problem (see Lecture 34):

$$u_t = c^2 \left(u_{xx} + u_{yy} \right), \quad 0 < x < a, \quad 0 < y < b, \quad t > 0, \quad c > 0 \quad (36.1)$$

$$u(x, y, 0) = f(x, y), \quad 0 < x < a, \quad 0 < y < b \quad (36.2)$$

$$u(x, 0, t) = 0, \quad u(x, b, t) = 0, \quad 0 < x < a, \quad t > 0 \quad (36.3)$$

$$u(0, y, t) = 0, \quad u(a, y, t) = 0, \quad 0 < y < b, \quad t > 0. \quad (36.4)$$

We shall find the solution of (36.1)–(36.4) by the method of separation of variables. For this, we assume that

$$u(x, y, t) = \phi(x, y) T(t) \neq 0 \quad (36.5)$$

so that

$$\left(\phi_{xx} + \phi_{yy} \right) T = \frac{1}{c^2} \phi T'$$

and hence on dividing by ϕT, we get

$$\left(\phi_{xx} + \phi_{yy} \right) \frac{1}{\phi} = \frac{T'}{c^2 T}.$$

Arguing as before the common value of the members of this equation must be a constant, which we take to be $-\lambda^2$. The equations that result are

$$\phi_{xx} + \phi_{yy} = -\lambda^2 \phi, \quad 0 < x < a, \quad 0 < y < b \quad (36.6)$$

$$T' + \lambda^2 c^2 T = 0, \quad t > 0. \quad (36.7)$$

R.P. Agarwal, D. O'Regan, *Ordinary and Partial Differential Equations*,
Universitext, DOI 10.1007/978-0-387-79146-3_36,
© Springer Science+Business Media, LLC 2009

It is also clear that (36.3) implies

$$\phi(x,0) = 0, \quad \phi(x,b) = 0, \tag{36.8}$$

whereas (36.4) implies

$$\phi(0,y) = 0, \quad \phi(a,y) = 0. \tag{36.9}$$

Next to solve (36.6), (36.8), (36.9) we assume $\phi(x,y) = X(x)Y(y)$, to obtain

$$\frac{X''(x)}{X(x)} + \frac{Y''(y)}{Y(y)} = -\lambda^2, \quad 0 < x < a, \quad 0 < y < b.$$

The sum of a function of x and a function of y can be constant only if these two functions are individually constant, i.e.,

$$\frac{X''}{X} = \text{constant}, \quad \frac{Y''}{Y} = \text{constant}.$$

We assume that

$$\frac{X''}{X} = -\mu^2 \quad \text{and} \quad \frac{Y''}{Y} = -\nu^2, \quad (\text{i.e., } \lambda^2 = \mu^2 + \nu^2)$$

so that

$$X'' + \mu^2 X = 0 \tag{36.10}$$

$$Y'' + \nu^2 Y = 0. \tag{36.11}$$

From (36.8) and (36.9) it also follows that

$$X(0) = 0, \quad X(a) = 0 \tag{36.12}$$

$$Y(0) = 0, \quad Y(b) = 0. \tag{36.13}$$

For (36.10), (36.12) the eigenvalues and eigenfunctions are

$$\mu_m^2 = \frac{m^2\pi^2}{a^2}, \quad X_m(x) = \sin\frac{m\pi x}{a}, \quad m = 1,2,\cdots. \tag{36.14}$$

Similarly, the eigenvalues and eigenfunctions of (36.11), (36.13) are

$$\nu_n^2 = \frac{n^2\pi^2}{b^2}, \quad Y_n(y) = \sin\frac{n\pi y}{b}, \quad n = 1,2,\cdots. \tag{36.15}$$

Notice that the indices n and m are independent. This means that ϕ will have a double index. Thus, a solution of (36.6), (36.8), (36.9) can be written as

$$\phi_{mn}(x,y) = X_m(x)Y_n(y), \quad \lambda_{mn}^2 = \mu_m^2 + \nu_n^2. \tag{36.16}$$

For $\lambda = \lambda_{mn}$, equation (36.7) takes the form

$$T'_{mn} + \lambda_{mn}^2 c^2 T_{mn} = 0,$$

which gives
$$T_{mn}(t) = \exp\left(-\lambda_{mn}^2 c^2 t\right). \tag{36.17}$$

Hence, in view of (36.5), (36.16), and (36.17), the solution $u(x,y,t)$ of (36.1) satisfying (36.3) and (36.4) can be written as

$$u(x,y,t) = \sum_{m=1}^{\infty}\sum_{n=1}^{\infty} a_{mn} \sin\frac{m\pi x}{a}\sin\frac{n\pi y}{b}\exp\left(-\lambda_{mn}^2 c^2 t\right), \tag{36.18}$$

where

$$\lambda_{mn}^2 = \frac{m^2\pi^2}{a^2} + \frac{n^2\pi^2}{b^2}. \tag{36.19}$$

Finally, this solution satisfies (36.2) if and only if

$$f(x,y) = u(x,y,0) = \sum_{m=1}^{\infty}\sum_{n=1}^{\infty} a_{mn}\sin\frac{m\pi x}{a}\sin\frac{n\pi y}{b}. \tag{36.20}$$

Multiplying (36.20) by $\sin(p\pi x/a)$ and integrating over $[0,a]$ gives

$$\int_0^a f(x,y)\sin\frac{p\pi x}{a}dx = \sum_{m=1}^{\infty}\sum_{n=1}^{\infty} a_{mn}\left(\int_0^a \sin\frac{m\pi x}{a}\sin\frac{p\pi x}{a}dx\right)\sin\frac{n\pi y}{b}.$$

However, since

$$\int_0^a \sin\frac{m\pi x}{a}\sin\frac{p\pi x}{a}dx = \begin{cases} a/2 & \text{if } m = p \\ 0 & \text{if } m \neq p \end{cases}$$

it follows that

$$\int_0^a f(x,y)\sin\frac{p\pi x}{a}dx = \sum_{n=1}^{\infty} a_{pn}\frac{a}{2}\sin\frac{n\pi y}{b}. \tag{36.21}$$

Now multiplying (36.21) by $\sin(q\pi y/b)$ and integrating over $[0,b]$, we find

$$\int_0^b\left(\int_0^a f(x,y)\sin\frac{p\pi x}{a}dx\right)\sin\frac{q\pi y}{b}dy = \sum_{n=1}^{\infty} a_{pn}\frac{a}{2}\int_0^b \sin\frac{n\pi y}{b}\sin\frac{q\pi y}{b}dy,$$

which is the same as

$$\int_0^a\int_0^b f(x,y)\sin\frac{p\pi x}{a}\sin\frac{q\pi y}{b}dxdy = \frac{a}{2}\cdot\frac{b}{2}a_{pq}$$

and hence

$$a_{mn} = \frac{4}{ab}\int_0^a\int_0^b f(x,y)\sin\frac{m\pi x}{a}\sin\frac{n\pi y}{b}dxdy. \tag{36.22}$$

Therefore, the solution of (36.1)–(36.4) appears as (36.18) where a_{mn} are given by (36.22).

In particular, we shall find the solution of (36.1)–(36.4) with $a = b = \pi$, $c = 1$ and $f(x, y) = xy$. Clearly, from (36.22) we have

$$
\begin{aligned}
a_{mn} &= \frac{4}{\pi^2} \int_0^\pi \int_0^\pi xy \sin mx \sin ny \, dx \, dy \\
&= \frac{4}{\pi^2} \cdot \frac{\pi(-1)^{m+1}}{m} \cdot \frac{\pi(-1)^{n+1}}{n} = \frac{4(-1)^{m+n}}{mn}.
\end{aligned}
$$

Hence, in view of (36.18), (36.19) the solution in this case is

$$
u(x, y, t) = \sum_{m=1}^\infty \sum_{n=1}^\infty \frac{4(-1)^{m+n}}{mn} \sin mx \sin ny \, e^{-(m^2+n^2)t}.
$$

Now we shall consider the heat equation (36.1) on a circular plate $0 < x^2 + y^2 < a^2$. From Problem 35.1 it follows that (36.1) in polar coordinates can be written as

$$
u_t = c^2 \left(u_{rr} + \frac{1}{r} u_r + \frac{1}{r^2} u_{\theta\theta} \right), \quad 0 < r < a, \quad -\pi < \theta \le \pi, \quad t > 0, \quad c > 0.
$$
(36.23)

We will find the solution of (36.23) subject to the initial and boundary conditions $u(r, \theta, 0) = f(r)$ and $u(a, \theta, t) = 0$. Since these two conditions are independent of θ, we must expect that u will also be independent of θ, i.e., $u = u(r, t)$. Thus, the problem we wish to solve is—

$$
u_t = c^2 \left(u_{rr} + \frac{1}{r} u_r \right), \quad 0 < r < a, \quad -\pi < \theta \le \theta, \quad t > 0, \quad c > 0 \quad (36.24)
$$

$$
u(r, 0) = f(r), \quad 0 < r < a \tag{36.25}
$$

$$
u(a, t) = 0, \quad t > 0 \tag{36.26}
$$

$$
|u(0, t)| < \infty, \quad t > 0. \tag{36.27}
$$

Let $u = u(r, t) = R(r)T(t) \ne 0$ in equation (36.24), to obtain

$$
\frac{T'}{c^2 T} = \frac{R'' + \frac{1}{r} R'}{R} = -\lambda^2,
$$

which leads to the ordinary DEs

$$
rR'' + R' + r\lambda^2 R = 0 \tag{36.28}
$$

and

$$
T' + c^2 \lambda^2 T = 0. \tag{36.29}
$$

Note that (36.28) is the Bessel equation of order zero (see Problem 9.5), and hence its solution is

$$R(r) = AJ_0(\lambda r) + BJ^0(\lambda r). \tag{36.30}$$

In view of (36.26) and (36.27) we need this solution to satisfy

$$R(a) = 0 \tag{36.31}$$

$$|R(0)| \quad \text{bounded.} \tag{36.32}$$

However, since $J^0(\lambda r) \to \infty$ as $r \to 0$, condition (36.32) implies that in (36.30) the constant B must be zero. Now the condition (36.31) is satisfied provided $AJ_0(\lambda a) = 0$, i.e., λa should be a root of the equation $J_0(\alpha) = 0$. The function $J_0(\alpha)$ has infinitely many positive zeros, which we write as α_n, $n = 1, 2, \cdots$. Thus, the solution of (36.28), (36.31), (36.32) can be written as

$$R(r) = J_0(\lambda_n r), \quad \lambda_n = \frac{\alpha_n}{a}, \quad n = 1, 2, \cdots. \tag{36.33}$$

Now with $\lambda^2 = \lambda_n^2$ the solution of equation (36.29) appears as

$$T_n(t) = e^{-\lambda_n^2 c^2 t}. \tag{36.34}$$

Hence, the general solution of (36.24), (36.26), (36.27) is

$$u(r, t) = \sum_{n=1}^{\infty} A_n e^{-\lambda_n^2 c^2 t} J_0(\lambda_n r). \tag{36.35}$$

This solution satisfies the condition (36.25) if and only if

$$f(r) = \sum_{n=1}^{\infty} A_n J_0(\lambda_n r). \tag{36.36}$$

To determine the unknowns A_n, $n = 1, 2, \cdots$ we recall the orthogonality of the Bessel functions. We multiply (36.36) by $J_0(\lambda_m r)r$ and integrate over 0 to a, to obtain

$$\int_0^a f(r)J_0(\lambda_m r)rdr = A_m \int_0^a J_0^2(\lambda_m r)rdr;$$

and hence in view of (13.7), we have

$$A_n = \frac{\int_0^a f(r)J_0(\lambda_n r)rdr}{\int_0^a J_0^2(\lambda_n r)rdr} = \frac{2\int_0^a f(r)J_0(\lambda_n r)rdr}{a^2 J_1^2(\lambda_n a)}, \quad n = 1, 2, \cdots. \tag{36.37}$$

In conclusion, the series (36.35) where A_n given by (36.37) is the solution of the initial-boundary value problem (36.24)–(36.27).

When $f(r) = u_0$, we can use Problem 9.2(v), to get

$$A_n = \frac{2u_0}{\lambda_n a J_1(\lambda_n a)};$$

and hence in this particular case the solution (36.35) reduces to

$$u(r,t) = 2u_0 \sum_{n=1}^{\infty} \frac{1}{\lambda_n a J_1(\lambda_n a)} e^{-\lambda_n^2 c^2 t} J_0(\lambda_n r).$$

Problems

36.1. Find the solution of the initial-boundary value problem (36.1)–(36.4) when

(i) $a = b = \pi$, $c = 1$ and $f(x,y) = \sin x \sin 2y$

(ii) $a = b = \pi$, $c = 1$ and $f(x,y) = x + y$.

36.2. Find the solution of the initial-boundary value problem (36.1), (36.2)

$$\begin{aligned}
u_y(x,0,t) = 0, \quad u_y(x,b,t) = 0, \quad 0 < x < a, \quad t > 0 \\
u_x(0,y,t) = 0, \quad u_x(a,y,t) = 0, \quad 0 < y < b, \quad t > 0.
\end{aligned} \qquad (36.38)$$

36.3. Find the solution of the initial-boundary value problem (36.1), (36.2)

$$\begin{aligned}
u(x,0,t) = 0, \quad u(x,b,t) = 0, \quad 0 < x < a, \quad t > 0 \\
u_x(0,y,t) = 0, \quad u_x(a,y,t) = 0, \quad 0 < y < b, \quad t > 0.
\end{aligned} \qquad (36.39)$$

36.4. Find the solution of the initial-boundary value problem (36.1), (36.2)

$$\begin{aligned}
u_y(x,0,t) = 0, \quad u_y(x,b,t) = 0, \quad 0 < x < a, \quad t > 0 \\
u(0,y,t) = 0, \quad u_x(a,y,t) = 0, \quad 0 < y < b, \quad t > 0.
\end{aligned} \qquad (36.40)$$

36.5. Find the solution of the initial-boundary value problem (36.1), (36.2)

$$\begin{aligned}
u(x,0,t) = 0, \quad u_y(x,b,t) = 0, \quad 0 < x < a, \quad t > 0 \\
u(0,y,t) = 0, \quad u_x(a,y,t) = 0, \quad 0 < y < b, \quad t > 0.
\end{aligned} \qquad (36.41)$$

36.6. Find the solution of the initial-boundary value problem (36.24), (36.25), (36.27), and $u_r(a,t) = 0$. In particular, show that when $f(r) = u_0$, then $u(r,t) \equiv u_0$.

36.7. Assume that a circular plate of radius a has insulated faces and heat capacity s calories per degree per square centimeter. Find $u(r,t)$ by solving (36.24)–(36.27) when

$$f(r) = \begin{cases} \dfrac{q_0}{s\pi\epsilon^2}, & 0 < r < \epsilon \\ 0, & \epsilon < r < a. \end{cases}$$

Further, use the fact that $J_1(x)/x \to 1/2$ as $x \to 0$ to find the limiting solution as $\epsilon \to 0$.

36.8. Find the solution of the initial–boundary value problem (36.24), (36.25), (36.27), and $hu(a,t) + ku_r(a,t) = 0$, $h > 0$, $k > 0$. In particular, find the solution when $f(r) = u_0$.

36.9. Let $u(r,t)$ be the temperature in a thin circular plate whose edge, $r = 1$ is kept at temperature $u = 0$, and whose initial temperature is $u = 1$, when there is surface heat transfer from the circular faces to surroundings at temperature zero. The heat equation can then be written as

$$u_t = u_{rr} + \frac{1}{r}u_r - hu,$$

where h is a positive constant. Find the series expansion of $u(r,t)$.

Answers or Hints

36.1. (i) $u(x,y,t) = e^{-5t}\sin x \sin 2y$ (ii) $u(x,y,t) = \sum_{m=1}^{\infty}\sum_{n=1}^{\infty}\frac{4}{\pi mn} \times e^{-(m^2+n^2)t}\{-(-1)^m - (-1)^n + 2(-1)^{n+m}\}\sin mx \sin ny$.

36.2. $u(x,y,t) = \sum_{m=0}^{\infty}\sum_{n=0}^{\infty} a_{mn}\cos\frac{m\pi x}{a}\cos\frac{n\pi y}{b}\exp\left(-\lambda_{mn}^2 c^2 t\right)$, $\lambda_{mn}^2 = \frac{m^2\pi^2}{a^2} + \frac{n^2\pi^2}{b^2}$.

36.3. $u(x,y,t) = \sum_{m=0}^{\infty}\sum_{n=1}^{\infty} a_{mn}\cos\frac{m\pi x}{a}\sin\frac{n\pi y}{b}\exp\left(-\lambda_{mn}^2 c^2 t\right)$, $\lambda_{mn}^2 = \frac{m^2\pi^2}{a^2} + \frac{n^2\pi^2}{b^2}$.

36.4. $u(x,y,t) = \sum_{m=0}^{\infty}\sum_{n=0}^{\infty} a_{mn}\sin\frac{(2m+1)\pi x}{2a}\cos\frac{n\pi y}{b}\exp\left(-\lambda_{mn}^2 c^2 t\right)$, $\lambda_{mn}^2 = \frac{(2m+1)^2\pi^2}{4a^2} + \frac{n^2\pi^2}{b^2}$.

36.5. $u(x,y,t) = \sum_{m=0}^{\infty}\sum_{n=0}^{\infty} a_{mn}\sin\frac{(2m+1)\pi x}{2a}\sin\frac{(2n+1)\pi y}{2b}\exp\left(-\lambda_{mn}^2 c^2 t\right)$, $\lambda_{mn}^2 = \frac{(2m+1)^2\pi^2}{4a^2} + \frac{(2n+1)^2\pi^2}{4b^2}$.

36.6. $u(r,t) = B_0 + \sum_{n=1}^{\infty} B_n e^{-\lambda_n^2 c^2 t} J_0(\lambda_n r)$, where $\lambda_n = \alpha_n/a$, α_n is a positive root of $J_0'(\alpha) = 0$, $B_0 = \frac{2}{a^2}\int_0^a rf(r)dr$, $B_n = \dfrac{2\int_0^a f(r)J_0(\lambda_n r)rdr}{a^2 J_0^2(\lambda_n a)}$, use $J_1(\alpha_n) = -J_0'(\alpha_n) = 0$.

36.7. $\frac{2q_0}{s\pi a^2} \sum_{n=1}^{\infty} \frac{1}{J_1^2(\lambda_n a)} \frac{J_1(\lambda_n \epsilon)}{(\lambda_n \epsilon)} e^{-\lambda_n^2 c^2 t} J_0(\lambda_n r) \longrightarrow \frac{q_0}{s\pi a^2} \sum_{n=1}^{\infty} \frac{1}{J_1^2(\lambda_n a)} \times$
$e^{-\lambda_n^2 c^2 t} J_0(\lambda_n r)$ where $\lambda_n = \alpha_n/a$ and α_n is a positive root of $J_0(\alpha) = 0$.

36.8. $u(r,t) = \sum_{n=1}^{\infty} C_n e^{-\lambda_n^2 c^2 t} J_0(\lambda_n r)$, where $\lambda_n = \alpha_n/a$, α_n is a positive root of $HJ_0(\alpha) + \alpha J_0'(\alpha) = 0$, $H = ah/k$, $C_n = \frac{2\lambda_n^2 \int_0^a f(r) J_0(\lambda_n r) r \, dr}{(\lambda_n^2 a^2 + H^2) J_0^2(\lambda_n a)}$,
$u(r,t) = 2u_0 \sum_{n=1}^{\infty} \frac{\lambda_n a J_1(\lambda_n a)}{(\lambda_n^2 a^2 + H^2) J_0^2(\lambda_n a)} e^{-\lambda_n^2 c^2 t} J_0(\lambda_n r)$.

36.9. $u(r,t) = 2e^{-ht} \sum_{j=1}^{\infty} \frac{J_0(\lambda_j r)}{\lambda_j J_1(\lambda_j)} e^{-\lambda_j^2 t}$, where λ_j are the positive roots of $J_0(\lambda) = 0$.

Lecture 37
Two-Dimensional Wave Equation

Using the method of separation of variables in this lecture we shall find vertical displacements of thin membranes occupying rectangular and circular regions.

The vertical displacement $u(x, y, t)$ of a thin rectangular membrane, which is homogeneous, perfectly flexible, maintained in a state of uniform tension, and occupying the plane region $0 \leq x \leq a$, $0 \leq y \leq b$ satisfies the two-dimensional wave equation

$$u_{tt} = c^2(u_{xx} + u_{yy}), \quad 0 < x < a, \quad 0 < y < b, \quad c > 0, \qquad (37.1)$$

where the positive constant c depends on the tension and physical properties of the membrane. If all four edges of the membrane are fixed, and it is given the initial shape $f(x, y)$ and released with velocity $g(x, y)$, then $u(x, y, t)$ satisfies the initial conditions (36.2),

$$u_t(x, y, 0) = g(x, y), \quad 0 < x < a, \quad 0 < y < b \qquad (37.2)$$

and the boundary conditions (36.3), (36.4).

Following exactly as in Lecture 36, the solution of the initial-boundary value problem (37.1), (36.2), (37.2), (36.3), (36.4) can be written as

$$u(x, y, t) = \sum_{m=1}^{\infty} \sum_{n=1}^{\infty} (a_{mn} \cos \lambda_{mn} ct + b_{mn} \sin \lambda_{mn} ct) \sin \frac{m\pi x}{a} \sin \frac{n\pi y}{b},$$
$$(37.3)$$

where λ_{mn} and a_{mn} are the same as in (36.19) and (36.22), and

$$b_{mn} = \frac{4}{abc\lambda_{mn}} \int_0^a \int_0^b g(x, y) \sin \frac{m\pi x}{a} \sin \frac{n\pi y}{b} dxdy. \qquad (37.4)$$

Now we shall consider the motion of a vibrating circular membrane that is clamped along its edge. We assume that the center of the membrane is at the origin of a polar coordinate system and the edge of the membrane lies on the circle $r = a$. Let $u(r, \theta, t)$ represent the displacement of a point (r, θ) of the membrane at time t. Again we assume that the membrane in thin,

R.P. Agarwal, D. O'Regan, *Ordinary and Partial Differential Equations*,
Universitext, DOI 10.1007/978-0-387-79146-3_37,
© Springer Science+Business Media, LLC 2009

homogeneous, perfectly flexible, maintained in a state of uniform tension, and subject to no external forces. Under these assumptions the equation of motion of the membrane is

$$\frac{1}{r}\frac{\partial}{\partial r}\left(r\frac{\partial u}{\partial r}\right)+\frac{1}{r^2}\frac{\partial^2 u}{\partial \theta^2}=\frac{1}{c^2}\frac{\partial^2 u}{\partial t^2}, \quad 0<r<a, \quad -\pi<\theta\leq\pi, \quad t>0. \quad (37.5)$$

Since the membrane is clamped along its edge, we have

$$u(a,\theta,t)=0 \qquad (37.6)$$

for all θ and positive t.

We assume that the membrane is set into motion by displacing its equilibrium position. Since there are no external forces, we can assume that there are possible modes of vibration in which the motion of each point is periodic. A *normal mode of vibration* is one in which all points of the membrane vibrate with the same period and pass through their equilibrium positions at the same time. We shall search for normal modes of vibration by considering possible displacement function of the form $u(r,\theta,t) = v(r,\theta)\cos(\omega t + d)$, where ω and d are some constants.

Since the membrane is circular, the function v must be periodic in θ with period 2π. For simplicity, we assume that $v(r,\theta) = R(r)\cos n\theta$, where n is a nonnegative integer. Thus, it follows that

$$u(r,\theta,t) = R(r)\cos(n\theta)\cos(\omega t + d).$$

A substitution of this choice of u into (37.5) and (37.6) yields

$$r^2\frac{d^2 R}{dr^2}+r\frac{dR}{dr}+\left[\left(\frac{\omega}{c}\right)^2 r^2-n^2\right]R=0, \quad R(a)=0. \qquad (37.7)$$

From the considerations in Lecture 9, the general solution of the Bessel DE in (37.7) can be written as

$$R(r) = AJ_n\left(\frac{\omega}{c}r\right)+BJ_{-n}\left(\frac{\omega}{c}r\right), \qquad (37.8)$$

where A and B are arbitrary constants. Clearly, from the physical reasons the displacement at the origin should be bounded; however, since $\lim_{r\to 0}|J_{-n}(\omega r/c)| \to \infty$, we must have $B = 0$. Finally, the condition $R(a) = 0$ is satisfied provided

$$0 = R(a) = AJ_n\left(\frac{\omega}{c}a\right).$$

Thus, the constant $\omega = cb_{n,p}/a$, where $b_{n,p}$ is a root of $J_n(x)$. Hence, any function of the form

$$u(r,\theta,t) = AJ_n\left(\frac{b_{n,p}}{a}r\right)\cos(n\theta)\cos\left(\frac{cb_{n,p}}{a}t+d\right)$$

gives a normal mode of vibration for the circular membrane.

Now we shall consider the vibrations of a circular membrane governed by the initial-boundary value problem (37.5),

$$u(r, \theta, 0) = f(r, \theta), \quad 0 < r < a, \quad -\pi < \theta \leq \pi \tag{37.9}$$

$$\frac{\partial u}{\partial t}(r, \theta, 0) = g(r, \theta), \quad 0 < r < a, \quad -\pi < \theta \leq \pi \tag{37.10}$$

$$u(a, \theta, t) = 0, \quad t > 0, \quad -\pi < \theta \leq \pi \tag{37.11}$$

$$|u(0, \theta, t)| < \infty, \quad t > 0, \quad -\pi < \theta \leq \pi \tag{37.12}$$

$$u(r, -\pi, t) = u(r, \pi, t), \quad 0 < r < a, \quad t > 0 \tag{37.13}$$

$$\frac{\partial u}{\partial \theta}(r, -\pi, t) = \frac{\partial u}{\partial \theta}(r, \pi, t), \quad 0 < r < a, \quad t > 0. \tag{37.14}$$

Clearly, this problem is a two-dimensional analog of (35.1)–(35.5).

We assume that $u(r, \theta, t)$ has the product form

$$u(r, \theta, t) = \phi(r, \theta)T(t) \neq 0, \tag{37.15}$$

which leads to the equations

$$\frac{1}{r}\frac{\partial}{\partial r}\left(r\frac{\partial \phi}{\partial r}\right) + \frac{1}{r^2}\frac{\partial^2 \phi}{\partial \theta^2} = -\lambda^2\phi, \quad 0 < r < a, \quad -\pi < \theta \leq \pi \tag{37.16}$$

and

$$T'' + \lambda^2 c^2 T = 0, \quad t > 0. \tag{37.17}$$

Next we assume that $\phi(r, \theta) = R(r)\Theta(\theta) \neq 0$, so that (37.16) takes the form

$$\frac{1}{r}(rR')'\Theta + \frac{1}{r^2}R\Theta'' = -\lambda^2 R\Theta.$$

In this equation the variables can be separated if we multiply by r^2 and divide it by $R\Theta$. Indeed, we get

$$\frac{r(rR')'}{R} + \lambda^2 r^2 = -\frac{\Theta''}{\Theta} = \mu^2,$$

which gives two differential equations

$$\Theta'' + \mu^2\Theta = 0, \quad -\pi < \theta \leq \pi \tag{37.18}$$

and

$$(rR')' - \frac{\mu^2}{r}R + \lambda^2 rR = 0, \quad 0 < r < a. \tag{37.19}$$

Clearly, in view of (37.13) and (37.14) we need to solve (37.18) with the boundary conditions

$$\Theta(-\pi) = \Theta(\pi) \tag{37.20}$$

$$\Theta'(-\pi) = \Theta'(\pi). \tag{37.21}$$

Further, equation (37.19) can be written as

$$rR'' + R' - \frac{\mu^2}{r}R + \lambda^2 rR = 0,$$

or

$$r^2 R'' + rR' + (\lambda^2 r^2 - \mu^2)R = 0. \tag{37.22}$$

Note that (37.22) is the Bessel equation (2.15). In view of (37.11) and (37.12) we need to solve (37.22) with the conditions

$$R(a) = 0 \tag{37.23}$$

$$|R(0)| \quad \text{bounded.} \tag{37.24}$$

For the problem (37.18), (37.20), (37.21) we know the eigenvalues and eigen-functions are

$$\begin{aligned} \mu_0^2 &= 0, \quad \Theta_0 = 1 \\ \mu_m^2 &= m^2, \quad \Theta_m(\theta) = \cos m\theta \quad \text{and} \quad \sin m\theta, \quad m = 1, 2, \cdots. \end{aligned} \tag{37.25}$$

From the considerations of Lecture 9 we note that for $\mu^2 = \mu_m^2 = m^2$ the solution of (37.22) can be written as

$$R_m(r) = \begin{cases} AJ_m(\lambda r) + BJ_{-m}(\lambda r) & \text{if} \quad m > 0 \\ AJ_0(\lambda r) + BJ^0(\lambda r) & \text{if} \quad m = 0. \end{cases}$$

However, since $J_{-m}(\lambda r)$ as well as $J^0(\lambda r) \to \infty$ as $r \to 0$, in $R_m(r)$ the constant B must be zero. So, we find that

$$R_m(r) = J_m(\lambda r), \quad m = 0, 1, \cdots. \tag{37.26}$$

Now this solution satisfies (37.23) if and only if

$$R_m(a) = J_m(\lambda a) = 0,$$

i.e., λa must be a root of the equation $J_m(\alpha) = 0$. However, we know that for each m, $J_m(\alpha) = 0$ has an infinite number of roots which we write as $\alpha_{m1}, \alpha_{m2}, \cdots, \alpha_{mn}, \cdots$.

In conclusion, the solution of (37.19), (37.23), (37.24) appears as

$$R(r) = J_m(\lambda_{mn} r), \tag{37.27}$$

where

$$\lambda_{mn} = \frac{\alpha_{mn}}{a}, \quad m = 0, 1, 2, \cdots, \quad n = 1, 2, \cdots. \tag{37.28}$$

From (37.25) and (37.27) it is clear that $\phi(r, \theta)$ takes the form

$$
\begin{cases}
J_m(\lambda_{mn}r)\cos m\theta, \quad J_m(\lambda_{mn}r)\sin m\theta, \quad m = 1, 2, \cdots, \quad n = 1, 2, \cdots \\
\text{and} \\
J_0(\lambda_{0n}r).
\end{cases}
$$

$$(37.29)$$

Now for $\lambda^2 = \lambda_{mn}^2$ equation (37.17) can be written as

$$
T_{mn}'' + \lambda_{mn}^2 c^2 T_{mn} = 0
$$

for which solutions are $\cos(\lambda_{mn}ct)$ and $\sin(\lambda_{mn}ct)$. Thus, the solutions of (37.5) satisfying (37.11) – (37.14) appear as

$$
\begin{array}{ll}
J_0(\lambda_{0n}r)\cos(\lambda_{0n}ct), & J_0(\lambda_{0n}r)\sin(\lambda_{0n}ct) \\
J_m(\lambda_{mn}r)\cos m\theta\cos(\lambda_{mn}ct), & J_m(\lambda_{mn}r)\cos m\theta\sin(\lambda_{mn}ct) \\
J_m(\lambda_{mn}r)\sin m\theta\cos(\lambda_{mn}ct), & J_m(\lambda_{mn}r)\sin m\theta\sin(\lambda_{mn}ct).
\end{array}
$$

Hence, the general solution of (37.5), (37.11) – (37.14) is

$$
\begin{aligned}
u(r, \theta, t) \;=\; & \sum_n a_{0n} J_0(\lambda_{0n}r)\cos(\lambda_{0n}ct) \\
& + \sum_{m,n} a_{mn} J_m(\lambda_{mn}r)\cos m\theta\cos(\lambda_{mn}ct) \\
& + \sum_{m,n} b_{m,n} J_m(\lambda_{mn}r)\sin m\theta\cos(\lambda_{mn}ct) \\
& + \sum_n A_{0n} J_0(\lambda_{0n}r)\sin(\lambda_{0n}ct) \\
& + \sum_{m,n} A_{mn} J_m(\lambda_{mn}r)\cos m\theta\sin(\lambda_{mn}ct) \\
& + \sum_{m,n} B_{mn} J_m(\lambda_{mn}r)\sin m\theta\sin(\lambda_{mn}ct).
\end{aligned}
$$

$$(37.30)$$

This solution satisfies the condition (37.9) if and only if

$$
\begin{aligned}
f(r, \theta) \;=\; & \sum_n a_{0n} J_0(\lambda_{0n}r) + \sum_{m,n} a_{mn} J_m(\lambda_{mn}r)\cos m\theta \\
& + \sum_{m,n} b_{mn} J_m(\lambda_{mn}r)\sin m\theta, \quad 0 < r < a, \quad -\pi < \theta \le \pi.
\end{aligned}
$$

$$(37.31)$$

Now recalling the orthogonality of the Bessel functions and the set $\{1, \cos m\theta, \sin n\theta\}$, we can find unknowns a_{0n}, a_{mn}, b_{mn} from the above relation. For example, if we multiply (37.31) by $r J_0(\lambda_{0p}r)$ and integrate over

0 to a with respect to r, and integrate over $-\pi$ to π with respect to θ, we obtain

$$\int_{-\pi}^{\pi} \int_0^a f(r,\theta)J_0(\lambda_{0p}r)rdrd\theta = a_{0p}2\pi \int_0^a J_0^2(\lambda_{0p}r)rdr$$

and hence

$$a_{0n} = \frac{\int_{-\pi}^{\pi} \int_0^a f(r,\theta)J_0(\lambda_{0n}r)rdrd\theta}{2\pi \int_0^a J_0^2(\lambda_{0n}r)rdr}, \quad n = 1,2,\cdots.$$

Finally, we remark that the constants A_{mn}, $m = 0,1,2,\cdots$, $n = 1,2,\cdots$ and B_{mn}, $m = 1,2,\cdots$, $n = 1,2,\cdots$ can be calculated by using the condition (37.10).

In the particular case when the initial displacements are functions of r alone, from the symmetry it follows that u will be independent of θ, and then the problem (37.5), (37.9)–(37.14) simplifies to

$$\frac{1}{r}\frac{\partial}{\partial r}\left(r\frac{\partial u}{\partial r}\right) = \frac{1}{c^2}\frac{\partial^2 u}{\partial t^2}, \quad 0 < r < a, \quad t > 0 \tag{37.32}$$

$$u(r,0) = f(r), \quad 0 < r < a \tag{37.33}$$

$$\frac{\partial u}{\partial t}(r,0) = g(r), \quad 0 < r < a \tag{37.34}$$

$$u(a,t) = 0, \quad t > 0 \tag{37.35}$$

$$|u(0,t)| < \infty, \quad t > 0. \tag{37.36}$$

From the above considerations the solution of the problem (37.32)–(37.36) appears as

$$u(r,t) = \sum_n a_{0n}J_0(\lambda_{0n}r)\cos(\lambda_{0n}ct) + \sum_n A_{0n}J_0(\lambda_{0n}r)\sin(\lambda_{0n}ct), \tag{37.37}$$

where

$$a_{0n} = \frac{2\int_0^a f(r)J_0(\lambda_{0n}r)rdr}{a^2 J_1^2(\lambda_{0n}a)}, \quad A_{0n} = \frac{2\int_0^a g(r)J_0(\lambda_{0n}r)rdr}{\lambda_{0n}ca^2 J_1^2(\lambda_{0n}a)}, \quad n = 1,2,\cdots.$$

Problems

37.1. Find the solution of the initial-boundary value problem (37.1), (36.2), (37.2), (36.3), (36.4) when $f(x,y) = Txy(x-a)(y-b)$ and $g(x,y) = 0$.

37.2. Find the solution of the initial–boundary value problem (37.1), (36.2), (37.2), (36.3), (36.4) when $a = 2$, $b = 3$, $c = 3$,

$$f(x,y) = \begin{cases} xy & 0 \le x < 1, \ 0 \le y < 3/2 \\ x(3-y) & 0 \le x < 1, \ 3/2 \le y \le 3 \\ (2-x)y & 1 \le x \le 2, \ 0 \le y < 3/2 \\ (2-x)(3-y) & 1 \le x \le 2, \ 3/2 \le y \le 3 \end{cases}$$

and $g(x,y) = 0$.

37.3. Find the solution of the initial-boundary value problem (37.1), (36.2), (37.2), (36.38).

37.4. Find the solution of the initial-boundary value problem (37.1), (36.2), (37.2), (36.39).

37.5. Find the solution of the initial-boundary value problem (37.1), (36.2), (37.2), (36.40).

37.6. Find the solution of the initial-boundary value problem (37.1), (36.2), (37.2), (36.41).

37.7. Find the solution of the initial-boundary value problem (37.32)–(37.36), when

(i) $a = 1$, $f(r) = \begin{cases} 1, & 0 < r < 1/2 \\ 0, & 1/2 < r < 1 \end{cases}$ and $g(r) = 0$

(ii) $a = 1$, $f(r) = 0$, $g(r) = 1$.

37.8. Find the solution of the initial-boundary value problem (37.32)–(37.36) when $f(r) = 0$ and

$$g(r) = \begin{cases} \dfrac{P_0}{\rho \pi \epsilon^2}, & 0 < r < \epsilon \\ 0, & \epsilon < r < a. \end{cases}$$

Further, use the fact that $J_1(x)/x \to 1/2$ as $x \to 0$ to find the limiting solution as $\epsilon \to 0$.

Answers or Hints

37.1. $\dfrac{64 T a^2 b^2}{\pi^6} \sum_{m=0}^{\infty} \sum_{n=0}^{\infty} \dfrac{1}{(2m+1)^3 (2n+1)^3} \cos \lambda_{2m+1,2n+1} ct \sin \dfrac{(2m+1)\pi x}{a}$
$\times \sin \dfrac{(2n+1)\pi y}{b}$, where $\lambda_{mn} = \left[\left(\dfrac{m\pi}{a} \right)^2 + \left(\dfrac{n\pi}{b} \right)^2 \right]^{1/2}$.

37.2. $u(x,y,t) = \dfrac{96}{\pi^4} \sum_{m=1}^{\infty} \sum_{n=1}^{\infty} \dfrac{1}{m^2 n^2} \sin \dfrac{m\pi}{3} \sin \dfrac{n\pi}{2} \cos(3\lambda_{mn}t) \sin \dfrac{m\pi y}{3}$
$\times \sin \dfrac{n\pi x}{2}$, where $\lambda_{mn}^2 = (m\pi/3)^2 + (n\pi/2)^2$.

37.3. $u(x,y,t) = \sum_{m=0}^{\infty} \sum_{n=0}^{\infty} (a_{mn} \cos \lambda_{mn} ct + b_{mn} \sin \lambda_{mn} ct) \cos \frac{m\pi x}{a}$
$\times \cos \frac{n\pi y}{b}$, $\lambda_{mn}^2 = \frac{m^2\pi^2}{a^2} + \frac{n^2\pi^2}{b^2}$.

37.4. $u(x,y,t) = \sum_{m=0}^{\infty} \sum_{n=1}^{\infty} (a_{mn} \cos \lambda_{mn} ct + b_{mn} \sin \lambda_{mn} ct) \cos \frac{m\pi x}{a}$
$\times \sin \frac{n\pi y}{b}$, $\lambda_{mn}^2 = \frac{m^2\pi^2}{a^2} + \frac{n^2\pi^2}{b^2}$.

37.5. $u(x,y,t) = \sum_{m=0}^{\infty} \sum_{n=0}^{\infty} (a_{mn} \cos \lambda_{mn} ct + b_{mn} \sin \lambda_{mn} ct)$
$\times \sin \frac{(2m+1)\pi x}{2a} \cos \frac{n\pi y}{b}$, $\lambda_{mn}^2 = \frac{(2m+1)^2\pi^2}{4a^2} + \frac{n^2\pi^2}{b^2}$.

37.6. $u(x,y,t) = \sum_{m=0}^{\infty} \sum_{n=0}^{\infty} (a_{mn} \cos \lambda_{mn} ct + b_{mn} \sin \lambda_{mn} ct)$
$\times \sin \frac{(2m+1)\pi x}{2a} \sin \frac{(2n+1)\pi y}{2b}$, $\lambda_{mn}^2 = \frac{(2m+1)^2\pi^2}{4a^2} + \frac{(2n+1)^2\pi^2}{4b^2}$.

37.7. (i) $u(r,t) = \sum_{n=1}^{\infty} \frac{J_1(\lambda_{0n}/2)}{\lambda_{0n} J_1^2(\lambda_{0n})} J_0(\lambda_{0n} r) \cos(\lambda_{0n} ct)$

(ii) $u(r,t) = \frac{2}{c} \sum_{n=1}^{\infty} \frac{\sin(\lambda_{0n} ct)}{\lambda_{0n}^2 J_1(\lambda_{0n})} J_0(\lambda_{0n} r)$.

37.8. $\frac{2P_0 c}{(\rho c^2)\pi a} \sum_{n=1}^{\infty} \frac{1}{(\lambda_{0n} a) J_1^2(\lambda_{0n} a)} \frac{J_1(\lambda_{0n}\epsilon)}{(\lambda_{0n}\epsilon)} J_0(\lambda_{0n} r) \sin(\lambda_{0n} ct) \longrightarrow$
$\frac{P_0 c}{(\rho c^2)\pi a} \sum_{n=1}^{\infty} \frac{1}{(\lambda_{0n} a) J_1^2(\lambda_{0n} a)} J_0(\lambda_{0n} r) \sin(\lambda_{0n} ct)$ where $\lambda_{0n} = \alpha_{0n}/a$ and α_{0n}
is a positive root of $J_0(\alpha) = 0$.

Lecture 38
Laplace Equation
in Three Dimensions

The three-dimensional Laplace equation occurs in problems such as gravitation, steady-state temperature, electrostatic potential, magnetostatics, fluid flow, and so on. In this lecture we shall use the method of separation of variables to find the solution of the Laplace equation in a three-dimensional box, and in a circular cylinder.

If the stream lines are curves in space, i.e., the heat flow is three dimensional, then instead of (34.3) we arrive at the equation

$$\frac{\partial u}{\partial t} = c^2 \left(\frac{\partial^2 u}{\partial x^2} + \frac{\partial^2 u}{\partial y^2} + \frac{\partial^2 u}{\partial z^2} \right). \tag{38.1}$$

In the steady state this equation reduces to

$$\Delta_3 u = u_{xx} + u_{yy} + u_{zz} = 0, \tag{38.2}$$

which is the three-dimensional Laplace equation. First, we shall find the solution of (38.2) in the three-dimensional box $D = \{0 < x < a,\ 0 < y < b,\ 0 < z < c\}$ satisfying the boundary conditions on the six sides

$$
\begin{aligned}
u(0,y,z) &= f_1(y,z), & u(a,y,z) &= f_2(y,z), & 0 < y < b, & \quad 0 < z < c \\
u(x,0,z) &= g_1(x,z), & u(x,b,z) &= g_2(x,z), & 0 < x < a, & \quad 0 < z < c \\
u(x,y,0) &= h_1(x,y), & u(x,y,c) &= h_2(x,y), & 0 < x < a, & \quad 0 < y < b.
\end{aligned}
\tag{38.3}
$$

Clearly, the solution of this problem can be obtained by summing the solutions of six problems of the type (38.2),

$$
\begin{aligned}
u(0,y,z) &= 0, & u(a,y,z) &= 0, & 0 < y < b, & \quad 0 < z < c \\
u(x,0,z) &= 0, & u(x,b,z) &= 0, & 0 < x < a, & \quad 0 < z < c \\
u(x,y,0) &= h_1(x,y), & u(x,y,c) &= 0, & 0 < x < a, & \quad 0 < y < b.
\end{aligned}
\tag{38.4}
$$

As such, Problem (38.2), (38.4) could occur in finding the potential function inside a rectangular parallelepiped in which four lateral faces and the top are at potential zero and the potential on the bottom is a given function of x and y (see Figure 38.1).

R.P. Agarwal, D. O'Regan, *Ordinary and Partial Differential Equations*,
Universitext, DOI 10.1007/978-0-387-79146-3_38,
© Springer Science+Business Media, LLC 2009

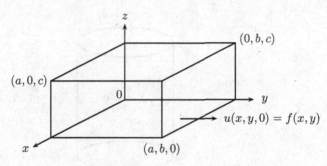

Figure 38.1

Following exactly as in Lectures 36 and 37, we separate the variables, i.e., assume that $u = u(x, y, z) = X(x)Y(y)Z(z)$, and use the homogeneous boundary conditions, to obtain

$$u(x, y, z) = \sum_{m=1}^{\infty} \sum_{n=1}^{\infty} a_{mn} \sinh \lambda_{mn}(c - z) \sin \frac{m\pi x}{a} \sin \frac{n\pi y}{b}, \qquad (38.5)$$

where λ_{mn} is the same as in (36.19), and

$$a_{mn} = \frac{4}{ab \sinh(c\lambda_{mn})} \int_0^a \int_0^b f(x, y) \sin \frac{m\pi x}{a} \sin \frac{n\pi y}{b} dx dy. \qquad (38.6)$$

In particular, when $a = b = c = \pi$, $f(x, y) = xy$ the solution of the problem (38.2), (38.4) can be written as

$$u(x, y, z) = \sum_{m=1}^{\infty} \sum_{n=1}^{\infty} \frac{4(-1)^{m+n}}{mn \sinh(\pi\lambda_{mn})} \sinh \lambda_{mn}(\pi - z) \sin mx \sin ny,$$

where $\lambda_{mn} = \sqrt{m^2 + n^2}$.

Now we shall find the steady-state temperature distribution in a solid cylinder made of homogeneous material. For this, we need to consider the Laplace equation in cylindrical coordinates (see Problem 39.6)

$$\frac{1}{r} \frac{\partial}{\partial r} \left(r \frac{\partial u}{\partial r} \right) + \frac{1}{r^2} \frac{\partial^2 u}{\partial \theta^2} + \frac{\partial^2 u}{\partial z^2} = 0, \quad 0 < r < a, \quad -\pi < \theta \leq \pi, \quad 0 < z < h \qquad (38.7)$$

with the boundary conditions

$$\begin{aligned}
u(r, \theta, 0) &= \alpha(r, \theta) && \text{bottom} \\
u(r, \theta, h) &= \beta(r, \theta) && \text{top} \\
u(a, \theta, z) &= \gamma(\theta, z) && \text{lateral side.}
\end{aligned} \qquad (38.8)$$

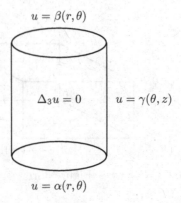

$$u = \beta(r, \theta)$$

$$\Delta_3 u = 0 \qquad u = \gamma(\theta, z)$$

$$u = \alpha(r, \theta)$$

Figure 38.2

Clearly, the solution of this problem can be obtained by summing the solutions of the following three problems:

$$(38.7), \quad u(r, \theta, 0) = 0, \quad u(r, \theta, h) = \beta(r, \theta), \quad u(a, \theta, z) = 0, \qquad (38.9)$$

$$(38.7), \quad u(r, \theta, 0) = \alpha(r, \theta), \quad u(r, \theta, h) = 0, \quad u(a, \theta, z) = 0 \quad (38.10)$$

and

$$(38.7), \quad u(r, \theta, 0) = 0, \quad u(r, \theta, h) = 0, \quad u(a, \theta, z) = \gamma(\theta, z). \qquad (38.11).$$

To find the solutions of these problems, we shall apply the method of separation of variables by assuming that u is a product of functions of r, θ, and z, i.e., $u = u(r, \theta, z) = R(r)\Theta(\theta)Z(z)$. Substituting the appropriate derivatives into the partial DE (38.7), we obtain

$$\frac{\Theta Z}{r} \frac{d}{dr}\left(r \frac{dR}{dr}\right) + \frac{RZ}{r^2} \frac{d^2\Theta}{d\theta^2} + R\Theta \frac{d^2 Z}{dz^2} = 0,$$

and now division of the above equation by $R\Theta Z/r^2$ yields

$$\frac{r}{R} \frac{d}{dr}\left(r \frac{dR}{dr}\right) + \frac{r^2}{Z} \frac{d^2 Z}{dz^2} = -\frac{1}{\Theta} \frac{d^2\Theta}{d\theta^2}.$$

Since the left-hand member of the last equation is independent of θ, the equation can be satisfied only if both members are equal to a constant. Hence,

$$-\frac{1}{\Theta} \frac{d^2\Theta}{d\theta^2} = m^2,$$

or

$$\Theta'' + m^2\Theta = 0. \qquad (38.12)$$

Note that we have chosen the separation constant m^2. This will force Θ (and u) to be periodic of period 2π in θ. This is in fact the desired situation in many applied problems. Clearly, the solutions of (38.12) are $\cos m\theta$ and $\sin m\theta$, $m = 0, 1, 2, \cdots$.

Now the second separation yields

$$\frac{1}{rR}\frac{d}{dr}\left(r\frac{dR}{dr}\right) - \frac{m^2}{r^2} = -\frac{1}{Z}\frac{d^2Z}{dz^2} = -\lambda.$$

Hence, we have

$$Z'' - \lambda Z = 0 \tag{38.13}$$

and

$$r^2 R'' + rR' + (\lambda r^2 - m^2)R = 0. \tag{38.14}$$

For the problem (38.9), we note that the condition $u(a, \theta, z) = 0$ leads to the boundary condition $R(a) = 0$. Further, as in Lecture 35 we need to impose at $r = 0$ the implicit boundary condition $|R(0)| < \infty$. The Bessel's equation (38.14) together with the boundary conditions

$$|R(0)| < \infty, \quad R(a) = 0 \tag{38.15}$$

has the eigenvalues $\lambda = \lambda_{mn}^2$ and the eigenfunctions $J_m(\lambda_{mn}r)$, $m = 0, 1, 2, \cdots$, $n = 1, 2, \cdots$, where λ_{mn} is the same as in (37.28). Now since $\lambda = \lambda_{mn}^2 > 0$, from the equation (38.13) and the condition $u(r, \theta, 0) = 0$ which implies $Z(0) = 0$, we have $Z(z) = \sinh(\lambda_{mn}z)$. Thus, by the principle of superposition the solution of (38.7) satisfying $u(r, \theta, 0) = 0$ and $u(a, \theta, z) = 0$ can be written as

$$\begin{aligned}
u(r, \theta, z) &= \sum_{m=0}^{\infty}\sum_{n=1}^{\infty} a_{mn}\sinh(\lambda_{mn}z)J_m(\lambda_{mn}r)\cos m\theta \\
&+ \sum_{m=1}^{\infty}\sum_{n=1}^{\infty} b_{mn}\sinh(\lambda_{mn}z)J_m(\lambda_{mn}r)\sin m\theta.
\end{aligned} \tag{38.16}$$

The unknowns a_{mn} and b_{mn} in (38.16) are determined by using the nonhomogeneous boundary condition $u(r, \theta, h) = \beta(r, \theta)$. For this, a Fourier series in θ and a Fourier–Bessel series in r are required.

The solution of the problem (38.10) can be obtained similarly, and appears as

$$\begin{aligned}
u(r, \theta, z) &= \sum_{m=0}^{\infty}\sum_{n=1}^{\infty} c_{mn}\sinh[\lambda_{mn}(h - z)]J_m(\lambda_{mn}r)\cos m\theta \\
&+ \sum_{m=1}^{\infty}\sum_{n=1}^{\infty} d_{mn}\sinh[\lambda_{mn}(h - z)]J_m(\lambda_{mn}r)\sin m\theta.
\end{aligned} \tag{38.17}$$

The unknowns c_{mn} and d_{mn} in (38.17) are obtained by using the nonhomogeneous boundary condition $u(r, \theta, 0) = \alpha(r, \theta)$.

Now we shall find the solution of the problem (38.11). We note that the conditions $u(r, \theta, 0) = u(r, \theta, h) = 0$ imply that $Z(0) = Z(h) = 0$, and hence the equation (38.13) has nontrivial solutions only when

$$\lambda = -\left(\frac{n\pi}{h}\right)^2, \quad n = 1, 2, \cdots \tag{38.18}$$

and $Z(z) = \sin n\pi z/h$. Substituting (38.18) in (39.14), we obtain a modified Bessel's DE

$$r^2 R'' + rR' - \left(\left(\frac{n\pi}{h}\right)^2 r^2 + m^2\right) R = 0. \tag{38.19}$$

This equation with the transformation $x = n\pi r/h$, $n = m$ is exactly the same as (9.15). Thus, the solutions of (38.19) are

$$I_m\left(\frac{n\pi}{h}r\right) \quad \text{and} \quad K_m\left(\frac{n\pi}{h}r\right).$$

Now since $|R(0)| < \infty$ and the solution K_m is singular at $r = 0$, we need to discard it. Hence, by the principle of superposition the solution of (38.7) satisfying $u(r, \theta, 0) = u(r, \theta, h) = 0$ can be written as

$$\begin{aligned}
u(r, \theta, z) &= \sum_{m=0}^{\infty} \sum_{n=1}^{\infty} e_{mn} I_m\left(\frac{n\pi}{h}r\right) \sin\frac{n\pi z}{h} \cos m\theta \\
&+ \sum_{m=1}^{\infty} \sum_{n=1}^{\infty} f_{mn} I_m\left(\frac{n\pi}{h}r\right) \sin\frac{n\pi z}{h} \sin m\theta.
\end{aligned} \tag{38.20}$$

Again the unknowns e_{mn} and f_{mn} in (38.20) are obtained by using the nonhomogeneous boundary condition $u(a, \theta, z) = \gamma(\theta, z)$.

In particular, if u is independent of z, then the temperature distribution in every circular cross section along the z-axis will be the same. In this case the problem is mathematically equivalent to the one we have discussed in Lecture 35. Similarly, if the temperature on the surface of the cylinder is prescribed in such a way that the functions α, β and γ are independent of θ, then the temperature inside the cylinder will also be independent of θ. In such a case, the problem (38.7), (38.8) reduces to

$$\frac{1}{r}\frac{\partial}{\partial r}\left(r\frac{\partial u}{\partial r}\right) + \frac{\partial^2 u}{\partial z^2} = 0, \quad 0 < r < a, \quad 0 < z < h \tag{38.21}$$

$$u(r, 0) = \alpha(r), \quad u(r, h) = \beta(r), \quad u(a, z) = \gamma(z);$$

and the solutions (38.16), (38.17) and (38.20), respectively, reduce to

$$u(r, z) = \sum_{n=1}^{\infty} a_n \sinh(\lambda_n z) J_0(\lambda_n r), \tag{38.22}$$

$$u(r, z) = \sum_{n=1}^{\infty} c_n \sinh[\lambda_n (h - z)] J_0(\lambda_n r) \qquad (38.23)$$

and

$$u(r, z) = \sum_{n=1}^{\infty} e_n I_0 \left(\frac{n\pi}{h} r\right) \sin \frac{n\pi z}{h}, \qquad (38.24)$$

where λ_n is the same as in (36.33).

In particular, when $\alpha(r) = u_0$, $\beta(r) = \gamma(r) = 0$, the solution (38.23) simplifies to

$$u(r, z) = \frac{2u_0}{a} \sum_{n=1}^{\infty} \frac{\sinh[\lambda_n (h - z)] J_0(\lambda_n r)}{\lambda_n J_1(\lambda_n a) \sinh(\lambda_n h)}.$$

Similarly, when $a = h = 1$, $\alpha(r) = 0$, $\beta(r) = 1 - r^2$, $\gamma(z) = 0$, the solution (38.22) becomes

$$u(r, z) = 8 \sum_{n=1}^{\infty} \frac{\sinh(\lambda_n z) J_0(\lambda_n r)}{\lambda_n^3 \sinh \lambda_n J_1(\lambda_n)}.$$

Lecture 39
Laplace Equation
in Three Dimensions (Cont'd.)

In this lecture we shall use the method of separation of variables to find the solutions of the Laplace equation in and outside a given sphere. We shall also discuss briefly Poisson's integral formulas.

From Problem 39.10 we know that the Laplace equation in spherical coordinates $x = r \sin\phi \cos\theta$, $y = r \sin\phi \sin\theta$, $z = r \cos\phi$, takes the form

$$\frac{1}{r^2} \frac{\partial}{\partial r} \left(r^2 \frac{\partial u}{\partial r} \right) + \frac{1}{r^2 \sin\phi} \frac{\partial}{\partial \phi} \left(\sin\phi \frac{\partial u}{\partial \phi} \right) + \frac{1}{r^2 \sin^2\phi} \frac{\partial^2 u}{\partial \theta^2} = 0. \qquad (39.1)$$

We assume that a solution of (39.1) can be written as $u = u(r, \theta, \phi) = R(r)\Theta(\theta)\Phi(\phi)$. Substituting the appropriate derivatives into the partial DE (39.1), we obtain

$$\frac{\Theta\Phi}{r^2} \frac{d}{dr} \left(r^2 \frac{dR}{dr} \right) + \frac{R\Theta}{r^2 \sin\phi} \frac{d}{d\phi} \left(\sin\phi \frac{d\Phi}{d\phi} \right) + \frac{R\Phi}{r^2 \sin^2\phi} \frac{d^2\Theta}{d\theta^2} = 0;$$

and now division of the above equation by $R\Theta\Phi/r^2 \sin^2\phi$ yields

$$\frac{\sin^2\phi}{R} \frac{d}{dr} \left(r^2 \frac{dR}{dr} \right) + \frac{\sin\phi}{\Phi} \frac{d}{d\phi} \left(\sin\phi \frac{d\Phi}{d\phi} \right) = -\frac{1}{\Theta} \frac{d^2\Theta}{d\theta^2}.$$

As earlier since the left-hand member of the last equation is independent of θ, the equation can be satisfied only if both members are equal to a constant. Hence,

$$-\frac{1}{\Theta} \frac{d^2\Theta}{d\theta^2} = m^2,$$

or

$$\Theta'' + m^2\Theta = 0. \qquad (39.2)$$

Note that again we have chosen the separation constant m^2. This will force Θ (and u) to be periodic of period 2π in θ. This is in fact the desired situation in many applied problems. Clearly, the solutions of (39.2) are $\cos m\theta$ and $\sin m\theta$, $m = 0, 1, 2, \cdots$.

The second separation yields

$$\frac{1}{R} \frac{d}{dr} \left(r^2 \frac{dR}{dr} \right) = -\left[\frac{1}{\Phi \sin\phi} \frac{d}{d\phi} \left(\sin\phi \frac{d\Phi}{d\phi} \right) - \frac{m^2}{\sin^2\phi} \right] = \lambda.$$

R.P. Agarwal, D. O'Regan, *Ordinary and Partial Differential Equations*,
Universitext, DOI 10.1007/978-0-387-79146-3_39,
© Springer Science+Business Media, LLC 2009

Hence, we have

$$\frac{d}{dr}\left(r^2\frac{dR}{dr}\right) - \lambda R = 0 \tag{39.3}$$

and

$$\frac{1}{\sin\phi\,d\phi}\frac{d}{d\phi}\left(\sin\phi\frac{d\Phi}{d\phi}\right) + \left(\lambda - \frac{m^2}{\sin^2\phi}\right)\Phi = 0. \tag{39.4}$$

Clearly, (39.3) can be written as a familiar Cauchy–Euler equation

$$r^2 R'' + 2rR - \lambda R = 0, \tag{39.5}$$

which has a solution $R = r^k$ provided $k^2 + k - \lambda = 0$. If we choose $k = n$, then $\lambda = n(n+1)$, and if we choose $k = -(n+1)$, then also $\lambda = n(n+1)$. Thus, with $\lambda = n(n+1)$, equation (39.5) has two linearly independent solutions r^n and $r^{-(n+1)}$.

In equation (39.4), we make the substitutions $x = \cos\phi$, $\Phi = y$, so that

$$\frac{d}{d\phi} = \frac{dx}{d\phi}\frac{d}{dx} = -\sin\phi\frac{d}{dx}$$

and hence

$$\begin{aligned}
\frac{d}{d\phi}\left(\sin\phi\frac{d\Phi}{d\phi}\right) &= -\sin\phi\frac{d}{dx}\left(\sin\phi\frac{dx}{d\phi}\frac{d\Phi}{dx}\right) \\
&= \sin\phi\frac{d}{dx}\left(\sin^2\phi\frac{dy}{dx}\right) \\
&= \sqrt{1-x^2}\frac{d}{dx}\left((1-x^2)\frac{dy}{dx}\right).
\end{aligned}$$

Thus equation (39.4) becomes

$$(1-x^2)y'' - 2xy' + \left[n(n+1) - \frac{m^2}{1-x^2}\right]y = 0, \tag{39.6}$$

which is exactly the same as Legendre's associated DE (7.20) with solutions $P_n^m(x)$ and $Q_n^m(x)$ defined in (7.21) and (7.22).

Now we shall find the solution of (39.1) in a sphere of radius a, satisfying the boundary condition (see Figure 39.1)

$$u(a,\theta,\phi) = f(\theta,\phi), \quad 0 < \theta < 2\pi, \quad 0 < \phi < \pi. \tag{39.7}$$

For this, as in earlier lectures discarding the solutions $r^{-(n+1)}$ and $Q_n^m(x)$, using Problem 7.12, and arranging the terms, we obtain

$$\begin{aligned}
u(r,\theta,\phi) = &\sum_{n=0}^{\infty}\left(\frac{r}{a}\right)^n\left[\frac{1}{2}a_{0n}P_n(\cos\phi)\right. \\
&\left. + \sum_{m=1}^{n}(a_{mn}\cos m\theta + b_{mn}\sin m\theta)P_n^m(\cos\phi)\right],
\end{aligned} \tag{39.8}$$

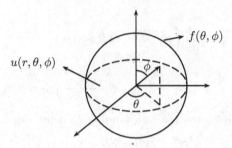

Figure 39.1

where

$$a_{mn} = \frac{(2n+1)(n-m)!}{2\pi(n+m)!} \int_0^{2\pi} \int_0^{\pi} f(\theta, \phi) P_n^m(\cos\phi) \cos m\theta \sin\phi \, d\phi \, d\theta,$$
(39.9)

$$b_{mn} = \frac{(2n+1)(n-m)!}{2\pi(n+m)!} \int_0^{2\pi} \int_0^{\pi} f(\theta, \phi) P_n^m(\cos\phi) \sin m\theta \sin\phi \, d\phi \, d\theta.$$
(39.10)

Next we shall consider the problem when $u(r, \theta, \phi) = u(r, \phi)$. This is not an unrealistic assumption, since many problems in electrostatics occur in this manner. In such a case, equation (39.1) reduces to

$$\frac{1}{r^2} \frac{\partial}{\partial r} \left(r^2 \frac{\partial u}{\partial r} \right) + \frac{1}{r^2 \sin\phi} \frac{\partial}{\partial \phi} \left(\sin\phi \frac{\partial u}{\partial \phi} \right) = 0.$$
(39.11)

We shall first consider (39.11) in a sphere of radius a, satisfying the boundary condition

$$u(a, \phi) = f(\cos\phi), \quad 0 < \phi < \pi.$$
(39.12)

From (39.8) the solution of the problem (39.11), (39.12) appears as

$$u(r, \phi) = \frac{1}{2} \sum_{n=0}^{\infty} \left(\frac{r}{a} \right)^n P_n(\cos\phi)(2n+1) \int_{-1}^{1} f(x) P_n(x) dx,$$
(39.13)

where in the integral we have used the substitution $x = \cos\phi$.

In particular, when

$$f(\cos\phi) = \begin{cases} V, & 0 < \phi < \pi/2 \ (0 < x < 1) \\ -V, & \pi/2 < \phi < \pi \ (-1 < x < 0) \end{cases}$$
(39.14)

we have

$$\int_{-1}^{1} f(x) P_n(x) dx = \begin{cases} 0 & \text{if } n \text{ is even} \\ 2V \int_0^1 P_n(x) dx & \text{if } n \text{ is odd,} \end{cases}$$

and hence (39.13) reduces to

$$u(r,\phi) = V\left[\frac{3}{2}\frac{r}{a}P_1(\cos\phi) - \frac{7}{8}\left(\frac{r}{a}\right)^3 P_3(\cos\phi) + \frac{11}{16}\left(\frac{r}{a}\right)^5 P_5(\cos\phi) + \cdots\right].$$

Similarly, to find the solution of (39.1) outside the sphere of radius a, satisfying the boundary condition (39.7), we discard the solutions r^n and $Q_n^m(x)$, of the equations (39.5) and (39.6), respectively, to obtain

$$u(r,\theta,\phi) = \sum_{n=0}^{\infty}\left(\frac{a}{r}\right)^{n+1}\left[\frac{1}{2}a_{0n}P_n(\cos\phi)\right.$$

$$\left. + \sum_{m=1}^{n}(a_{mn}\cos m\theta + b_{mn}\sin m\theta)P_n^m(\cos\phi)\right],\tag{39.15}$$

where the constants a_{mn} and b_{mn} remain exactly the same as in (39.9) and (39.10).

Thus, the solution of the problem (39.11), (39.12) outside the sphere of radius a, can be written as

$$u(r,\phi) = \frac{1}{2}\sum_{n=0}^{\infty}\left(\frac{a}{r}\right)^{n+1}P_n(\cos\phi)(2n+1)\int_{-1}^{1}f(x)P_n(x)dx.\tag{39.16}$$

This solution in particular, when $f(\cos\phi)$ is given by (39.14) becomes

$$u(r,\phi) = V\left[\frac{3}{2}\left(\frac{a}{r}\right)^2 P_1(\cos\phi) - \frac{7}{8}\left(\frac{a}{r}\right)^4 P_3(\cos\phi)\right.$$

$$\left. + \frac{11}{16}\left(\frac{a}{r}\right)^6 P_5(\cos\phi) + \cdots\right].$$

By adjusting the terms it can be shown that the solution (39.8) can be equivalently written as

$$u(r,\theta,\phi) = \frac{a(a^2 - r^2)}{4\pi}\int_0^{2\pi}\int_0^{\pi}\frac{f(\overline{\theta},\overline{\phi})\sin\overline{\phi}\,d\overline{\phi}\,d\overline{\theta}}{(r^2 + a^2 - 2ra\cos\psi)^{3/2}}, \quad r < a,\tag{39.17}$$

where

$$\cos\psi = \cos\phi\cos\overline{\phi} + \sin\phi\sin\overline{\phi}\cos(\theta - \overline{\theta}).$$

This is *Poisson's integral formula* in three dimensions for the interior problem (39.1), (39.7). In this formula $(r^2 + a^2 - 2ra\cos\psi)^{1/2}$ is the distance from the point (r,θ,ϕ) to the point $(a,\overline{\theta},\overline{\phi})$.

Similarly, the solution (39.15) of the exterior problem (39.1), (39.7) can be written as *Poisson's integral formula*:

$$u(r,\theta,\phi) = \frac{a(r^2 - a^2)}{4\pi}\int_0^{2\pi}\int_0^{\pi}\frac{f(\overline{\theta},\overline{\phi})\sin\overline{\phi}\,d\overline{\phi}\,d\overline{\theta}}{(r^2 + a^2 - 2ra\cos\psi)^{3/2}}, \quad r > a.\tag{39.18}$$

From (39.8) as well as (39.17) it is clear that

$$u(0, \theta, \phi) = \frac{1}{4\pi} \int_0^{2\pi} \int_0^\pi f(\overline{\theta}, \overline{\phi}) \sin \overline{\phi} d\overline{\phi} d\overline{\theta}.$$

Thus, the value of u at the center of a sphere is the mean value of $u(a, \theta, \phi)$ over the surface of the sphere. This fact is called the *mean value theorem* and holds for all functions that satisfy Laplace's equation on the sphere.

Similarly, from (39.15) as well as (39.18) it follows that when $r \to \infty$,

$$u(r, \theta, \phi) = \frac{a}{4\pi r} \int_0^{2\pi} \int_0^\pi f(\overline{\theta}, \overline{\phi}) \sin \overline{\phi} d\overline{\phi} d\overline{\theta} + O\left(\frac{1}{r^2}\right).$$

Finally, we consider a more general boundary condition for Laplace's equation (39.1) in the sphere $r < a$,

$$\cos \alpha \, \frac{\partial u}{\partial r}(a, \theta, \phi) + \sin \alpha \, u(a, \theta, \phi) = f(\theta, \phi). \tag{39.19}$$

Here α may assume values between 0 and $\pi/2$. Clearly, for $\alpha = \pi/2$, (39.19) reduces to the *Dirichlet condition* (39.7), and for $\alpha = 0$ it reduces to *Neumann condition* (39.20) which is considered in Problem 39.11. An intermediate value of α corresponds to a *mixed condition*. Physically this occurs when we have free radiation of heat according to Newton's law of cooling.

For the mixed problem also the solution remains the same as (39.8), and it satisfies the boundary condition (39.19) if and only if

$$
\begin{aligned}
f(\theta, \phi) &= \sum_{n=0}^{\infty} \left(\sin \alpha + \frac{n}{a} \cos \alpha\right) \left[\frac{1}{2} a_{0n} P_n(\cos \phi) \right. \\
&\quad + \left. \sum_{m=1}^{n} (a_{mn} \cos m\theta + b_{mn} \sin m\theta) P_n^m(\cos \phi)\right].
\end{aligned}
$$

Thus, the constants a_{mn} and b_{mn} remain exactly the same as in (39.9) and (39.10), except that each need to be divided by the factor $\left(\sin \alpha + \frac{n}{a} \cos \alpha\right)$.

In particular, when $\alpha = \pi/4$, $f(\theta, \phi) = f(\cos \phi) = (1/\sqrt{2}) \cos \phi$, the solution of the mixed problem (39.11), (39.19) in view of (39.13) can be written as

$$
\begin{aligned}
u(r, \phi) &= \frac{1}{2} \sum_{n=0}^{\infty} \left(\frac{r}{a}\right)^n P_n(\cos \phi) \frac{2n+1}{\frac{1}{\sqrt{2}} + \frac{n}{a}\frac{1}{\sqrt{2}}} \int_{-1}^1 \frac{1}{\sqrt{2}} x P_n(x) dx \\
&= \frac{1}{2} \frac{r}{a} P_1(\cos \phi) \frac{3}{1+\frac{1}{a}} \frac{2}{3} \\
&= \frac{r \cos \phi}{1+a}.
\end{aligned}
$$

Problems

39.1. Show that

$$u(x, y, z) = \int_0^{2\pi} f(x \cos t + y \sin t + iz, t)dt$$

is a solution of Laplace's equation (38.2).

39.2. Show that the gravitational potential due to the attraction of n particles

$$\mu \sum_{i=1}^n \frac{m_i}{r_i},$$

where m_i is the mass of the particle, (a_i, b_i, c_i) its coordinates, and $r_i^2 = (x - a_i)^2 + (y - b_i)^2 + (z - c_i)^2$, satisfies Laplace's equation (38.2).

39.3. Show that the potential of a body at an exterior point (x, y, z),

$$\mu \iiint \frac{\rho(a, b, c)da\,db\,dc}{[(x - a)^2 + (y - b)^2 + (z - c)^2]^{1/2}},$$

where ρ is the density and the integral is extended over the body, satisfies Laplace's equation (38.2).

39.4. Find the solution of the boundary value problem (38.2), (38.4) when $a = b = c = \pi$, $h_1(x, y) = \sin x \sin^3 y$.

39.5. Find the solution of the following initial-boundary value problem

$$u_t = u_{xx} + u_{yy} + u_{zz}, \quad 0 < x < \pi, \ 0 < y < \pi, \ 0 < z < \pi, \ t > 0$$
$$u(x, y, z, 0) = f(x, y, z)$$
$$u_x(0, y, z, t) = 0, \quad u_x(\pi, y, z, t) = 0$$
$$u_y(x, 0, z, t) = 0, \quad u_y(x, \pi, z, t) = 0$$
$$u_z(x, y, 0, t) = 0, \quad u_z(x, y, \pi, t) = 0.$$

In particular, find the solution when $f(x, y, z) = xyz$.

39.6. Show that in cylindrical coordinates, $x = r \cos \theta$, $y = r \sin \theta$, $z = z$, Laplace's equation (38.2) becomes

$$\frac{1}{r} \frac{\partial}{\partial r} \left(r \frac{\partial u}{\partial r} \right) + \frac{1}{r^2} \frac{\partial^2 u}{\partial \theta^2} + \frac{\partial^2 u}{\partial z^2} = 0.$$

39.7. Find the steady-state, bounded temperature distribution in the interior of a solid cylinder of radius a and height h, given that the temperature of the curved lateral surface is kept at zero, the base is insulated, and the top is kept at u_0.

39.8. Find the steady-state, bounded temperature distribution in the interior of a solid cylinder of radius 1 and height π, given that the temperature of the curved lateral surface is $\gamma(z) = z$, and the base and top are insulated.

39.9. Find the solution of the following problem:

$$u_{rr} + (1/r)u_r + u_{zz} = 0, \quad 0 < r < a, \quad z > 0$$
$$u(r,0) = \alpha(r), \quad 0 < r < a$$
$$|u(r,z)| < \infty, \quad 0 < r < a, \quad z > 0$$
$$u_r(a,z) + ku(a,z) = 0, \quad z > 0, \quad k > 0.$$

39.10. Show that in spherical coordinates, $x = r\sin\phi\cos\theta$, $y = r\sin\phi\sin\theta$, $z = r\cos\phi$, Laplace's equation (38.2) becomes

$$\frac{1}{r^2}\frac{\partial}{\partial r}\left(r^2\frac{\partial u}{\partial r}\right) + \frac{1}{r^2\sin\phi}\frac{\partial}{\partial\phi}\left(\sin\phi\frac{\partial u}{\partial\phi}\right) + \frac{1}{r^2\sin^2\phi}\frac{\partial^2 u}{\partial\theta^2} = 0.$$

39.11. Find the solution of (39.1) in a sphere of radius 1, satisfying the boundary condition

$$\frac{\partial u}{\partial r}(1,\theta,\phi) = f(\theta,\phi), \quad 0 < \theta < 2\pi, \quad 0 < \phi < \pi \qquad (39.20)$$

where $\int_0^{2\pi}\int_0^\pi f(\theta,\phi)\sin\phi d\phi d\theta = 0$.

39.12. A solid hemisphere of radius a has its plane face perfectly insulated, while the temperature of its curved surface is given by $f(\cos\phi)$. Find the steady-state, bounded temperature at any point in the interior of the hemisphere.

39.13. The temperature on the surface of a solid homogeneous sphere of radius a is prescribed by

$$u(a,\phi) = \begin{cases} V, & 0 < \phi < \pi/2 \\ 0, & \pi/2 < \phi < \pi. \end{cases}$$

Find the steady-state temperature distribution in and outside the sphere.

39.14. Find the steady-state temperature distribution $u(r,\phi)$ in a hollow sphere with its inner surface $r = a$ is kept at temperature $u(a,\phi) = f(\cos\phi)$, and its outer surface $r = b$ at $u(b,\phi) = 0$.

39.15. The Laplace equation can be used in mathematical modeling of the growth of a spherical tumor as follows: Assume that the tumor is initially the shape of a sphere of radius a, and as it grows remains a sphere of radius $R(t)$. Let $p(r,t)$ be the pressure within the tumor at distance r from its center and at time t. Then, p satisfies

$$\frac{1}{r^2}\frac{\partial^2(rp)}{\partial r^2} = S, \quad r < R(t),$$

where S is the rate of volume change per unit volume. Let $u(r,t)$ be the nutrient concentration outside the tumor and u_0 be that inside the tumor. Then, u satisfies

$$\frac{1}{r^2}\frac{\partial^2(ru)}{\partial r^2} = 0, \quad r > R(t).$$

The above equations are subject to the following conditions:

(a) $p(0,t)$ is bounded

(b) $p(R(t),t) = \beta/R(t)$

(c) $u_r - \mu(u - u_0)^{1/2} = 0$ for $r = R(t)$

(d) $\lim_{r\to\infty} u(r,t) = u_\infty$,

where S, β, μ, u_0 and u_∞ are constants.

(i)   Use (a) and (b) to find $p(r,t)$, $r < R(t)$.

(ii) Use (c) and (d) to find $u(r,t)$, $r > R(t)$.

Now we know $p(r,t)$ and $u(r,t)$ in terms of the unknown expanding radius of the tumor $R(t)$. However, $R(t)$ can be determined by solving the initial value problem

$$R'(t) = -\frac{\partial p}{\partial r} + \lambda(u - u_0)^{1/2} \quad \text{for} \quad r = R(t), \quad R(0) = a.$$

(iii) Find the limiting size of the tumor, i.e., R when $R'(t) = 0$.

Problems in which one of the boundary is unknown and changes with time are known as *moving boundary value problems*.

39.16.   The Schrödinger equation for a single particle of mass m moving in a potential field $V(x,y,z)$ is

$$\frac{ih}{2\pi}\psi_t = -\frac{h^2}{8m\pi^2}\Delta_3\psi + V(x,y,z)\psi. \tag{39.21}$$

(i)   Show that separating out the t dependence as $\psi(x,y,z,t) = \phi(x,y,z)e^{-2\pi iEt/h}$ results in the partial DE

$$\Delta_3\phi + \frac{8m\pi^2}{h^2}(E - V)\phi = 0; \tag{39.22}$$

here h is Planck's constant, and the total energy E is assumed to be a constant.

(ii) If $V \equiv 0$ in (39.22), the resulting equation is called Helmholtz's equation. For this equation set $\phi = X(x)Y(y)Z(z)$ to determine the DEs for X, Y, and Z.

39.17. The time-independent Schrödinger equation for a single particle of mass m moving in a potential field V in spherical coordinates is given by

$$\frac{1}{r^2}\frac{\partial}{\partial r}\left(r^2\frac{\partial \psi}{\partial r}\right) + \frac{1}{r^2 \sin\phi}\frac{\partial}{\partial \phi}\left(\sin\phi \frac{\partial \psi}{\partial \phi}\right) + \frac{1}{r^2 \sin^2\phi}\frac{\partial^2 \psi}{\partial \theta^2} + \frac{8m\pi^2}{h^2}(E-V)\psi = 0,$$

where in some applications $V = V(r)$. For this equation set $\psi = R(r)\Phi(\phi)\Theta(\theta)$ to determine the DEs for R, Φ and Θ.

Answers or Hints

39.1. Verify directly.

39.2. Verify directly.

39.3. Verify directly.

39.4. $\frac{3}{4}\sin x \sin y \frac{\sinh\sqrt{2}(\pi-z)}{\sinh\sqrt{2}\pi} - \frac{1}{4}\sin x \sin 3y \frac{\sinh\sqrt{10}(\pi-z)}{\sinh\sqrt{10}\pi}$.

39.5. $u(x,y,z,t) = \frac{1}{8}a_{000} + \frac{1}{4}\sum_{\ell} a_{\ell 00}e^{-\ell^2 t}\cos\ell x + \frac{1}{4}\sum_{m} a_{0m0}e^{-m^2 t}\cos my$
$+ \frac{1}{4}\sum_{n} a_{00n}e^{-n^2 t}\cos nz + \frac{1}{2}\sum_{\ell m} a_{\ell m0}e^{-(\ell^2+m^2)t}\cos\ell x \cos my$
$+ \frac{1}{2}\sum_{mn} a_{0mn}e^{-(m^2+n^2)t}\cos my \cos nz + \frac{1}{2}\sum_{\ell n} a_{\ell 0n}e^{-(\ell^2+n^2)t}\cos\ell x \cos nz$
$+ \sum_{\ell mn} a_{\ell mn}e^{-(\ell^2+m^2+n^2)t}\cos\ell x \cos my \cos nz$, where
$a_{\ell mn} = \frac{8}{\pi^3}\int_0^\pi \int_0^\pi \int_0^\pi f(x,y,z)\cos\ell x \cos my \cos nz\,dxdydz$. $a_{000} = \pi^3$,
$a_{\ell 00} = -2\pi[1-(-1)^\ell]/\ell^2$, $a_{\ell m0} = 4[1-(-1)^\ell][1-(-1)^m]/(\pi\ell^2 m^2)$,
$a_{\ell mn} = -8[1-(-1)^\ell][1-(-1)^m][1-(-1)^n]/(\pi^2\ell^2 m^2 n^2)$.

39.6. Verify directly.

39.7. $u(r,z) = \frac{2u_0}{a}\sum_{n=1}^{\infty}\frac{\cosh(\lambda_n z)J_0(\lambda_n r)}{\lambda_n J_1(\lambda_n a)\cosh(\lambda_n h)}$.

39.8. $u(r,z) = \frac{\pi}{2} - \frac{4}{\pi}\sum_{n=1}^{\infty}\frac{I_0[(2n-1)r]}{(2n-1)^2 I_0(2n-1)}\cos(2n-1)z$.

39.9. $u(r,z) = \frac{2}{a^2}\sum_{n=1}^{\infty}\frac{\lambda_n^2 \exp(-\lambda_n z)J_0(\lambda_n r)}{(\lambda_n^2+k^2)J_0^2(\lambda_n a)}\int_0^a r\alpha(r)J_0(\lambda_n r)dr$, where λ_n
are the positive roots of $\lambda_n J_0'(\lambda_n a) + kJ_0(\lambda_n a) = 0$.

39.10. Verify directly.

39.11. $u(r,\theta,\phi) = \sum_{n=1}^{\infty}\frac{r^n}{n}\left[\frac{1}{2}a_{0n}P_n(\cos\phi) + \sum_{m=1}^{n}(a_{mn}\cos m\theta + b_{mn}\sin m\theta)P_n^m(\cos\phi)\right]$.

39.12. We need to find the solution of (39.11) where $0 < r < a$, $0 < \phi < \pi/2$ with the boundary conditions $u_z(r, \pi/2) = 0$, $0 < r < a$, $u(a, \phi) = f(\cos\phi)$, $0 < \phi < \pi/2$, $|u(r, \phi)| < \infty$, $0 < r < a$, $0 < \phi < \pi/2$. Since $z = r\cos\phi$, $\frac{\partial u}{\partial \phi} = \frac{\partial u}{\partial z}\frac{\partial z}{\partial \phi} = -r\sin\phi\frac{\partial u}{\partial z}$, and hence at $\phi = \pi/2$ we have $\frac{\partial u}{\partial z} = -\frac{1}{r}\frac{\partial u}{\partial \phi}$ and now the condition $u_z = 0$ implies that $u_\phi = 0$ at $\phi = \pi/2$.
$u(r, \phi) = \sum_{n=0}^{\infty}(4n+1)\left(\frac{r}{a}\right)^{2n} P_{2n}(\cos\phi)\int_0^1 f(x)P_{2n}(x)dx$.

39.13. Use Problems 7.2, 7.8(i) to show that (39.13) and (39.16) reduce to
$u(r, \phi) = \frac{V}{2}\left[1 + \sum_{n=0}^{\infty}(-1)^n\left(\frac{4n+3}{2n+2}\right)\frac{(2n)!}{2^{2n}(n!)^2}\left(\frac{r}{a}\right)^{2n+1} P_{2n+1}(\cos\phi)\right]$,
$u(r, \phi) = \frac{V}{2}\left[\frac{a}{r} + \sum_{n=0}^{\infty}(-1)^n\left(\frac{4n+3}{2n+2}\right)\frac{(2n)!}{2^{2n}(n!)^2}\left(\frac{a}{r}\right)^{2n+2} P_{2n+1}(\cos\phi)\right]$.

39.14. $u(r, \phi) = \frac{1}{2}\sum_{n=0}^{\infty}\frac{b^{2n+1}-r^{2n+1}}{b^{2n+1}-a^{2n+1}}(2n+1)\left(\frac{a}{r}\right)^{n+1} P_n(\cos\phi)$
$\times \int_{-1}^1 f(x)P_n(x)dx$.

39.15. Solve directly.

39.16. (i) Verify directly (ii) $X'' + \mu^2 X = 0$, $Y'' + \nu^2 Y = 0$, $Z'' + \lambda^2 Z = 0$ where $\mu^2 + \nu^2 + \lambda^2 = \frac{8m\pi^2 E}{h^2}$.

39.17. $\frac{d^2\Theta}{d\theta^2} = -m_\ell^2\Theta$ (Azimuthal equation),
$\frac{1}{r^2}\frac{d}{dr}\left(r^2\frac{dR}{dr}\right) + \frac{8m\pi^2}{h^2}\left[E - V - \frac{h^2}{8m\pi^2}\frac{\ell(\ell+1)}{r^2}\right] R = 0$ (radial equation),
$\frac{1}{\sin\phi}\frac{d}{d\phi}\left(\sin\phi\frac{d\Phi}{d\phi}\right) + \left[\ell(\ell+1) - \frac{m_\ell^2}{\sin^2\phi}\right]\Phi = 0$ (angular equation),
where the constants ℓ and m_ℓ, respectively, are the orbital angular momentum and magnetic quantum numbers.

Lecture 40
Nonhomogeneous Equations

In Lectures 30–39 we employed the Fourier method to solve homogeneous partial DEs together with appropriate initial and boundary conditions. However, often in applications the governing partial DE is nonhomogeneous. In this lecture we shall demonstrate how the Fourier method can be employed to solve nonhomogeneous problems. Here our approach is similar to that of we have discussed in Lecture 24.

When a heat source is present within the domain of interest, instead of (30.1), the governing partial DE is nonhomogeneous:

$$u_t - c^2 u_{xx} = q(x,t), \quad 0 < x < a, \quad t > 0, \quad c > 0. \tag{40.1}$$

We shall consider (40.1) with the initial condition (30.2), and the homogeneous Robin's conditions

$$a_0 u(0,t) - a_1 u_x(0,t) = 0, \quad t > 0, \quad a_0^2 + a_1^2 > 0 \tag{40.2}$$

$$d_0 u(a,t) + d_1 u_x(a,t) = 0, \quad t > 0, \quad d_0^2 + d_1^2 > 0. \tag{40.3}$$

Let λ_n, $X_n(x)$, $n \geq 1$ be the eigenvalues and eigenfunctions of the problem (30.5),

$$\begin{aligned} a_0 X(0) - a_1 X'(0) &= 0 \\ d_0 X(a) + d_1 X'(a) &= 0. \end{aligned} \tag{40.4}$$

We seek a solution of (40.1)–(40.3), (30.2) in the form

$$u(x,t) = \sum_{n=1}^{\infty} T_n(t) X_n(x), \tag{40.5}$$

where the time-dependent coefficients $T_n(t)$ have to be determined. Assuming that termwise differentiation is permitted, we obtain

$$u_t = \sum_{n=1}^{\infty} T_n'(t) X_n(x) \tag{40.6}$$

and

$$u_{xx} = \sum_{n=1}^{\infty} T_n(t) X_n''(x) = \frac{1}{c^2} \sum_{n=1}^{\infty} \lambda_n T_n(t) X_n(x). \tag{40.7}$$

R.P. Agarwal, D. O'Regan, *Ordinary and Partial Differential Equations*,
Universitext, DOI 10.1007/978-0-387-79146-3_40,
© Springer Science+Business Media, LLC 2009

Substituting these expressions (40.6) and (40.7) in equation (40.1), we obtain

$$\sum_{n=1}^{\infty} [T_n'(t) - \lambda_n T_n(t)] X_n(x) = q(x, t). \tag{40.8}$$

For a fixed value of t, equation (40.8) represents a Fourier series representation of the function $q(x, t)$, with Fourier coefficients

$$T_n' - \lambda_n T_n = \frac{\int_0^a q(x, t) X_n(x) dx}{\int_0^a X_n^2(x) dx} = Q_n(t), \quad \text{say}, \quad n = 1, 2, \cdots. \tag{40.9}$$

Assuming $\lambda_n \neq 0$, then for each n, (40.9) is a first-order linear ordinary DE with general solution

$$T_n(t) = \left[c_n + \int_0^t Q_n(s) e^{-\lambda_n s} ds \right] e^{\lambda_n t}, \quad n = 1, 2, \cdots. \tag{40.10}$$

Substituting (40.10) in (40.5), we find the formal solution

$$u(x, t) = \sum_{n=1}^{\infty} \left[c_n + \int_0^t Q_n(s) e^{-\lambda_n s} ds \right] e^{\lambda_n t} X_n(x). \tag{40.11}$$

Finally, to determine the constants c_n, $n = 1, 2, \cdots$ we set $t = 0$ in (40.11) and use the prescribed initial condition (30.2), to get

$$u(x, 0) = f(x) = \sum_{n=1}^{\infty} c_n X_n(x),$$

which immediately gives

$$c_n = \frac{\int_0^a f(x) X_n(x) dx}{\int_0^a X_n^2(x) dx}, \quad n = 1, 2, \cdots. \tag{40.12}$$

In particular, we consider the initial-boundary value problem

$$\begin{aligned} u_t - u_{xx} &= -(1 - x) \cos at, \quad 0 < x < 1, \quad t > 0 \\ u(0, t) &= 0, \quad u(1, t) = 0, \quad u(x, 0) = 0. \end{aligned} \tag{40.13}$$

For (40.13) it is clear that $\lambda_n = -n^2 \pi^2$, $X_n(x) = \sin n\pi x$, $c_n = 0$, and hence from (40.11), we have

$$\begin{aligned} u(x, t) &= \sum_{n=1}^{\infty} \left[\int_0^t \left\{ \frac{\int_0^1 (\tau - 1) \cos as \sin n\pi\tau d\tau}{\int_0^1 \sin^2 n\pi\tau d\tau} \right\} e^{n^2\pi^2 s} ds \right] e^{-n^2\pi^2 t} \sin n\pi x \\ &= \sum_{n=1}^{\infty} \left[\int_0^t \left\{ \frac{-2 \cos as}{n\pi} \right\} e^{n^2\pi^2 s} ds \right] e^{-n^2\pi^2 t} \sin n\pi x \\ &= \sum_{n=1}^{\infty} \frac{2}{n\pi(a^2 + n^4\pi^4)} \left[n^2\pi^2 \left(e^{-n^2\pi^2 t} - \cos at \right) - a \sin at \right] \sin n\pi x. \end{aligned}$$

In the derivation of the wave equation (32.1) if there is a force $q(x,t)$ per unit length acting at right angles to the string and m is the mass density per unit length, then the resulting partial DE is nonhomogeneous

$$u_{tt} - c^2 u_{xx} = \frac{1}{m} q(x,t), \quad 0 < x < a, \quad t > 0. \tag{40.14}$$

We shall find the solution of (40.14) subject to the initial and boundary conditions (32.2)–(32.5). Again, we assume that the solution can be written as (40.5), where now $X_n(x)$, $n \geq 1$ are the eigenfunctions of the problem (32.14), i.e.,

$$u(x,t) = \sum_{n=1}^{\infty} T_n(t) \sin \frac{n\pi x}{a}. \tag{40.15}$$

Now substituting (40.15) in (40.14), we obtain

$$\sum_{n=1}^{\infty} \left[T_n''(t) + \frac{n^2 \pi^2 c^2}{a^2} T_n(t) \right] \sin \frac{n\pi x}{a} = \frac{1}{m} q(x,t)$$

and hence

$$T_n'' + \frac{n^2 \pi^2 c^2}{a^2} T_n = \frac{2}{ma} \int_0^a q(x,t) \sin \frac{n\pi x}{a} dx = R_n(t), \quad \text{say}, \quad n = 1, 2, \cdots. \tag{40.16}$$

Using the methods presented in Lecture 2, the general solution of (40.16) can be written as

$$T_n(t) = a_n \cos \frac{n\pi ct}{a} + b_n \sin \frac{n\pi ct}{a} + \frac{a}{n\pi c} \int_0^t \sin \frac{n\pi c}{a}(t-s) R_n(s) ds.$$

Substituting this in (40.15), we get

$$u(x,t) = \sum_{n=1}^{\infty} \left[a_n \cos \frac{n\pi ct}{a} + b_n \sin \frac{n\pi ct}{a} + \frac{a}{n\pi c} \int_0^t \sin \frac{n\pi c}{a}(t-s) R_n(s) ds \right]$$
$$\times \sin \frac{n\pi x}{a}. \tag{40.17}$$

Finally, this solution satisfies the initial conditions (32.2) and (32.3) if and only if a_n and b_n are the same as in (32.19) and (32.21).

In particular, if the string is vibrating under the force of gravity, the forcing function is given by $q(x,t) = -g$, and then we have

$$R_n(t) = \frac{2}{ma} \int_0^a -g \sin \frac{n\pi x}{a} dx = -\frac{2g}{mn\pi} \left(1 - (-1)^n \right).$$

Now we assume

$$f(x) = \begin{cases} x, & 0 < x < a/2 \\ a - x, & a/2 < x < a \end{cases}$$

and $g(x) = 0$, so that

$$a_n = \frac{2}{a}\left[\int_0^{a/2} x\sin\frac{n\pi x}{a}\,dx + \int_{a/2}^a (a-x)\sin\frac{n\pi x}{a}\,dx\right] = \frac{4a}{n^2\pi^2}\sin\frac{n\pi}{2}$$

and $b_n = 0$. Thus, in this case the required solution is

$$
\begin{aligned}
u(x,t) &= \sum_{n=1}^{\infty}\left[\frac{4a}{n^2\pi^2}\sin\frac{n\pi}{2}\cos\frac{n\pi ct}{a}\right.\\
&\qquad \left. -\frac{2g}{mn\pi}(1-(-1)^n)\int_0^t \sin\frac{nc\pi}{a}(t-s)ds\right]\sin\frac{n\pi x}{a}\\
&= \sum_{n=1}^{\infty}\left[\left\{\frac{4a(-1)^{n+1}}{(2n-1)^2\pi^2} + \frac{4ga^2}{m(2n-1)^3\pi^3 c^2}\right\}\cos\frac{(2n-1)\pi ct}{a}\right.\\
&\qquad \left. -\frac{4ga^2}{m(2n-1)^3\pi^3 c^2}\right]\sin\frac{n\pi x}{a}.
\end{aligned}
$$

Next we shall consider the nonhomogeneous Laplace equation known as *Poisson's equation*:

$$\Delta_2 u = u_{xx} + u_{yy} = q(x,y), \quad 0 < x < a, \quad 0 < y < b. \tag{40.18}$$

This equation appears in electrostatics theory. We shall find the solution of the boundary value problem (40.18), (34.5)–(34.8). We assume that the solution can be written as

$$u(x,y) = \sum_{n=1}^{\infty} X_n(x)Y_n(y) = \sum_{n=1}^{\infty}\sin\frac{n\pi x}{a}Y_n(y), \tag{40.19}$$

where $X_n(x) = \sin(n\pi x/a)$ are the eigenfunctions of the problem (34.9), (34.10). Substituting (40.19) in (40.18), we obtain

$$\sum_{n=1}^{\infty}\left[Y_n''(y) - \frac{n^2\pi^2}{a^2}Y_n(y)\right]\sin\frac{n\pi x}{a} = q(x,y)$$

and hence

$$Y_n'' - \frac{n^2\pi^2}{a^2}Y_n = \frac{2}{a}\int_0^a q(x,y)\sin\frac{n\pi x}{a}\,dx = S_n(y), \quad \text{say}, \quad n = 1,2,\cdots. \tag{40.20}$$

Now remembering that we will have to satisfy the boundary conditions (34.5) and (34.6), we write the solution of (40.20) in terms of Green's function (see Problem 15.9) and substitute it in (40.19), to obtain

$$
\begin{aligned}
u(x,y) = \sum_{n=1}^{\infty} \Bigg(\frac{1}{\sinh \frac{n\pi b}{a}} \Bigg\{ & \left[a_n \sinh \frac{n\pi(b-y)}{a} + b_n \sinh \frac{n\pi y}{a} \right] \\
& - \frac{a}{n\pi} \left[\sinh \frac{n\pi(b-y)}{a} \int_0^y S_n(\tau) \sinh \frac{n\pi\tau}{a} d\tau \right. \\
& + \left. \sinh \frac{n\pi y}{a} \int_y^b S_n(\tau) \sinh \frac{n\pi(b-\tau)}{a} d\tau \right] \Bigg\} \Bigg) \sin \frac{n\pi x}{a}.
\end{aligned}
$$
$$(40.21)$$

Clearly, the conditions (34.5) and (34.6) now easily determine

$$
a_n = \frac{2}{a} \int_0^a f(x) \sin \frac{n\pi x}{a} dx, \quad b_n = \frac{2}{a} \int_0^a g(x) \sin \frac{n\pi x}{a} dx. \qquad (40.22)
$$

In the particular case, when $q(x,y) = -1$, $f(x) = g(x) = 0$, we find $a_n = b_n = 0$,

$$
S_n(y) = -\frac{2}{n\pi} (1 - (-1)^n)
$$

and the solution (40.21) becomes

$$
u(x,y) = \frac{4a^2}{\pi^3} \sum_{n \text{ odd}}^{\infty} \frac{1}{n^3} \left[1 - \frac{\sinh \frac{n\pi y}{a} + \sinh \frac{n\pi(b-y)}{a}}{\sinh \frac{n\pi b}{a}} \right] \sin \frac{n\pi x}{a}.
$$

As in Lecture 34 we also note that if $u_1(x,y)$ is the solution of the problem (40.18), (34.5)–(34.8), and $u_2(x,y)$ is the solution of the problem (34.4), (34.18)–(34.21), then $u(x,y) = u_1(x,y) + u_2(x,y)$ is the solution of the Poisson equation (40.18) satisfying the boundary conditions (34.5), (34.6), (34.20), (34.21).

Problems

40.1. Consider the nonhomogeneous partial DE

$$
\frac{\partial}{\partial x} \left(k(x) \frac{\partial u}{\partial x} \right) = \rho(x) c(x) \frac{\partial u}{\partial t} - q(x), \quad \alpha < x < \beta, \quad t > 0
$$

together with the initial and boundary conditions (31.2)–(31.4). Show that its solution can be written as (31.21), where now $v(x)$ is the solution of the nonhomogeneous equation

$$
\frac{d}{dx} \left(k(x) \frac{dv}{dx} \right) = -q(x), \quad \alpha < x < \beta
$$

satisfying the same boundary conditions (31.6). In particular, find the solution of the initial-boundary value problem

$$u_t - c^2 u_{xx} = q(x), \quad 0 < x < a, \quad t > 0, \quad c > 0$$

satisfying (30.2)–(30.4), when

(i) $q(x) = \mu \sin \nu x$

(ii) $q(x) = \mu e^{-\nu x}$ (occurs in radioactive decay)

(iii) $q(x) = \mu x e^{-\nu x}$.

40.2. Find the solution of the initial-boundary value problem (40.1), (30.2)–(30.4) when

(i) $a = 1, \ c = 1, \ f(x) = 0, \ q(x,t) = 2x - 5 \sin t$

(ii) $a = 1, \ c = 1, \ f(x) = 2x, \ q(x,t) = x + t$

(iii) $a = \pi, \ c = 2, \ f(x) = (1/2)x(\pi - x)^2, \ q(x,t) = xt^2$.

40.3. Find the solution of the nonhomogeneous partial DE

$$u_{tt} - c^2 u_{xx} = q(x), \quad 0 < x < a, \quad t > 0$$

satisfying the initial and boundary conditions (32.2), (32.3), (33.20).

40.4. Find the solution of the initial-boundary value problem (40.14), (32.2)–(32.5) when

(i) $a = \pi, \ f(x) = 0, \ g(x) = 0, \ q(x,t) = mq_0 c^2 \sin \omega t$

(ii) $a = \pi, \ c = 1, \ f(x) = 0, \ g(x) = x, \ q(x,t) = mxt$

(iii) $a = 1, \ c = 1, \ f(x) = x(1 - x), \ g(x) = 0, \ q(x,t) = m(x + t)$

(iv) $a = 2\pi, \quad c = 1, \quad f(x) = x(x - 2\pi), \quad g(x) = 0, \quad q(x,t) =$
$$m \begin{cases} x, \ 0 \le x < \pi \\ 2\pi - x, \ \pi \le x \le 2\pi. \end{cases}$$

40.5. Find the solution of the initial-boundary value problem (40.14), (32.2)–(32.4), $u_x(a, t) = 0, \ t > 0$.

40.6. Find the solution of the initial-boundary value problem (40.14), (32.2), (32.3), $u_x(0, t) = 0, \ t > 0$, (32.5).

40.7. Find the solution of the boundary value problem (39.18), (34.5)–(34.8) when

(i) $a = \pi, \ b = \pi, \ f(x) = 0, \ g(x) = T_0, \ q(x,y) = -xy$

(ii) $a = \pi, \ b = \pi, \ f(x) = 1, \ g(x) = 1, \ q(x,y) = -\sin 3y$.

40.8. Find the solution of the following boundary value problem:

$$u_{rr} + \frac{1}{r}u_r + u_{zz} = -q(r,z), \quad 0 < r < 1, \quad 0 < z < h$$

$$u(r,0) = 0, \quad u(r,h) = 0, \quad u(1,z) = f(z).$$

Answers or Hints

40.1. $v(x) + \frac{2}{a}\sum_{n=1}^{\infty}\left(\int_0^a (f(x) - v(x))\sin\frac{n\pi x}{a}dx\right)e^{-(n^2\pi^2 c^2/a^2)t}\sin\frac{n\pi x}{a}$
where $v(x) = \frac{1}{ac^2}\left[(a-x)\int_0^x sq(s)ds + x\int_x^a (a-s)q(s)ds\right]$ (i) $v(x) = \frac{\mu}{ac^2\nu^2}(a\sin\nu x - x\sin\nu a)$ (ii) $v(x) = \frac{\mu}{ac^2\nu^2}[a(1-e^{-\nu x}) - x(1-e^{-\nu a})]$
(iii) $v(x) = \frac{\mu}{ac^2\nu^3}[2a(1-e^{-\nu x}) - 2x(1-e^{-\nu a}) - ax\nu(e^{-\nu x} - e^{-\nu a})]$.

40.2. (i) $\sum_{n=1}^{\infty}\left[\frac{4}{n^3\pi^3}(-1)^{n-1}(1 - e^{-n^2\pi^2 t}) - \frac{10}{n\pi}\frac{(1-(-1)^n)}{1+n^2\pi^2}(n^2\pi^2\sin t\right.$
$\left. - \cos t + e^{-n^2\pi^2 t})\right]\sin n\pi x$ (ii) $\sum_{n=1}^{\infty}\frac{2}{n^5\pi^5}\left[2(-1)^{n+1}n^4\pi^4 e^{-n^2\pi^2 t}\right.$
$\left. -((-1)^n - 1)n^2\pi^2 t + (e^{-n^2\pi^2 t} - 1)(1 - (-1)^n + (-1)^n n^2\pi^2)\right]\sin n\pi x$
(iii) $\sum_{n=1}^{\infty}\left[\frac{2}{n^3}(2 + (-1)^n)e^{-4n^2 t} + \frac{(-1)^{n+1}}{16n^7}(8n^4 t^2 - 4n^2 t + 1 - e^{-4n^2 t})\right]$
$\times \sin nx$.

40.3. $v(x) + \sum_{n=1}^{\infty}\left(a_n\cos\frac{n\pi ct}{a} + b_n\sin\frac{n\pi ct}{a}\right)\sin\frac{n\pi x}{a}$ where
$a_n = \frac{2}{a}\int_0^a (f(x) - v(x))\sin\frac{n\pi x}{a}dx$, $b_n = \frac{2}{n\pi c}\int_0^a (f(x) - v(x))\sin\frac{n\pi x}{a}dx$
and $v(x) = \frac{(a-x)}{a}A + \frac{x}{a}B + \frac{1}{ac^2}\left[(a-x)\int_0^x sq(s)ds + x\int_x^a (a-s)q(s)ds\right]$.

40.4. (i) $\frac{4q_0 c}{\pi}\sum_{n\,\text{odd}}^{\infty}\frac{\sin nx}{n^2(\omega^2 - n^2 c^2)}(\omega\sin nct - nc\sin\omega t)$ (ii) $\sum_{n=1}^{\infty}\frac{2(-1)^{n+1}}{n^2}$
$\times\left[\left(1 - \frac{1}{n^2}\right)\sin nt + \frac{t}{n}\right]\sin nx$ (iii) $\sum_{n=1}^{\infty}\frac{2}{n^3\pi^3}\left[(1 - (-1)^n)(2\cos n\pi t + t\right.$
$\left. - \frac{1}{n\pi}\sin n\pi t) + (-1)^n(\cos n\pi t - 1)\right]\sin n\pi x$ (iv) $\sum_{n=1}^{\infty}\left[a_n\cos\frac{nt}{2} + \right.$
$\left. \frac{32}{\pi n^4}\sin\frac{n\pi}{2}\right]\sin\frac{nx}{2}$, where $a_n = -\frac{32}{\pi n^4}\sin\frac{n\pi}{2} + \frac{16}{\pi n^3}((-1)^n - 1)$.

40.5. See Problem 33.4.

40.6. See Problem 33.5.

40.7. (i) $\frac{4T_0}{\pi}\sum_{n=1}^{\infty}\frac{\sinh(2n-1)y}{(2n-1)\sinh(2n-1)\pi}\sin(2n-1)x + \sum_{m=1}^{\infty}\sum_{n=1}^{\infty}\frac{4(-1)^{m+n}}{mn(m^2+n^2)}$
$\times\sin mx\sin ny$ (ii) $\frac{4}{\pi}\sum_{n=1}^{\infty}\frac{\sinh(2n-1)y + \sinh(2n-1)(\pi-y)}{(2n-1)\sinh(2n-1)\pi}\sin(2n-1)x$
$+ \frac{4\sin 3y}{\pi}\sum_{m=1}^{\infty}\frac{\sin(2m-1)x}{(2m-1)[(2m-1)^2+9]}$.

40.8. $u(r,z) = \sum_{n=1}^{\infty}a_n I_0\left(\frac{n\pi r}{h}\right)\sin\frac{n\pi z}{h} + \sum_{m=1}^{\infty}\sum_{n=1}^{\infty}b_{mn}J_0(\lambda_n r)\sin\frac{m\pi z}{h}$,
where $J_0(\lambda_n) = 0$, $n = 1, 2, \cdots$, $a_n I_0\left(\frac{n\pi}{h}\right) = \frac{2}{h}\int_0^h f(z)\sin\frac{n\pi z}{h}dz$, and
$b_{mn} = \frac{4}{hJ_1^2(\lambda_n)\left(\frac{m^2\pi^2}{h^2}+\lambda_n^2\right)}\int_0^1\int_0^h q(r,z)rJ_0(\lambda_n r)\sin\frac{m\pi z}{h}dzdr$.

Lecture 41
Fourier Integral and Transforms

The Fourier integral is a natural extension of Fourier trigonometric series in the sense that it represents a piecewise smooth function whose domain is semi–infinite or infinite. In this lecture we shall develop the Fourier integral with an intuitive approach, and then discuss Fourier cosine and sine integrals which are extensions of Fourier cosine and sine series, respectively. This leads to Fourier cosine and sine transform pairs.

Let $f_p(x)$ be a periodic function of period $2p$ that can be represented by a Fourier trigonometric series

$$f_p(x) = \frac{a_0}{2} + \sum_{n=1}^{\infty} (a_n \cos \omega_n x + b_n \sin \omega_n x), \quad \omega_n = \frac{n\pi}{p},$$

where

$$a_n = \frac{1}{p} \int_{-p}^{p} f_p(t) \cos \omega_n t\, dt, \quad n \geq 0$$

$$b_n = \frac{1}{p} \int_{-p}^{p} f_p(t) \sin \omega_n t\, dt, \quad n \geq 1.$$

The problem we shall consider is what happens to the above series when $p \to \infty$. For this we insert a_n and b_n, to obtain

$$f_p(x) = \frac{1}{2p} \int_{-p}^{p} f_p(t) dt + \frac{1}{p} \sum_{n=1}^{\infty} \left[\cos \omega_n x \int_{-p}^{p} f_p(t) \cos \omega_n t\, dt \right.$$

$$\left. + \sin \omega_n x \int_{-p}^{p} f_p(t) \sin \omega_n t\, dt \right].$$

We now set

$$\Delta \omega = \omega_{n+1} - \omega_n = \frac{(n+1)\pi}{p} - \frac{n\pi}{p} = \frac{\pi}{p}.$$

Then, $1/p = \Delta\omega/\pi$, and we may write the Fourier series in the form

$$f_p(x) = \frac{1}{2p} \int_{-p}^{p} f_p(t) dt + \frac{1}{\pi} \sum_{n=1}^{\infty} \left[(\cos \omega_n x) \Delta\omega \int_{-p}^{p} f_p(t) \cos \omega_n t\, dt \right.$$

$$\left. + (\sin \omega_n x) \Delta\omega \int_{-p}^{p} f_p(t) \sin \omega_n t\, dt \right]. \tag{41.1}$$

R.P. Agarwal, D. O'Regan, *Ordinary and Partial Differential Equations*,
Universitext, DOI 10.1007/978-0-387-79146-3_41,
© Springer Science+Business Media, LLC 2009

This representation is valid for any fixed p, arbitrarily large, but fixed.

We now let $p \to \infty$ and assume that the resulting nonperiodic function $f(x) = \lim_{p \to \infty} f_p(x)$ is absolutely integrable on the x-axis, i.e., $\int_{-\infty}^{\infty} |f(x)| dx < \infty$. Then, $1/p \to 0$, and the value of the first term on the right side of (41.1) approaches zero. Also, $\Delta \omega = \pi/p \to 0$ and the infinite series in (41.1) becomes an integral from 0 to ∞, which represents $f(x)$, i.e.,

$$f(x) = \frac{1}{\pi} \int_0^{\infty} \left[\cos \omega x \int_{-\infty}^{\infty} f(t) \cos \omega t \, dt + \sin \omega x \int_{-\infty}^{\infty} f(t) \sin \omega t \, dt \right] d\omega.$$

(41.2)

Now if we introduce the notations

$$A(\omega) = \frac{1}{\pi} \int_{-\infty}^{\infty} f(t) \cos \omega t \, dt, \quad B(\omega) = \frac{1}{\pi} \int_{-\infty}^{\infty} f(t) \sin \omega t \, dt, \quad (41.3)$$

then (41.2) can be written as

$$f(x) = \int_0^{\infty} [A(\omega) \cos \omega x + B(\omega) \sin \omega x] d\omega. \quad (41.4)$$

This representation of $f(x)$ is called *Fourier integral*.

The following result gives precise conditions for the existence of the integral in (41.4), and the meaning of the equality.

Theorem 41.1. Let $f(x)$, $-\infty < x < \infty$ be piecewise continuous on each finite interval, and $\int_{-\infty}^{\infty} |f(x)| < \infty$, i.e., f is absolutely integrable on $(-\infty, \infty)$. Then, $f(x)$ can be represented by a Fourier integral (41.4). Further, at each x,

$$\int_0^{\infty} [A(\omega) \cos \omega x + B(\omega) \sin \omega x] d\omega = \frac{1}{2}[f(x + 0) + f(x - 0)].$$

Example 41.1. We shall find the Fourier integral representation of the single pulse function

$$f(x) = \begin{cases} 1 & \text{if } |x| < 1 \\ 0 & \text{if } |x| > 1. \end{cases}$$

From (41.3), we have

$$A(\omega) = \frac{1}{\pi} \int_{-\infty}^{\infty} f(t) \cos \omega t \, dt = \frac{1}{\pi} \int_{-1}^{1} \cos \omega t \, dt = \frac{2 \sin \omega}{\pi \omega}$$

$$B(\omega) = \frac{1}{\pi} \int_{-1}^{1} \sin \omega t \, dt = 0.$$

Thus, (41.4) gives the representation

$$f(x) = \frac{2}{\pi} \int_0^\infty \frac{\cos \omega x \sin \omega}{\omega} d\omega. \qquad (41.5)$$

Now from Theorem 41.1 it is clear that

$$\int_0^\infty \frac{\cos \omega x \sin \omega}{\omega} d\omega = \begin{cases} \dfrac{\pi}{2} & \text{if } |x| < 1 \\[2mm] \dfrac{\pi}{2} \dfrac{(1+0)}{2} = \dfrac{\pi}{4} & \text{if } x = \pm 1 \\[2mm] 0 & \text{if } |x| > 1. \end{cases} \qquad (41.6)$$

This integral is called *Dirichlet's discontinuous factor.*

From (41.6) it is clear that

$$\int_0^\infty \frac{\sin \omega}{\omega} d\omega = \frac{\pi}{2}. \qquad (41.7)$$

For an even or odd function, the Fourier integral becomes simpler. Indeed, if $f(x)$ is an even function, then $B(\omega) = 0$ in (41.3) and

$$A(\omega) = \frac{2}{\pi} \int_0^\infty f(t) \cos \omega t \, dt \qquad (41.8)$$

and the Fourier integral (41.4) reduces to the *Fourier cosine integral,*

$$f(x) = \int_0^\infty A(\omega) \cos \omega x \, d\omega. \qquad (41.9)$$

Similarly, if $f(x)$ is odd, then in (41.3) we have $A(\omega) = 0$ and

$$B(\omega) = \frac{2}{\pi} \int_0^\infty f(t) \sin \omega t \, dt$$

and the Fourier integral (41.4) reduces to the *Fourier sine integral*

$$f(x) = \int_0^\infty B(\omega) \sin \omega x \, d\omega. \qquad (41.10)$$

Example 41.2. We shall find the Fourier cosine and sine integrals of the function $f(x) = e^{-ax}$, $x > 0$, $a > 0$. Since

$$\begin{aligned} A(\omega) &= \frac{2}{\pi} \int_0^\infty e^{-at} \cos \omega t \, dt \\ &= \frac{2}{\pi} \times -\frac{a}{a^2 + \omega^2} e^{-at} \left(-\frac{\omega}{a} \sin \omega t + \cos \omega t \right) \Big|_0^\infty = \frac{2a/\pi}{a^2 + \omega^2} \end{aligned}$$

it follows that

$$f(x) = e^{-ax} = \frac{2a}{\pi} \int_0^\infty \frac{\cos \omega x}{a^2 + \omega^2} d\omega.$$

From this representation it is clear that

$$\int_0^\infty \frac{\cos \omega x}{a^2 + \omega^2} d\omega = \frac{\pi}{2a} e^{-ax}. \qquad (41.11)$$

Further, since

$$B(\omega) = \frac{2}{\pi} \int_0^\infty e^{-at} \sin \omega t \, dt = \frac{2\omega/\pi}{a^2 + \omega^2},$$

we have

$$f(x) = e^{-ax} = \frac{2}{\pi} \int_0^\infty \frac{\omega \sin \omega x}{a^2 + \omega^2} d\omega$$

and hence

$$\int_0^\infty \frac{\omega \sin \omega x}{a^2 + \omega^2} d\omega = \frac{\pi}{2} e^{-ax}. \qquad (41.12)$$

The integrals (41.11) and (41.12) are known as *Laplace integrals*.

Now for an even function $f(x)$ the Fourier integral is the Fourier cosine integral (41.9) where $A(\omega)$ is given by (41.8). We set $A(\omega) = \sqrt{2/\pi} \, F_c(\omega)$, where c indicates cosine. Then, replacing t by x, we get

$$F_c(\omega) = \sqrt{\frac{2}{\pi}} \int_0^\infty f(x) \cos \omega x \, dx \qquad (41.13)$$

and

$$f(x) = \sqrt{\frac{2}{\pi}} \int_0^\infty F_c(\omega) \cos \omega x \, d\omega. \qquad (41.14)$$

Formula (41.13) gives from $f(x)$ a new function $F_c(\omega)$ called the *Fourier cosine transform* of $f(x)$, whereas (41.14) gives back $f(x)$ from $F_c(\omega)$, and we call it the *inverse Fourier cosine transform* of $F_c(\omega)$. Relations (41.13) and (41.14) together form a Fourier cosine transform pair.

Similarly, for an odd function $f(x)$, the *Fourier sine transform* is

$$F_s(\omega) = \sqrt{\frac{2}{\pi}} \int_0^\infty f(x) \sin \omega x \, dx \qquad (41.15)$$

and the *inverse Fourier sine transform* is

$$f(x) = \sqrt{\frac{2}{\pi}} \int_0^\infty F_s(\omega) \sin \omega x \, d\omega. \qquad (41.16)$$

Relations (41.15) and (41.16) together form a Fourier sine transform pair.

Example 41.3. We shall find the Fourier cosine and sine transforms of the function

$$f(x) = \begin{cases} k, & 0 < x < a \\ 0, & x > a. \end{cases}$$

Clearly, we have

$$F_c(\omega) = \sqrt{\frac{2}{\pi}} k \int_0^a \cos \omega x \, dx = \sqrt{\frac{2}{\pi}} k \left[\frac{\sin a\omega}{\omega} \right].$$

$$F_s(\omega) = \sqrt{\frac{2}{\pi}} k \int_0^a \sin \omega x \, dx = \sqrt{\frac{2}{\pi}} k \left[\frac{1 - \cos a\omega}{\omega} \right].$$

Example 41.4. We shall find the Fourier cosine transform of the function e^{-x}. Clearly, we have

$$
\begin{aligned}
F_c(\omega) &= \sqrt{\frac{2}{\pi}} \int_0^\infty e^{-x} \cos \omega x \, dx \\
&= \sqrt{\frac{2}{\pi}} \frac{e^{-x}}{1+\omega^2} (-\cos \omega x + \omega \sin \omega x) \Big|_0^\infty = \frac{\sqrt{2/\pi}}{1+\omega^2}.
\end{aligned}
$$

If $f(x)$ is absolutely integrable on the positive x-axis and piecewise continuous on every finite interval, then the Fourier cosine and sine transforms of f exist. Furthermore, it is clear that F_c and F_s are linear operators, i.e.,

$$F_c(af + bg) = aF_c(f) + bF_c(g)$$

and

$$F_s(af + bg) = aF_s(f) + bF_s(g).$$

Theorem 41.2. Let $f(x)$ be continuous and absolutely integrable on the x-axis, let $f'(x)$ be piecewise continuous on each finite interval, and let $f(x) \to 0$ as $x \to \infty$. Then,

(i) $F_c(f'(x)) = \omega F_s(f(x)) - \sqrt{\frac{2}{\pi}} f(0)$, and

(ii) $F_s(f'(x)) = -\omega F_c(f(x))$.

Proof. To show (i), we integrate by parts, to obtain

$$
\begin{aligned}
F_c(f'(x)) &= \sqrt{\frac{2}{\pi}} \int_0^\infty f'(x) \cos \omega x \, dx \\
&= \sqrt{\frac{2}{\pi}} \left[f(x) \cos \omega x \Big|_0^\infty + \omega \int_0^\infty f(x) \sin \omega x \, dx \right] \\
&= -\sqrt{\frac{2}{\pi}} f(0) + \omega F_s(f(x)).
\end{aligned}
$$

The proof of (ii) is similar. ∎

Similarly, we can show that

$$F_c(f''(x)) = -\omega^2 F_c(f(x)) - \sqrt{\frac{2}{\pi}} f'(0),$$

$$F_s(f''(x)) = -\omega^2 F_s(f(x)) + \sqrt{\frac{2}{\pi}} \omega f(0).$$

Fourier Cosine Transforms

	$f(x)$	$F_c(f)$
1.	$\begin{cases} 1, & 0 < x < a \\ 0, & \text{otherwise} \end{cases}$	$\sqrt{\dfrac{2}{\pi}} \dfrac{\sin a\omega}{\omega}$
2.	$x^{a-1}, \; 0 < a < 1$	$\sqrt{\dfrac{2}{\pi}} \dfrac{\Gamma(a)}{\omega^a} \cos\dfrac{a\omega}{2}$
3.	$e^{-ax}, \; a > 0$	$\sqrt{\dfrac{2}{\pi}} \left(\dfrac{a}{a^2 + \omega^2} \right)$
4.	$e^{-ax^2}, \; a > 0$	$\dfrac{1}{\sqrt{2a}} e^{-\omega^2/4a}$
5.	$x^n e^{-ax}, \; a > 0$	$\sqrt{\dfrac{2}{\pi}} \dfrac{\cdot\, n!}{(a^2 + \omega^2)^{n+1}} \operatorname{Re}(a + i\omega)^{n+1}$
6.	$\begin{cases} \cos x, & 0 < x < a \\ 0, & \text{otherwise} \end{cases}$	$\dfrac{1}{\sqrt{2\pi}} \left[\dfrac{\sin a(1 - \omega)}{1 - \omega} + \dfrac{\sin a(1 + \omega)}{1 + \omega} \right]$
7.	$\cos ax^2, \; a > 0$	$\dfrac{1}{\sqrt{2a}} \cos\left(\dfrac{\omega^2}{4a} - \dfrac{\pi}{4} \right)$
8.	$\sin ax^2, \; a > 0$	$\dfrac{1}{\sqrt{2a}} \cos\left(\dfrac{\omega^2}{4a} + \dfrac{\pi}{4} \right)$
9.	$\dfrac{\sin ax}{x}, \; a > 0$	$\begin{cases} \sqrt{\dfrac{\pi}{2}}, & \omega < a \\ 0, & \omega > a \end{cases}$
10.	$\dfrac{e^{-x} \sin x}{x}$	$\dfrac{1}{\sqrt{2\pi}} \tan^{-1} \dfrac{2}{\omega^2}$
11.	$J_0(ax), \; a > 0$	$\begin{cases} \sqrt{\dfrac{2}{\pi}} \dfrac{1}{\sqrt{a^2 - \omega^2}}, & \omega < a \\ 0, & \omega > a \end{cases}$
12.	$\operatorname{sech} ax, \; a > 0$	$\sqrt{\dfrac{\pi}{2}} \dfrac{1}{a} \operatorname{sech} \dfrac{\pi\omega}{2a}.$

Fourier Sine Transforms

	$f(x)$	$F_s(f)$
1.	$\begin{cases} 1, & 0 < x < a \\ 0, & \text{otherwise} \end{cases}$	$\sqrt{\dfrac{2}{\pi}} \left[\dfrac{1 - \cos a\omega}{\omega} \right]$
2.	$\dfrac{1}{\sqrt{x}}$	$\dfrac{1}{\sqrt{\omega}}$
3.	$\dfrac{1}{x^{3/2}}$	$2\sqrt{\omega}$
4.	$x^{a-1}, \quad 0 < a < 1$	$\sqrt{\dfrac{2}{\pi}} \dfrac{\Gamma(a)}{\omega^a} \sin \dfrac{a\pi}{2}$
5.	e^{-x}	$\sqrt{\dfrac{2}{\pi}} \left(\dfrac{\omega}{1 + \omega^2} \right)$
6.	$\dfrac{e^{-ax}}{x}, \quad a > 0$	$\sqrt{\dfrac{2}{\pi}} \tan^{-1} \dfrac{x}{a}$
7.	$x^n e^{-ax}, \quad a > 0$	$\sqrt{\dfrac{2}{\pi}} \dfrac{n!}{(a^2 + \omega^2)^{n+1}} \operatorname{Im}(a + i\omega)^{n+1}$
8.	$xe^{-ax^2}, \quad a > 0$	$\dfrac{\omega}{(2a)^{3/2}} e^{-\omega^2/4a}$
9.	$\begin{cases} \sin x, & 0 < x < a \\ 0, & \text{otherwise} \end{cases}$	$\dfrac{1}{\sqrt{2\pi}} \left[\dfrac{\sin a(1 - \omega)}{1 - \omega} - \dfrac{\sin a(1 + \omega)}{1 + \omega} \right]$
10.	$\dfrac{\cos ax}{x}, \quad a > 0$	$\begin{cases} \sqrt{\dfrac{\pi}{2}}, & a < \omega \\ 0, & a > \omega \end{cases}$
11.	$\tan^{-1} \dfrac{2a}{x}, \quad a > 0$	$\sqrt{2\pi} \dfrac{\sinh a\omega}{\omega} e^{-a\omega}$
12.	$\operatorname{cosech} ax, \quad a > 0$	$\sqrt{\dfrac{\pi}{2}} \dfrac{1}{a} \tanh \dfrac{\pi\omega}{2a}.$

Lecture 42
Fourier Integral and Transforms (Cont'd.)

In this lecture we shall introduce the complex Fourier integral and the Fourier transform pair, and find the Fourier transform of the derivative of a function. Then, we shall state and prove the Fourier convolution theorem, which is an important result.

We note that (41.2) is the same as

$$
\begin{aligned}
f(x) &= \frac{1}{\pi} \int_0^\infty \int_{-\infty}^\infty f(t)[\cos \omega t \cos \omega x + \sin \omega t \sin \omega x] dt d\omega \\
&= \frac{1}{\pi} \int_0^\infty \left[\int_{-\infty}^\infty f(t) \cos(\omega x - \omega t) dt \right] d\omega.
\end{aligned}
\tag{42.1}
$$

The integral in brackets is an even function of ω; we denote it by $\tilde{F}(\omega)$. Since $\cos(\omega x - \omega t)$ is an even function of ω, the function f does not depend on ω, and we integrate with respect to t (not ω), the integral of $\tilde{F}(\omega)$ from $\omega = 0$ to ∞ is $1/2$ times the integral of $\tilde{F}(\omega)$ from $-\infty$ to ∞. Thus,

$$
f(x) = \frac{1}{2\pi} \int_{-\infty}^\infty \left[\int_{-\infty}^\infty f(t) \cos(\omega x - \omega t) dt \right] d\omega.
\tag{42.2}
$$

From the above argument it is clear that

$$
\frac{1}{2\pi} \int_{-\infty}^\infty \left[\int_{-\infty}^\infty f(t) \sin(\omega x - \omega t) dt \right] d\omega = 0.
\tag{42.3}
$$

A combination of (42.2) and (42.3) gives

$$
f(x) = \frac{1}{2\pi} \int_{-\infty}^\infty \int_{-\infty}^\infty f(t) e^{i\omega(x-t)} dt d\omega.
\tag{42.4}
$$

This is called the *complex Fourier integral*.

From the above representation of $f(x)$, we have

$$
f(x) = \frac{1}{\sqrt{2\pi}} \int_{-\infty}^\infty \left[\frac{1}{\sqrt{2\pi}} \int_{-\infty}^\infty f(t) e^{-i\omega t} dt \right] e^{i\omega x} d\omega.
\tag{42.5}
$$

R.P. Agarwal, D. O'Regan, *Ordinary and Partial Differential Equations*,
Universitext, DOI 10.1007/978-0-387-79146-3_42,
© Springer Science+Business Media, LLC 2009

The expression in brackets is a function of ω, is denoted by $F(\omega)$ or $F(f)$, and is called the *Fourier transform* of f. Now writing x for t, we get

$$F(\omega) = \frac{1}{\sqrt{2\pi}} \int_{-\infty}^{\infty} f(x) e^{-i\omega x} dx \qquad (42.6)$$

and with this (42.5) becomes

$$f(x) = \frac{1}{\sqrt{2\pi}} \int_{-\infty}^{\infty} F(\omega) e^{i\omega x} d\omega. \qquad (42.7)$$

The representation (42.7) is called the *inverse Fourier transform* of $F(\omega)$.

Finally, as in Theorem 41.1, if $f(x)$, $-\infty < x < \infty$ is piecewise continuous on each finite interval, and $\int_{-\infty}^{\infty} |f(x)| < \infty$. Then, the Fourier transform (42.6) of $f(x)$ exists. Further, at each x,

$$\frac{1}{\sqrt{2\pi}} \int_{-\infty}^{\infty} F(\omega) e^{i\omega x} d\omega = \frac{1}{2}[f(x+0) + f(x-0)].$$

Example 42.1. We shall find the Fourier transform of

$$f(x) = \begin{cases} k, & 0 < x < a \\ 0, & \text{otherwise.} \end{cases}$$

Clearly, we have

$$F(\omega) = \frac{1}{\sqrt{2\pi}} \int_0^a k e^{-i\omega x} dx = \frac{k(1 - e^{-ia\omega})}{i\omega\sqrt{2\pi}}.$$

Example 42.2. We shall find the Fourier transform of $f(x) = e^{-ax^2}$, $a > 0$. We have

$$
\begin{aligned}
F(\omega) &= \frac{1}{\sqrt{2\pi}} \int_{-\infty}^{\infty} e^{[-ax^2 - i\omega x]} dx \\
&= \frac{1}{\sqrt{2\pi}} \int_{-\infty}^{\infty} \exp\left[-\left(\sqrt{a}x + \frac{i\omega}{2\sqrt{a}}\right)^2 + \left(\frac{i\omega}{2\sqrt{a}}\right)^2\right] dx \\
&= \frac{1}{\sqrt{2\pi}} \exp\left(-\frac{\omega^2}{4a}\right) \int_{-\infty}^{\infty} \exp\left[-\left(\sqrt{a}x + \frac{i\omega}{2\sqrt{a}}\right)^2\right] dx \\
&= \frac{1}{\sqrt{2\pi}} \exp\left(-\frac{\omega^2}{4a}\right) \int_{-\infty}^{\infty} \exp\left(-v^2\right) \frac{dv}{\sqrt{a}}, \quad \sqrt{a}x + \frac{i\omega}{2\sqrt{a}} = v \\
&= \frac{1}{\sqrt{2\pi}} \exp\left(-\frac{\omega^2}{4a}\right) \frac{1}{\sqrt{a}} \sqrt{\pi} = \frac{1}{\sqrt{2a}} e^{-\omega^2/4a}.
\end{aligned}
$$

Example 42.3. We shall find the Fourier transform of the square wave function

$$f(x) = \begin{cases} 0, & x < a \\ 1, & a \le x \le b \\ 0, & x > b. \end{cases}$$

From (42.6), we have

$$F(\omega) = \frac{1}{\sqrt{2\pi}} \int_{-\infty}^{\infty} f(x)e^{-i\omega x} dx = \frac{1}{\sqrt{2\pi}} \int_{a}^{b} e^{-i\omega x} dx$$

$$= \begin{cases} \dfrac{e^{-i\omega b} - e^{-i\omega a}}{-\sqrt{2\pi} i\omega}, & \omega \ne 0 \\ \dfrac{b-a}{\sqrt{2\pi}}, & \omega = 0. \end{cases}$$

Further, it follows that

$$\frac{1}{\sqrt{2\pi}} \int_{-\infty}^{\infty} \frac{e^{-i\omega b} - e^{-i\omega a}}{-\sqrt{2\pi} i\omega} e^{i\omega x} d\omega = \begin{cases} 0, & x < a \\ 1/2, & x = a \\ 1, & a < x < b \\ 1/2, & x = b \\ 0, & x > b. \end{cases}$$

Example 42.4. We shall find the Fourier transform of the function

$$f(x) = \begin{cases} x^2 e^{-x}, & x > 0 \\ 0, & x < 0. \end{cases}$$

We have

$$F(\omega) = \frac{1}{\sqrt{2\pi}} \int_{0}^{\infty} f(x)e^{-i\omega x} dx = \frac{1}{\sqrt{2\pi}} \int_{0}^{\infty} x^2 e^{-x} e^{-i\omega x} dx$$

$$= \frac{1}{\sqrt{2\pi}} \int_{0}^{\infty} x^2 e^{-x(1+i\omega)} dx = \sqrt{\frac{2}{\pi}} \frac{1}{(1+i\omega)^3}.$$

The Fourier transform is a linear operation. We state this property in the following result.

Theorem 42.1 (Linearity Property). The Fourier transform is a linear operation, i.e., for arbitrary functions $f(x)$ and $g(x)$ whose Fourier transforms exist and arbitrary constants a and b,

$$F(af + bg) = aF(f) + bF(g).$$

Now we state and prove the following two results, which will be used repeatedly in the next two lectures.

Theorem 42.2 (Transform of the Derivative). Let $f(x)$ be continuous on the x-axis and $f(x) \to 0$ as $|x| \to \infty$. Furthermore, let $f'(x)$ be absolutely integrable on the x-axis. Then

$$F(f'(x)) = i\omega F(f).$$

Proof. Integrating by parts and using $f(x) \to 0$ as $|x| \to \infty$, we obtain

$$
\begin{aligned}
F(f'(x)) &= \frac{1}{\sqrt{2\pi}} \int_{-\infty}^{\infty} f'(x)e^{-i\omega x} dx \\
&= \frac{1}{\sqrt{2\pi}} \left[f(x)e^{-i\omega x} \Big|_{-\infty}^{\infty} - (-i\omega) \int_{-\infty}^{\infty} f(x)e^{-i\omega x} dx \right] \\
&= i\omega F(f(x)). \quad \blacksquare
\end{aligned}
$$

It is clear that

$$F(f'') = i\omega F(f') = (i\omega)^2 F(f) = -\omega^2 F(f).$$

Example 42.5. Clearly, we have

$$
\begin{aligned}
F\left(xe^{-x^2}\right) &= F\left(-\frac{1}{2}(e^{-x^2})'\right) = -\frac{1}{2}F\left((e^{-x^2})'\right) \\
&= -\frac{1}{2}i\omega F\left(e^{-x^2}\right) = -\frac{1}{2}i\omega \frac{1}{\sqrt{2}} e^{-\omega^2/4} \\
&= -\frac{i\omega}{2\sqrt{2}} e^{-\omega^2/4}.
\end{aligned}
$$

Theorem 42.3 (Convolution Theorem). Suppose that $f(x)$ and $g(x)$ are piecewise continuous, bounded, and absolutely integrable functions on the x-axis. Then

$$F(f * g) = \sqrt{2\pi}F(f)F(g), \tag{42.8}$$

where $f * g$ is the convolution of functions f and g defined as

$$(f * g)(x) = \int_{-\infty}^{\infty} f(t)g(x - t)dt = \int_{-\infty}^{\infty} f(x - t)g(t)dt. \tag{42.9}$$

Proof. By the definition and an interchange of the order of integration, we have

$$
\begin{aligned}
F(f * g) &= \frac{1}{\sqrt{2\pi}} \int_{-\infty}^{\infty} \int_{-\infty}^{\infty} f(\tau)g(x - \tau)e^{-i\omega x} d\tau dx \\
&= \frac{1}{\sqrt{2\pi}} \int_{-\infty}^{\infty} \int_{-\infty}^{\infty} f(\tau)g(x - \tau)e^{-i\omega x} dx d\tau.
\end{aligned}
$$

Now we make the substitution $x - \tau = \nu$, so that $x = \tau + \nu$ and

$$
\begin{aligned}
F(f * g) &= \frac{1}{\sqrt{2\pi}} \int_{-\infty}^{\infty} \int_{-\infty}^{\infty} f(\tau) g(\nu) e^{-i\omega(\tau+\nu)} d\nu d\tau \\
&= \frac{1}{\sqrt{2\pi}} \int_{-\infty}^{\infty} f(\tau) e^{-i\omega\tau} d\tau \int_{-\infty}^{\infty} g(\nu) e^{-i\omega\nu} d\nu \\
&= \sqrt{2\pi} F(f) G(g). \quad \blacksquare
\end{aligned}
$$

By taking the inverse Fourier transform on both sides of (42.8) and writing $F(f) = \hat{f}(\omega)$ and $F(g) = \hat{g}(\omega)$, and noting that $\sqrt{2\pi}$ and $1/\sqrt{2\pi}$ cancel each other, we obtain

$$
(f * g)(x) = \int_{-\infty}^{\infty} \hat{f}(\omega) \hat{g}(\omega) e^{i\omega x} d\omega. \tag{42.10}
$$

Fourier Transforms

	$f(x)$	$F(\omega)$
1.		
2.	$\begin{cases} 1, & -b < x < b \\ 0, & \text{otherwise} \end{cases}$	$\sqrt{\dfrac{2}{\pi}} \dfrac{\sin b\omega}{\omega}$
3.	$\begin{cases} 1, & b < x < c \\ 0, & \text{otherwise} \end{cases}$	$\dfrac{e^{-ib\omega} - e^{-ic\omega}}{i\omega\sqrt{2\pi}}$
4.	$\dfrac{1}{x^2 + a^2}, \quad a > 0$	$\sqrt{\dfrac{\pi}{2}} \dfrac{e^{-a|\omega|}}{a}$
5.	$\begin{cases} x, & 0 < x < b \\ 2x - a, & b < x < 2b \\ 0, & \text{otherwise} \end{cases}$	$\dfrac{-1 + 2e^{ib\omega} - e^{-2ib\omega}}{\sqrt{2\pi}\omega^2}$
6.	$\begin{cases} e^{-ax}, & x > 0 \\ 0, & \text{otherwise} \end{cases}$	$\dfrac{1}{\sqrt{2\pi}(a + i\omega)}$
7.	$\begin{cases} e^{ax}, & b < x < c \\ 0, & \text{otherwise} \end{cases}$	$\dfrac{e^{(a-i\omega)c} - e^{(a-i\omega)b}}{\sqrt{2\pi}(a - i\omega)}$
8.	$\begin{cases} e^{iax}, & -b < x < b \\ 0, & \text{otherwise} \end{cases}$	$\sqrt{\dfrac{2}{\pi}} \dfrac{\sin b(\omega - a)}{\omega - a}$
9.	$\begin{cases} e^{iax}, & b < x < c \\ 0, & \text{otherwise} \end{cases}$	$\dfrac{i}{\sqrt{2\pi}} \dfrac{e^{ib(a-\omega)} - e^{ic(a-\omega)}}{a - \omega}$
10.	$\dfrac{\sin ax}{x}, \quad a > 0$	$\begin{cases} \sqrt{\dfrac{\pi}{2}}, & |\omega| < a \\ 0, & |\omega| > a. \end{cases}$

Problems

42.1. Find the Fourier integral representation of the following functions:

(i) $f(x) = \begin{cases} -1, & -1 < x < 0 \\ 1, & 0 < x < 1 \\ 0, & \text{otherwise} \end{cases}$

(ii) $f(x) = e^{-|x|}$

(iii) $f(x) = \begin{cases} 1 - x^2, & |x| < 1 \\ 0, & |x| > 1. \end{cases}$

42.2. Use Fourier integral representation to show that

(i) $\displaystyle\int_0^\infty \frac{\sin x\omega}{\omega}\,d\omega = \begin{cases} \pi/2, & x > 0 \\ 0, & x = 0 \\ -\pi/2, & x < 0 \end{cases}$

(ii) $\displaystyle\int_0^\infty \frac{\cos x\omega + \omega \sin x\omega}{1 + \omega^2}\,d\omega = \begin{cases} 0, & x < 0 \\ \pi/2, & x = 0 \\ \pi e^{-x}, & x > 0 \end{cases}$

(iii) $\displaystyle\int_0^\infty \frac{\omega^3 \sin x\omega}{\omega^4 + 4}\,d\omega = \frac{\pi}{2} e^{-x}\cos x \ \text{ if } \ x > 0$

(iv) $\displaystyle\int_0^\infty \frac{1 - \cos \pi\omega}{\omega}\sin x\omega\,d\omega = \begin{cases} \pi/2, & 0 < x < \pi \\ 0, & x > \pi \end{cases}$

(v) $\displaystyle\int_0^\infty \frac{\cos(\pi\omega/2)\cos x\omega}{1 - \omega^2}\,d\omega = \begin{cases} (\pi/2)\cos x, & |x| < \pi/2 \\ 0, & |x| > \pi/2 \end{cases}$

(vi) $\displaystyle\int_0^\infty \frac{\sin \pi\omega \sin x\omega}{1 - \omega^2}\,d\omega = \begin{cases} (\pi/2)\sin x, & 0 \le x \le \pi \\ 0, & x > \pi. \end{cases}$

42.3. Show that

(i) $\cos ax^2 = \dfrac{1}{\sqrt{\pi a}}\displaystyle\int_0^\infty \cos\left(\frac{\omega^2}{4a} - \frac{\pi}{4}\right)\cos \omega x\,d\omega, \quad a > 0$

(ii) $\sin ax^2 = \dfrac{1}{\sqrt{\pi a}}\displaystyle\int_0^\infty \cos\left(\frac{\omega^2}{4a} + \frac{\pi}{4}\right)\cos \omega x\,d\omega, \quad a > 0$

(iii) $e^{-x}\cos x = \dfrac{2}{\pi}\displaystyle\int_0^\infty \frac{\omega^2 + 2}{\omega^4 + 4}\cos \omega x\,d\omega, \quad x > 0.$

42.4. Show that

(i) $e^{-x}\cos x = \dfrac{2}{\pi}\displaystyle\int_0^\infty \frac{\omega^3}{\omega^4 + 4}\sin \omega x\,d\omega, \quad x > 0$

(ii) $\tan^{-1}\dfrac{2a}{x} = 2\displaystyle\int_0^\infty \frac{e^{-a\omega}}{\omega}\sinh a\omega \sin \omega x\,d\omega, \quad a > 0, \ x > 0$

(iii) $\cosh ax = \dfrac{1}{a} \displaystyle\int_0^\infty \tanh \dfrac{\pi\omega}{2a} \sin \omega x d\omega, \quad a > 0, \quad x > 0.$

42.5.   Find Fourier transforms of the following functions:

(i) $f(x) = \begin{cases} e^{-x} \sin x, & x > 0 \\ 0, & x < 0 \end{cases}$

(ii) $f(x) = \begin{cases} e^{-x}, & x > 0 \\ e^{2x}, & x < 0 \end{cases}$

(iii) $f(x) = e^{-a|x|}, \quad a > 0.$

42.6.   Parseval's equality for the Fourier transform (42.6) and its inverse (42.7) becomes

$$\int_{-\infty}^{\infty} |f(x)|^2 dx = \int_{-\infty}^{\infty} |F(\omega)|^2 d\omega.$$

Use Example 42.3 to show that

$$\frac{1}{2\pi} \int_{-\infty}^{\infty} \frac{|e^{-i\omega b} - e^{-i\omega a}|^2}{\omega^2} d\omega = b - a.$$

42.7.   Show that the solution of the integral equation

$$\int_0^\infty f(\omega) \sin x\omega d\omega = g(x),$$

where $g(x) = 1$ when $0 < x < \pi$, $g(x) = 0$ when $x > \pi$ is

$$f(\omega) = \frac{2}{\pi} \frac{1 - \cos \pi\omega}{\omega}, \quad \omega > 0.$$

42.8.   Show that the integral equation

$$\int_0^\infty f(\omega) \cos x\omega d\omega = e^{-x},$$

has the solution

$$f(\omega) = \frac{2}{\pi} \frac{1}{1 + \omega^2}, \quad \omega > 0.$$

42.9.   Find the solution of the integral equation

$$f(x) = g(x) + \int_{-\infty}^{\infty} h(t) f(x - t) dt,$$

where the functions f, g and h satisfy the conditions of Theorem 41.1. In particular, solve the integral equation

$$\int_{-\infty}^{\infty} \frac{f(t)dt}{(x-t)^2 + a^2} = \frac{1}{x^2 + b^2}, \quad 0 < a < b.$$

42.10.   Evaluate the following integrals

(i) $\displaystyle\int_{-\infty}^{\infty} \frac{dx}{(x^2 + a^2)(x^2 + b^2)}, \quad a > 0, \quad b > 0$

(ii) $\displaystyle\int_{-\infty}^{\infty} \frac{x^2\, dx}{(x^2 + a^2)(x^2 + b^2)}, \quad a > 0, \quad b > 0.$

Answers or Hints

42.1. (i) $\frac{2}{\pi}\int_0^{\infty} \left(\frac{1-\cos\omega}{\omega}\right) \sin\omega x d\omega$ (ii) $\frac{2}{\pi}\int_0^{\infty} \frac{\cos\omega x}{1+\omega^2} d\omega$
(iii) $\frac{4}{\pi}\int_0^{\infty} \left(\frac{\sin\omega - \omega\cos\omega}{\omega^3}\right) \cos\omega x d\omega.$

42.5. (i) $\frac{1}{\sqrt{2\pi}[(1+i\omega)^2+1]}$ (ii) $\frac{3}{\sqrt{2\pi}[2+i\omega+\omega^2]}$ (iii) $\sqrt{\frac{2}{\pi}} \frac{a}{(a^2+\omega^2)}.$

42.9. $f(x) = \frac{1}{\sqrt{2\pi}} \int_{-\infty}^{\infty} \frac{F(g)}{1-\sqrt{2\pi}F(h)} e^{i\omega x} d\omega, \quad \frac{a(b-a)}{b\pi(x^2+(b-a)^2)}.$

42.10. Put $x = 0$ in (4.9) and (4.10). (i) $\frac{\pi}{ab(a+b)}$ (ii) $\frac{\pi}{(a+b)}.$

Lecture 43
Fourier Transform Method for Partial DEs

Here and in the next lecture we shall consider problems in infinite domains which can be effectively solved by finding the Fourier transform or the Fourier sine or cosine transform of the unknown function. For such problems usually the method of separation of variables does not work because the Fourier series are not adequate to yield complete solutions. This is due to the fact that often these problems require a continuous superposition of separated solutions. We shall illustrate the method by considering several examples.

Example 43.1. We will show how the Fourier transform applies to the heat equation. We consider the heat flow problem of an infinitely long thin bar insulated on its lateral surface, which is modeled by the following initial-value problem

$$u_t = c^2 u_{xx}, \quad -\infty < x < \infty, \quad t > 0, \quad c > 0$$

$$u \quad \text{and} \quad u_x \quad \text{finite as} \quad |x| \to \infty, \quad t > 0 \tag{43.1}$$

$$u(x,0) = f(x), \quad -\infty < x < \infty,$$

where the function f is piecewise smooth and absolutely integrable in $(-\infty, \infty)$.

Let $U(\omega, t)$ be the Fourier transform of $u(x,t)$. Thus, from the Fourier transform pair, we have

$$u(x,t) = \frac{1}{\sqrt{2\pi}} \int_{-\infty}^{\infty} U(\omega,t) e^{i\omega x} d\omega$$

$$U(\omega,t) = \frac{1}{\sqrt{2\pi}} \int_{-\infty}^{\infty} u(x,t) e^{-i\omega x} dx.$$

Assuming that the derivatives can be taken under the integral, we get

$$\frac{\partial u}{\partial t} = \frac{1}{\sqrt{2\pi}} \int_{-\infty}^{\infty} \frac{\partial U(\omega,t)}{\partial t} e^{i\omega x} dx$$

$$\frac{\partial u}{\partial x} = \frac{1}{\sqrt{2\pi}} \int_{-\infty}^{\infty} U(\omega,t)(i\omega) e^{i\omega x} d\omega$$

$$\frac{\partial^2 u}{\partial x^2} = \frac{1}{\sqrt{2\pi}} \int_{-\infty}^{\infty} U(\omega,t)(i\omega)^2 e^{i\omega x} d\omega.$$

R.P. Agarwal, D. O'Regan, *Ordinary and Partial Differential Equations*,
Universitext, DOI 10.1007/978-0-387-79146-3_43,
© Springer Science+Business Media, LLC 2009

In order for $u(x,t)$ to satisfy the heat equation, we must have

$$0 = \frac{\partial u}{\partial t} - c^2 \frac{\partial^2 u}{\partial x^2} = \frac{1}{\sqrt{2\pi}} \int_{-\infty}^{\infty} \left[\frac{\partial U(\omega,t)}{\partial t} + c^2 \omega^2 U(\omega,t) \right] e^{i\omega x} d\omega.$$

Thus, U must be a solution of the ordinary differential equation

$$\frac{dU}{dt} + c^2 \omega^2 U = 0.$$

The initial condition is determined by

$$
\begin{aligned}
U(\omega,0) &= \frac{1}{\sqrt{2\pi}} \int_{-\infty}^{\infty} u(x,0) e^{-i\omega x} dx \\
&= \frac{1}{\sqrt{2\pi}} \int_{-\infty}^{\infty} f(x) e^{-i\omega x} dx = F(\omega).
\end{aligned}
$$

Therefore, we have

$$U(\omega,t) = F(\omega) e^{-\omega^2 c^2 t},$$

and hence

$$u(x,t) = \frac{1}{\sqrt{2\pi}} \int_{-\infty}^{\infty} F(\omega) e^{-\omega^2 c^2 t} e^{i\omega x} d\omega. \tag{43.2}$$

Now since

$$\frac{1}{\sqrt{2\pi}} \int_{-\infty}^{\infty} e^{-\omega^2 c^2 t} e^{i\omega x} d\omega = \sqrt{2\pi} \frac{e^{-x^2/(4c^2 t)}}{\sqrt{4\pi c^2 t}},$$

if in (42.10) we denote $F(\omega) = \hat{f}(\omega)$ and $\hat{g}(\omega) = e^{-\omega^2 c^2 t}$, then from (43.2) it follows that

$$
\begin{aligned}
u(x,t) &= \frac{1}{\sqrt{2\pi}} \int_{-\infty}^{\infty} f(\mu) \sqrt{2\pi} \frac{e^{-(x-\mu)^2/4c^2 t}}{\sqrt{4\pi c^2 t}} d\mu \\
&= \int_{-\infty}^{\infty} f(\mu) \frac{e^{-(x-\mu)^2/4c^2 t}}{\sqrt{4\pi c^2 t}} d\mu.
\end{aligned}
\tag{43.3}
$$

This formula is due to Gauss and Weierstrass.

For each μ the function $(x,t) \to e^{-(x-\mu)^2/4c^2 t}/\sqrt{4\pi c^2 t}$ is a solution of the heat equation and is called the *fundamental solution*. Thus, (43.3) gives a representation of the solution as a continuous superposition of the fundamental solution.

We recall that the *standard normal distribution function* Φ is defined as

$$\Phi(c) = \frac{1}{\sqrt{2\pi}} \int_{-\infty}^{c} e^{-z^2/2} dz.$$

This is a continuous increasing function with $\Phi(-\infty) = 0$, $\Phi(0) = 1/2$, $\Phi(\infty) = 1$. If $a < b$, then we can write

$$\int_a^b \frac{e^{-(x-\mu)^2/4c^2t}}{\sqrt{4\pi c^2 t}} d\mu = \frac{1}{\sqrt{2\pi}} \int_{(x-b)/\sqrt{2c^2t}}^{(x-a)/\sqrt{2c^2t}} e^{-z^2/2} dz, \qquad \frac{(x-\mu)}{\sqrt{2c^2t}} = z$$

$$= \frac{1}{\sqrt{2\pi}} \int_{-\infty}^{\frac{(x-a)}{\sqrt{2c^2t}}} e^{-z^2/2} dz - \frac{1}{\sqrt{2\pi}} \int_{-\infty}^{\frac{(x-b)}{\sqrt{2c^2t}}} e^{-z^2/2} dz$$

$$= \Phi\left(\frac{x-a}{\sqrt{2c^2t}}\right) - \Phi\left(\frac{x-b}{\sqrt{2c^2t}}\right).$$

(43.4)

From (43.3) and (43.4) it is clear that the solution of the problem

$$u_t = c^2 u_{xx}, \quad -\infty < x < \infty, \quad t > 0$$

$$u(x,0) = f(x) = \begin{cases} 0, & x < a \\ L, & a \le x \le b \\ 0, & x > b \end{cases}$$

can be written as

$$u(x,t) = L\Phi\left(\frac{x-a}{\sqrt{2c^2t}}\right) - L\Phi\left(\frac{x-b}{\sqrt{2c^2t}}\right).$$

Now using the properties $\Phi(-\infty) = 0$, $\Phi(0) = 1/2$, $\Phi(\infty) = 1$ we can verify that

$$\lim_{t \to 0} u(x,t) = \begin{cases} 0, & x < a \\ L/2, & x = a \\ L, & a < x < b \\ L/2, & x = b \\ 0, & x > b. \end{cases}$$

Example 43.2. Consider the problem

$$u_t = c^2 u_{xx}, \quad x > 0, \quad t > 0$$
$$u(0,t) = 0, \quad t > 0$$
$$u \text{ and } u_x \text{ finite as } x \to \infty, \quad t > 0$$
$$u(x,0) = f(x), \quad x > 0,$$

(43.5)

which appears in heat flow in a semi–infinite region. In (43.5) the function f is piecewise smooth and absolutely integrable in $[0, \infty)$.

We define the odd function

$$\bar{f}(x) = \begin{cases} f(x), & x > 0 \\ -f(-x), & x < 0. \end{cases}$$

Then from (43.3), we have

$$u(x,t) = \int_{-\infty}^{\infty} \frac{e^{-(x-\mu)^2/4c^2t}}{\sqrt{4\pi c^2 t}} \bar{f}(\mu)d\mu$$

$$= \int_{-\infty}^{0} \frac{e^{-(x-\mu)^2/4c^2t}}{\sqrt{4\pi c^2 t}} \bar{f}(\mu)d\mu + \int_{0}^{\infty} \frac{e^{-(x-\mu)^2/4c^2t}}{\sqrt{4\pi c^2 t}} \bar{f}(\mu)d\mu.$$

In the first integral we change μ to $-\mu$ and use the oddness of $\bar{f}$, to obtain

$$\int_{-\infty}^{0} \frac{e^{-(x-\mu)^2/4c^2t}}{\sqrt{4\pi c^2 t}} \bar{f}(\mu)d\mu = -\int_{0}^{\infty} \frac{e^{-(x+\mu)^2/4c^2t}}{\sqrt{4\pi c^2 t}} f(\mu)d\mu.$$

Thus, the solution of the problem (43.5) can be written as

$$u(x,t) = \int_{0}^{\infty} \frac{e^{-(x-\mu)^2/4c^2t} - e^{-(x+\mu)^2/4c^2t}}{\sqrt{4\pi c^2 t}} f(\mu)d\mu. \qquad (43.6)$$

The above procedure to find the solution of (43.5) is called *the method of images*.

In an analogous way it can be shown that the solution of the problem

$$u_t = c^2 u_{xx}, \quad x > 0, \quad t > 0$$
$$u_x(0,t) = 0, \quad t > 0$$
$$u \quad \text{and} \quad u_x \quad \text{finite as} \quad x \to \infty, \quad t > 0 \qquad (43.7)$$
$$u(x,0) = f(x), \quad x > 0$$

can be written as

$$u(x,t) = \int_{0}^{\infty} \frac{e^{-(x-\mu)^2/4c^2t} + e^{-(x+\mu)^2/4c^2t}}{\sqrt{4\pi c^2 t}} f(\mu)d\mu. \qquad (43.8)$$

Here, of course, we need to extend $f(x)$ to an even function

$$\bar{f}(x) = \begin{cases} f(x), & x > 0 \\ f(-x), & x < 0. \end{cases}$$

In (43.7) the physical significance of the condition $u_x(0,t) = 0$ is that there is a perfect insulation, i.e., there is no heat flux across the surface.

Example 43.3. Consider the initial-value problem for the wave equation

$$u_{tt} = c^2 u_{xx}, \quad -\infty < x < \infty, \quad t > 0, \quad c > 0$$
$$u \quad \text{and} \quad u_x \quad \text{finite as} \quad |x| \to \infty, \quad t > 0$$
$$u(x,0) = f_1(x), \quad -\infty < x < \infty \qquad (43.9)$$
$$u_t(x,0) = f_2(x), \quad -\infty < x < \infty,$$

where the functions f_1 and f_2 are piecewise smooth and absolutely integrable in $(-\infty, \infty)$.

To find the solution of this problem, we introduce the Fourier transforms

$$F_j(\omega) = \frac{1}{\sqrt{2\pi}} \int_{-\infty}^{\infty} e^{-i\omega x} f_j(x) dx, \quad j = 1, 2$$

and its inversion formulas

$$f_j(x) = \frac{1}{\sqrt{2\pi}} \int_{-\infty}^{\infty} e^{i\omega x} F_j(\omega) d\omega, \quad j = 1, 2.$$

We also need the Fourier representation of the solution $u(x, t)$,

$$u(x, t) = \frac{1}{\sqrt{2\pi}} \int_{-\infty}^{\infty} U(\omega, t) e^{i\omega x} d\omega,$$

where $U(\omega, t)$ is an unknown function, which we will now determine. For this, we substitute this into the differential equation (43.9), to obtain

$$0 = \frac{1}{\sqrt{2\pi}} \int_{-\infty}^{\infty} \left[\frac{\partial^2 U(\omega, t)}{\partial t^2} + c^2 \omega^2 U(\omega, t) \right] e^{i\omega x} d\omega.$$

Thus, U must be a solution of the ordinary differential equation

$$\frac{d^2 U}{dt^2} + c^2 \omega^2 U = 0,$$

whose solution can be written as

$$U(\omega, t) = c_1(\omega) \cos \omega c t + c_2(\omega) \sin \omega c t.$$

To find $c_1(\omega)$ and $c_2(\omega)$, we note that

$$f_1(x) = u(x, 0) = \frac{1}{\sqrt{2\pi}} \int_{-\infty}^{\infty} c_1(\omega) e^{i\omega x} d\omega$$

$$f_2(x) = \frac{\partial u(x, 0)}{\partial t} = \frac{1}{\sqrt{2\pi}} \int_{-\infty}^{\infty} \omega c \, c_2(\omega) e^{i\omega x} d\omega$$

and hence $F_1(\omega) = c_1(\omega)$ and $F_2(\omega) = \omega c \, c_2(\omega)$.

Therefore, it follows that

$$U(\omega, t) = F_1(\omega) \cos \omega c t + \frac{F_2(\omega)}{\omega c} \sin \omega c t.$$

and hence the Fourier representation of the solution is

$$u(x,t) = \frac{1}{\sqrt{2\pi}} \int_{-\infty}^{\infty} \left[F_1(\omega)\cos\omega ct + \frac{F_2(\omega)}{\omega c}\sin\omega ct \right] e^{i\omega x} d\omega. \quad (43.10)$$

Now since $\cos\theta = (e^{i\theta} + e^{-i\theta})/2$, $\sin\theta = (e^{i\theta} - e^{-i\theta})/2i$, we have

$$\frac{1}{\sqrt{2\pi}} \int_{-\infty}^{\infty} F_1(\omega)(\cos\omega ct)e^{i\omega x} dx$$

$$= \frac{1}{2}\frac{1}{\sqrt{2\pi}} \int_{-\infty}^{\infty} F_1(\omega)\left(e^{i\omega ct} + e^{-i\omega ct}\right) e^{i\omega x} d\omega$$

$$= \frac{1}{2}\frac{1}{\sqrt{2\pi}} \int_{-\infty}^{\infty} F_1(\omega)\left(e^{i\omega(x+ct)} + e^{i\omega(x-ct)}\right) d\omega$$

$$= \frac{1}{2}[f_1(x+ct) + f_1(x-ct)].$$

Similarly,

$$\frac{1}{\sqrt{2\pi}} \int_{-\infty}^{\infty} F_2(\omega)\frac{\sin\omega ct}{\omega c}e^{i\omega x} d\omega$$

$$= \frac{1}{2}\frac{1}{\sqrt{2\pi}} \int_{-\infty}^{\infty} F_2(\omega)\frac{e^{i\omega ct} - e^{-i\omega ct}}{i\omega c}e^{i\omega x} d\omega$$

$$= \frac{1}{2}\frac{1}{\sqrt{2\pi}} \int_{-\infty}^{\infty} F_2(\omega)\frac{e^{i\omega(x+ct)} - e^{i\omega(x-ct)}}{i\omega c} d\omega$$

$$= \frac{1}{2c}\frac{1}{\sqrt{2\pi}} \int_{-\infty}^{\infty} F_2(\omega)\left(\int_{x-ct}^{x+ct} e^{i\omega\xi}d\xi\right) d\omega$$

$$= \frac{1}{2c}\int_{x-ct}^{x+ct} \left[\frac{1}{\sqrt{2\pi}} \int_{-\infty}^{\infty} e^{i\omega\xi}F_2(\omega)d\omega\right] d\xi$$

$$= \frac{1}{2c}\int_{x-ct}^{x+ct} f_2(\xi)d\xi.$$

Putting these together yields d'Alembert's formula

$$u(x,t) = \frac{1}{2}[f_1(x+ct) + f_1(x-ct)] + \frac{1}{2c}\int_{x-ct}^{x+ct} f_2(\xi)d\xi, \quad (43.11)$$

which is also obtained in Problem 33.10.

Lecture 44
Fourier Transform Method for Partial DEs (Cont'd.)

In this lecture we shall employ the Fourier transform, or the Fourier sine or cosine transform to find solutions of the Laplace equation in infinite domains. Here we shall also introduce finite Fourier sine and cosine transforms, and demonstrate how easily these can be used directly to solve some finite domain problems.

Example 44.1. We shall find the solution of the following problem involving the Laplace equation in a half-plane:

$$u_{xx} + u_{yy} = 0, \quad -\infty < x < \infty, \quad y > 0$$
$$u(x, 0) = f(x), \quad -\infty < x < \infty \tag{44.1}$$
$$|u(x, y)| \le M, \quad -\infty < x < \infty, \quad y > 0,$$

where the function f is piecewise smooth and absolutely integrable in $(-\infty, \infty)$. If $f(x) \to 0$ as $|x| \to \infty$, then we also have the implied boundary conditions $\lim_{|x| \to \infty} u(x, y) = 0$, $\lim_{y \to +\infty} u(x, y) = 0$.

For this, as in Lecture 43, we let

$$f(x) = \frac{1}{\sqrt{2\pi}} \int_{-\infty}^{\infty} F(\omega)e^{i\omega x}d\omega, \quad F(\omega) = \frac{1}{\sqrt{2\pi}} \int_{-\infty}^{\infty} f(x)e^{-i\omega x}dx$$

and

$$u(x, y) = \frac{1}{\sqrt{2\pi}} \int_{-\infty}^{\infty} U(\omega, y)e^{i\omega x}d\omega.$$

We find that

$$0 = u_{xx} + u_{yy} = \frac{1}{\sqrt{2\pi}} \int_{-\infty}^{\infty} \left[-\omega^2 U(\omega, y) + \frac{\partial^2 U(\omega, y)}{\partial y^2} \right] e^{i\omega x}d\omega.$$

Thus, U must satisfy the ordinary differential equation

$$\frac{d^2 U}{dy^2} = \omega^2 U$$

and the initial condition $U(\omega, 0) = F(\omega)$ for each ω.

R.P. Agarwal, D. O'Regan, *Ordinary and Partial Differential Equations*, Universitext, DOI 10.1007/978-0-387-79146-3_44,
© Springer Science+Business Media, LLC 2009

The general solution of the ordinary differential equation is $c_1 e^{\omega y} + c_2 e^{-\omega y}$. If we impose the initial condition and the boundedness condition, the solution becomes

$$U(\omega, y) = \left\{ \begin{array}{ll} F(\omega)e^{-\omega y}, & \omega \geq 0 \\ F(\omega)e^{\omega y}, & \omega < 0 \end{array} \right\} = F(\omega)e^{-|\omega| y}.$$

Thus, the desired Fourier representation of the solution is

$$u(x, y) = \frac{1}{\sqrt{2\pi}} \int_{-\infty}^{\infty} F(\omega)e^{-|\omega| y} e^{i\omega x} d\omega.$$

To obtain an explicit representation, we insert the formula for $F(\omega)$ and formally interchange the order of integration, to obtain

$$\begin{aligned} u(x, y) &= \frac{1}{2\pi} \int_{-\infty}^{\infty} \left(\int_{-\infty}^{\infty} f(\xi)e^{-i\omega \xi} d\xi \right) e^{-|\omega| y} e^{i\omega x} d\omega \\ &= \frac{1}{2\pi} \int_{-\infty}^{\infty} \left(\int_{-\infty}^{\infty} e^{i\omega(x-\xi)} e^{-|\omega| y} d\omega \right) f(\xi) d\xi. \end{aligned}$$

Now the inner integral is

$$\begin{aligned} \int_{-\infty}^{\infty} e^{i\omega(x-\xi)} e^{-|\omega| y} d\omega &= 2\mathrm{Re} \int_{0}^{\infty} e^{i\omega(x-\xi)} e^{-\omega y} d\omega \\ &= 2\mathrm{Re}\frac{1}{y - i(x - \xi)} = \frac{2y}{y^2 + (x - \xi)^2}. \end{aligned}$$

Therefore, the solution $u(x, y)$ can be explicitly written as

$$u(x, y) = \frac{1}{\pi} \int_{-\infty}^{\infty} \frac{y}{y^2 + (x - \xi)^2} f(\xi) d\xi. \tag{44.2}$$

This representation is known as *Poisson's integral formula*.

In particular, for

$$u(x, 0) = f(x) = \left\{ \begin{array}{ll} 1, & a < x < b \\ 0, & \text{otherwise} \end{array} \right.$$

(44.2) becomes

$$u(x, y) = \frac{1}{\pi} \int_{a}^{b} \frac{y}{y^2 + (x - \xi)^2} d\xi = \frac{1}{\pi} \int_{a}^{b} \frac{d\xi/y}{\frac{(x-\xi)^2}{y^2} + 1}.$$

Using the substitution $v = (\xi - x)/y$, we have $d\xi = y \, dv$, so that

$$\begin{aligned} u(x, y) &= \frac{1}{\pi} \int_{(a-x)/y}^{(b-x)/y} \frac{1}{1 + v^2} dv \\ &= \frac{1}{\pi} \left(\tan^{-1}\frac{b - x}{y} - \tan^{-1}\frac{a - x}{y} \right) = \frac{1}{\pi}(\theta_b - \theta_a), \end{aligned}$$

where θ_a and θ_b are as in Figure 44.1.

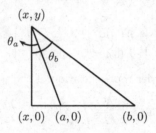

Figure 44.1

Example 44.2. We shall employ the Fourier sine transform to find the solution of the following problem involving the Laplace equation in a semi-infinite strip:

$$\begin{aligned}
&u_{xx} + u_{yy} = 0, \quad 0 < x < \infty, \quad 0 < y < b \\
&u(x,0) = f(x), \quad 0 < x < \infty \\
&u(0,y) = 0, \quad 0 < y < b \\
&u(x,b) = 0, \quad 0 < x < \infty,
\end{aligned} \tag{44.3}$$

where the function f is piecewise smooth and absolutely integrable in $[0,\infty)$. We shall also need the boundary conditions $\lim_{x\to\infty} u(x,y) = 0$ and $\lim_{x\to\infty} u_x(x,y) = 0$.

For this, we let

$$f(x) = \sqrt{\frac{2}{\pi}} \int_0^\infty F_s(\omega) \sin\omega x\, d\omega, \quad F_s(\omega) = \sqrt{\frac{2}{\pi}} \int_0^\infty f(x) \sin\omega x\, dx$$

and

$$u(x,y) = \sqrt{\frac{2}{\pi}} \int_0^\infty U_s(\omega,y) \sin\omega x\, d\omega.$$

This, as in Example 44.1, leads to the same ordinary DE $U_s'' = \omega^2 U_s$, and hence

$$U_s(\omega,y) = c_1(\omega)\cosh\omega y + c_2(\omega)\sinh\omega y.$$

Now the boundary condition $U_s(\omega,b) = 0$ yields

$$c_1(\omega) = -c_2(\omega)\frac{\sinh\omega b}{\cosh\omega b}.$$

Thus, we have

$$U_s(\omega,y) = -c_2(\omega)\frac{\sinh\omega b}{\cosh\omega b}\cosh\omega y + c_2(\omega)\sinh\omega y = c_2(\omega)\frac{\sinh\omega(y-b)}{\cosh\omega b}.$$

Now since $U_s(\omega, 0) = F_s(\omega)$, we find

$$c_2(\omega) = -F_s(\omega)\frac{\cosh \omega b}{\sinh \omega b},$$

and therefore

$$U_s(\omega, y) = F_s(\omega)\frac{\sinh \omega(b-y)}{\sinh \omega b}.$$

This gives the solution

$$
\begin{aligned}
u(x,y) &= \sqrt{\frac{2}{\pi}}\int_0^\infty F_s(\omega)\frac{\sinh \omega(b-y)}{\sinh \omega b}\sin \omega x\, d\omega \\
&= \frac{2}{\pi}\int_0^\infty \int_0^\infty f(t)\sin \omega t\, \frac{\sinh \omega(b-y)}{\sinh \omega b}\sin \omega x\, dt\, d\omega.
\end{aligned}
$$

Next we shall introduce finite Fourier sine and cosine transforms.

Definition 44.1. The *finite Fourier sine transform* of $f(x)$, $0 < x < L$ is defined as

$$F_s(n) = \int_0^L f(x)\sin \frac{n\pi x}{L}dx, \tag{44.4}$$

where $n \geq 1$ is an integer. The function $f(x)$ is then called the *inverse finite Fourier sine transform* of $F_s(n)$ and is given by

$$f(x) = \frac{2}{L}\sum_{n=1}^\infty F_s(n)\sin \frac{n\pi x}{L}. \tag{44.5}$$

Definition 44.2. The *finite Fourier cosine transform* of $f(x)$, $0 < x < L$ is defined as

$$F_c(n) = \int_0^L f(x)\cos \frac{n\pi x}{L}dx, \tag{44.6}$$

where $n \geq 0$ is an integer. The function $f(x)$ is then called the *inverse finite Fourier cosine transform* of $F_c(n)$ and is given by

$$f(x) = \frac{1}{L}F_c(0) + \frac{2}{L}\sum_{n=1}^\infty F_c(n)\cos \frac{n\pi x}{L}. \tag{44.7}$$

Finite Fourier transforms are useful in solving partial differential equations. For this, we note that

$$\int_0^L \frac{\partial u(x,t)}{\partial x}\sin \frac{n\pi x}{L}dx = u(x,t)\sin \frac{n\pi x}{L}\Big|_0^L - \frac{n\pi}{L}\int_0^L u(x,t)\cos \frac{n\pi x}{L}dx$$

and hence

$$F_s\left(\frac{\partial u}{\partial x}\right) = -\frac{n\pi}{L}F_c(n) \tag{44.8}$$

and, similarly,

$$F_c\left(\frac{\partial u}{\partial x}\right) = \frac{n\pi}{L}F_s(n) - [u(0,t) - u(L,t)\cos n\pi], \qquad (44.9)$$

$$
\begin{aligned}
F_s\left(\frac{\partial^2 u}{\partial x^2}\right) &= -\frac{n\pi}{L}F_c\left(\frac{\partial u}{\partial x}\right) \\
&= -\frac{n^2\pi^2}{L^2}F_s(n) + \frac{n\pi}{L}[u(0,t) - u(L,t)\cos n\pi],
\end{aligned}
\qquad (44.10)
$$

$$F_c\left(\frac{\partial^2 u}{\partial x^2}\right) = -\frac{n^2\pi^2}{L^2}F_c(n) - [u_x(0,t) - u_x(L,t)\cos n\pi]. \qquad (44.11)$$

Example 44.3. We will use finite Fourier sine transform to find the solution of the problem

$$
\begin{aligned}
\frac{\partial u}{\partial t} &= \frac{\partial^2 u}{\partial x^2}, \quad 0 < x < 4, \quad t > 0 \\
u(x,0) &= 2x, \quad 0 < x < 4 \\
u(0,t) &= u(4,t) = 0, \quad t > 0.
\end{aligned}
\qquad (44.12)
$$

Taking the finite Fourier sine transform with $L = 4$ of both sides of the partial differential equation gives

$$\int_0^4 \frac{\partial u}{\partial t}\sin\frac{n\pi x}{4}dx = \int_0^4 \frac{\partial^2 u}{\partial x^2}\sin\frac{n\pi x}{4}dx.$$

Writing U for $F_s(n)$ and using (44.10) with $u(0,t) = 0,\ u(4,t) = 0$ leads to

$$\frac{dU(n,t)}{dt} = -\frac{n^2\pi^2}{16}U,$$

which can be solved to obtain

$$U(n,t) = ce^{-n^2\pi^2 t/16}.$$

Now taking the finite Fourier sine transform of the condition $u(x,0) = 2x$, we have

$$
\begin{aligned}
U(n,0) &= \int_0^4 2x\sin\frac{n\pi x}{4}dx \\
&= \left[2x\left(-\frac{\cos n\pi x/4}{n\pi/4}\right) - 2\left(-\frac{\sin n\pi x/4}{n^2\pi^2/16}\right)\right]\Bigg|_0^4 = -\frac{32}{n\pi}\cos n\pi.
\end{aligned}
$$

Since $c = U(n,0)$ it follows that

$$U(n,t) = -\frac{32}{n\pi}\cos n\pi\, e^{-n^2\pi^2 t/16}.$$

Thus, from (44.5) we get

$$u(x, t) = -\frac{16}{\pi} \sum_{n=1}^{\infty} \frac{\cos n\pi}{n} e^{-n^2\pi^2 t/16} \sin \frac{n\pi x}{4}.$$

Problems

44.1. Use the Fourier transform to find the solution of the boundary value problem

$$y'' + ay' + by = f(x)$$
$$y(x) \to 0, \quad y'(x) \to 0 \quad \text{as} \quad |x| \to \infty,$$

where the function f is piecewise smooth and absolutely integrable in $(-\infty, \infty)$. In particular, find the solution when $a = 0$, $b = -1$, $f(x) = e^{-|x|}$.

44.2. Use the Fourier sine transform to find the solution of the boundary value problem

$$y'' - k^2 y = f(x), \quad 0 < x < \infty, \quad k > 0$$
$$y(0) = 1, \quad y(x) \to 0, \quad y'(x) \to 0 \quad \text{as} \quad x \to \infty,$$

where the function f is piecewise smooth and absolutely integrable in $[0, \infty)$.

44.3. Use the Fourier cosine transform to find the solution of the boundary value problem

$$y'' - k^2 y = f(x), \quad 0 < x < \infty, \quad k > 0$$
$$y'(0) = 1, \quad y(x) \to 0, \quad y'(x) \to 0 \quad \text{as} \quad x \to \infty,$$

where the function f is piecewise smooth and absolutely integrable in $[0, \infty)$.

44.4. Use the Fourier transform to find solutions of the following ordinary DEs satisfying $y(x) \to 0$, $y'(x) \to 0$ as $|x| \to \infty$:

(i) $xy'' + y' + xy = 0$ (Bessel DE of order zero)
(ii) $xy'' + y' - xy = 0$ (modified Bessel DE of order zero)
(iii) $y'' + y' + xy = 0$.

44.5. Show that the solution (43.3) can be written as

$$u(x, t) = \frac{1}{\sqrt{\pi}} \int_{-\infty}^{\infty} e^{-w^2} f(x + 2\sqrt{c^2 t}\, w) dw.$$

44.6.   Show that the solution (43.6) can be written as

$$u(x,t) = \frac{2}{\pi} \int_0^\infty \int_0^\infty f(\mu)e^{-c^2\omega^2 t} \sin\omega\mu \sin\omega x \, d\omega d\mu.$$

44.7.   Show that the solution (43.8) can be written as

$$u(x,t) = \frac{2}{\pi} \int_0^\infty \int_0^\infty f(\mu)e^{-c^2\omega^2 t} \cos\omega\mu \cos\omega x \, d\omega d\mu.$$

44.8.   Find the solution of (43.1) when the initial temperature distribution in the rod is given by

$$f(x) = e^{-x^2/2}, \quad -\infty < x < \infty.$$

44.9.   Find the solution of (43.5) when the initial temperature distribution in the rod is given by

$$f(x) = xe^{-x^2/4a^2}, \quad x > 0.$$

44.10.   Use the Fourier cosine transform to solve the following problem:

$$u_t = u_{xx}, \quad x > 0, \quad t > 0$$
$$u(x,0) = 0, \quad x > 0$$
$$u(x,t) \to 0 \text{ as } x \to \infty, \quad t > 0$$
$$u_x(0,t) = f(t), \quad t > 0$$

where the function f is piecewise smooth and absolutely integrable in $[0, \infty)$.

44.11.   Use the Fourier transform to solve the following problem for a heat equation with transport term

$$u_t = c^2 u_{xx} + k u_x, \quad -\infty < x < \infty, \quad t > 0, \quad c > 0, \quad k > 0$$
$$u(x,0) = f(x), \quad -\infty < x < \infty$$
$$u(x,t) \text{ and } u_x(x,t) \text{ finite as } |x| \to \infty, \quad t > 0$$

where the function f is piecewise smooth and absolutely integrable in $(-\infty, \infty)$.

44.12.   Use the Fourier transform to solve the following nonhomogeneous problem

$$u_t = c^2 u_{xx} + q(x,t), \quad -\infty < x < \infty, \quad t > 0$$
$$u(x,0) = f(x), \quad -\infty < x < \infty$$
$$u(x,t) \to 0, \quad u_x(x,t) \to 0 \text{ as } |x| \to \infty, \quad t > 0$$

where the function f is piecewise smooth and absolutely integrable in $(-\infty, \infty)$.

44.13.   Find the solution of the problem (43.9) when

(i) $f_1(x) = 3\sin 2x$, $f_2(x) = 0$
(ii) $f_1(x) = e^{-|x|}$, $f_2(x) = 0$
(iii) $f_1(x) = 0$, $f_2(x) = 4\cos 5x$
(iv) $f_1(x) = 0$, $f_2(x) = e^{-|x|}$.

44.14.   Use the Fourier sine transform to solve the following problem:

$$u_{tt} = u_{xx}, \quad x > 0, \quad t > 0$$
$$u(x, 0) = 0, \quad x > 0$$
$$u_t(x, 0) = 0, \quad x > 0$$
$$u(0, t) = f(t), \quad t > 0$$
$$u(x, t) \quad \text{and} \quad u_x(x, t) \to 0 \quad \text{as} \quad x \to \infty, \quad t > 0,$$

where the function f is piecewise smooth and absolutely integrable in $[0, \infty)$.

44.15.   Find the solution of the wave equation

$$u_{tt} = c^2 u_{xx} - ku, \quad -\infty < x < \infty, \quad t > 0, \quad c > 0, \quad k > 0$$

satisfying the same conditions as in (43.9).

44.16.   Show that the solution of the following Neumann problem:

$$u_{xx} + u_{yy} = 0, \quad -\infty < x < \infty, \quad y > 0$$
$$u_y(x, 0) = f(x), \quad -\infty < x < \infty$$
$$u(x, y) \quad \text{and} \quad u_y(x, y) \to 0 \quad \text{as} \quad (x^2 + y^2) \to \infty,$$

where the function f is piecewise smooth and absolutely integrable in $(-\infty, \infty)$, can be written as

$$u(x, y) = c + \frac{1}{2\pi} \int_{-\infty}^{\infty} f(\xi) \ln[y^2 + (x - \xi)^2] d\xi,$$

where c is an arbitrary constant.

44.17.   Find the solution of the following problem:

$$u_{xx} + u_{yy} = 0, \quad 0 < x < \infty, \quad 0 < y < b$$
$$u(x, 0) = f(x), \quad 0 < x < \infty$$
$$u_x(0, y) = 0, \quad 0 < y < b$$
$$u_y(x, b) = 0, \quad 0 < x < \infty,$$

where the function f is piecewise smooth and absolutely integrable in $[0, \infty)$.

44.18.   Find the bounded solution of the following problem:

$$u_{xx} + u_{yy} = 0, \quad 0 < x < c, \quad 0 < y < \infty$$
$$u_y(x, 0) = 0, \quad 0 < x < c$$
$$u(0, y) = 0, \quad 0 < y < \infty$$
$$u_x(c, y) = f(y), \quad 0 < y < \infty,$$

where the function f is piecewise smooth and absolutely integrable in $[0, \infty)$.

44.19.   Use the finite Fourier cosine transform to find the solution of the following problem:

$$u_t = c^2 u_{xx}, \quad 0 < x < a, \quad t > 0, \quad c > 0$$
$$u(x, 0) = f(x), \quad 0 < x < a$$
$$u_x(0, t) = u_x(a, t) = 0, \quad t > 0.$$

44.20.   Use the finite Fourier sine transform to find the solution of the following problem:

$$u_{tt} = c^2 u_{xx}, \quad 0 < x < a, \quad t > 0, \quad c > 0$$
$$u(x, 0) = f(x), \quad 0 < x < a$$
$$u_t(x, 0) = 0, \quad 0 < x < a$$
$$u(0, t) = u(a, t) = 0, \quad t > 0.$$

Answers or Hints

44.1. $y(x) = -\int_{-\infty}^{\infty} f(t)g(x - t)dt$, where $g(x) = \frac{1}{\sqrt{a^2 - 4b}}$
$\times \exp\left[-\frac{1}{2}(ax + \sqrt{a^2 - 4b}\, |x|)\right]$. $y(x) = \frac{1}{2}\begin{cases} -e^{-x}(1 + x), & x \geq 0 \\ xe^x, & x < 0. \end{cases}$

44.2. $y(x) = e^{-kx} - \sqrt{\frac{2}{\pi}} \int_0^{\infty} \frac{\sin \omega x}{\omega^2 + k^2} F(\omega)d\omega$.

44.3. $y(x) = -\frac{1}{k}e^{-kx} - \sqrt{\frac{2}{\pi}} \int_0^{\infty} \frac{\cos \omega x}{\omega^2 + k^2} F(\omega)d\omega$.

44.4. (i) $y(x) = \frac{c}{\pi} \int_0^{\infty} \frac{1}{\omega} \sin\left(\frac{1}{\omega} + \omega x\right) d\omega$ (ii) $y(x) = \frac{c}{\pi} \int_0^{\infty} \frac{\cos \omega x}{\sqrt{1 + \omega^2}} d\omega$
(iii) $y(x) = \frac{c}{\pi} \int_0^{\infty} e^{-\omega^2/2} \cos\left(\frac{\omega^3}{3} - \omega x\right) d\omega$.

44.5. Use the substitution $\mu = x + 2\sqrt{c^2 t}\, w$.

44.6. Use $e^{-ax^2} = \frac{1}{\sqrt{\pi a}} \int_0^\infty e^{-\omega^2/4a} \cos \omega x d\omega$.

44.7. Use the same identity as in Problem 44.2.

44.8. $u(x,t) = e^{-x^2/(2+4c^2t)}/\sqrt{1+2c^2t}$.

44.9. $u(x,t) = xe^{-x^2/4(a^2+c^2t)}/(1+c^2t/a^2)^{3/2}$.

44.10. $u(x,t) = -\frac{1}{\sqrt{\pi}} \int_0^t \frac{1}{\sqrt{t-\mu}} f(\mu) \exp\left(-\frac{x^2}{4(t-\mu)}\right) d\mu$.

44.11. $u(x,t) = \int_{-\infty}^\infty f(\mu) \frac{e^{-(x-\mu+kt)^2/4c^2t}}{\sqrt{4\pi c^2 t}} d\mu$.

44.12. $u(x,t) = \int_{-\infty}^\infty f(\mu) \frac{e^{-(x-\mu)^2/4c^2t}}{\sqrt{4\pi c^2 t}} d\mu$
$+ \int_0^t \int_{-\infty}^\infty q(\mu,\tau) \frac{e^{-(x-\mu)^2/4c^2(t-\tau)}}{\sqrt{4\pi c^2 (t-\tau)}} d\mu d\tau$.

44.13. (i) $3 \sin 2x \cos 2ct$ (ii) $\frac{2}{\pi} \int_0^\infty \frac{\cos \omega ct \cos \omega x}{1+\omega^2} d\omega$ (iii) $\frac{4}{5c} \cos 5x \sin 5ct$
(iv) $\frac{2}{\pi c} \int_0^\infty \frac{\sin \omega ct \cos \omega x}{\omega(1+\omega^2)} d\omega$.

44.14. $u(x,t) = \frac{2}{\pi} \int_0^\infty \int_0^t f(\mu) \sin \omega(t-\mu) \sin \omega x d\mu d\omega$.

44.15. $u(x,t) = \frac{1}{\sqrt{2\pi}} \int_{-\infty}^\infty \left[F_1(\omega) \cos t\sqrt{k+\omega^2 c^2} + F_2(\omega) \frac{\sin t\sqrt{k+\omega^2 c^2}}{\sqrt{k+\omega^2 c^2}} \right] e^{i\omega x} d\omega$.

44.16. $z = u_y$ satisfies the Dirichlet problem (44.1).

44.17. Use the Fourier cosine transform
$u(x,y) = \frac{2}{\pi} \int_0^\infty \int_0^\infty f(t) \cos \omega t \frac{\cosh \omega(b-y)}{\cosh \omega b} \cos \omega x dt d\omega$.

44.18. Use the Fourier cosine transform
$u(x,y) = \frac{2}{\pi} \int_0^\infty \int_0^\infty f(t) \cos \omega t \frac{\sinh \omega x}{\omega \cosh \omega c} \cos \omega y dt d\omega$.

44.19. $u(x,t) = \frac{1}{a} F_c(0) + \frac{2}{a} \sum_{n=1}^\infty F_c(n) e^{-(n^2\pi^2 c^2/a^2)t} \cos \frac{n\pi x}{a}$,
$F_c(n) = \int_0^a f(x) \cos \frac{n\pi x}{a} dx$.

44.20. $u(x,t) = \frac{2}{a} \sum_{n=1}^\infty F_s(n) \cos \frac{n\pi ct}{a} \sin \frac{n\pi x}{a}$,
$F_s(n) = \int_0^a f(x) \sin \frac{n\pi x}{a} dx$.

Lecture 45
Laplace Transforms

The method of Laplace transforms has the advantage of directly giving the solutions of differential equations with given initial and boundary conditions without the necessity of first finding the general solution and then evaluating from it the arbitrary constants. Moreover, the ready table of Laplace transforms reduces the problem of solving differential equations to mere algebraic manipulations. In this lecture we shall introduce some basic concepts of Laplace transform theory.

We begin with the following definition of Laplace transform.

Definition 45.1. The *Laplace transform* of a function $f(x)$, $0 \leq x < \infty$ is defined by the improper integral

$$\mathcal{L}[f(x)] = F(s) = \int_0^\infty e^{-sx} f(x) dx, \tag{45.1}$$

where it is assumed that the integral converges for at least one value of s, say, $s = s_0$. Clearly, then the integral converges for all $s > s_0$.

Thus, the Laplace transform is an operator which transforms the function $f(x)$ into its image $F(s)$. The original function $f(x)$ in (45.1) is called the *inverse transform*, or *inverse* of $F(s)$ and will be denoted by $\mathcal{L}^{-1}[F]$; i.e., we shall write

$$f(x) = \mathcal{L}^{-1}[F]. \tag{45.2}$$

Of course, s may be a complex number whose real part is sufficiently large to make (45.1) convergent, but in the early part of the theory, it is more definite to think of s as a real positive number.

When evaluating the Laplace transform of some function $f(x)$, we actually use $f(x)$ only for $0 \leq x < \infty$. Hence it should be irrelevant, from the mathematical point of view, if and how $f(x)$ is defined for $x < 0$. However, some properties of Laplace transform, particularly those which reflect a relationship with the Fourier integral, can be better understood if $f(x)$ is assigned the value zero for $-\infty < x < 0$.

Definition 45.2. The *Heaviside function* $H(x)$, defined by

$$H(x) = \begin{cases} 0, & x < 0 \\ 1, & x > 0, \end{cases} \tag{45.3}$$

R.P. Agarwal, D. O'Regan, *Ordinary and Partial Differential Equations*,
Universitext, DOI 10.1007/978-0-387-79146-3_45,
© Springer Science+Business Media, LLC 2009

is a discontinuous function that is important in certain applications. It is discontinuous at $x = 0$.

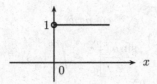

Figure 45.1

Example 45.1. To find Laplace transform of $H(x)$, we note that

$$\int_0^\infty e^{-sx} H(x)dx = \int_0^\infty e^{-sx}dx = \lim_{b\to\infty} \int_0^b e^{-sx}dx$$

$$= \lim_{b\to\infty} \left[-\frac{e^{-sx}}{s}\right]\Big|_0^b = \lim_{b\to\infty} \left[\frac{1}{s} - \frac{e^{-bs}}{s}\right].$$

Thus, if $s > 0$ the above limit exists and, we obtain

$$\mathcal{L}[H] = \mathcal{L}[1] = \frac{1}{s}. \tag{45.4}$$

Example 45.2.

$$\mathcal{L}[e^{ax}] = \int_0^\infty e^{ax}e^{-sx}dx = \int_0^\infty e^{(a-s)x}dx$$

$$= \frac{1}{a-s}e^{(a-s)x}\Big|_0^\infty = \frac{1}{s-a}, \quad s > a. \tag{45.5}$$

Example 45.3.

$$\mathcal{L}[x^{n+1}] = \int_0^\infty e^{-sx}x^{n+1}dx = -\frac{1}{s}e^{-sx}x^{n+1}\Big|_0^\infty + \frac{(n+1)}{s}\int_0^\infty e^{-sx}x^n dx$$

$$= \frac{(n+1)}{s}\int_0^\infty e^{-sx}x^n dx = \frac{(n+1)}{s}\mathcal{L}[x^n].$$

Thus, in view of (45.4), we have

$$\mathcal{L}[x] = \frac{1}{s}\mathcal{L}[1] = \frac{1}{s^2}, \quad \mathcal{L}[x^2] = \frac{2}{s}\mathcal{L}[x] = \frac{2!}{s^3}, \cdots, \mathcal{L}[x^n] = \frac{n!}{s^{n+1}}. \tag{45.6}$$

Example 45.4. Let $a > 0$ be a number. Then,

$$\mathcal{L}[x^a] = \int_0^\infty e^{-sx}x^a dx = \int_0^\infty e^{-\tau}\left(\frac{\tau}{s}\right)^a \frac{d\tau}{s} \quad \text{(using } x = \tau/s)$$

$$= \frac{1}{s^{a+1}}\int_0^\infty e^{-\tau}\tau^a d\tau = \frac{\Gamma(a+1)}{s^{a+1}}. \tag{45.7}$$

Recall that $\Gamma(n+1) = n!$, so that (45.6) follows from (45.7).

Example 45.5.

$$
\begin{aligned}
\mathcal{L}[\cos ax] &= \int_0^\infty e^{-sx} \cos ax\, dx \\
&= \left. e^{-sx} \frac{\sin ax}{a} \right|_0^\infty - \int_0^\infty (-s) e^{-sx} \frac{\sin ax}{a} dx \\
&= \frac{s}{a} \int_0^\infty e^{-sx} \sin ax\, dx \\
&= \left. \frac{s}{a} e^{-sx} \frac{\cos ax}{-a} \right|_0^\infty - \frac{s}{a} \int_0^\infty (-s) e^{-sx} \frac{\cos ax}{-a} dx \\
&= \frac{s}{a^2} - \frac{s^2}{a^2} \mathcal{L}[\cos ax], \quad s > 0
\end{aligned}
$$

and hence

$$
\mathcal{L}[\cos ax] = \frac{s}{s^2 + a^2}, \quad s > 0. \tag{45.8}
$$

Similarly, we have

$$
\mathcal{L}[\sin ax] = \frac{a}{s^2 + a^2}, \quad s > 0. \tag{45.9}
$$

Theorem 45.1 (Linearity Property). Let $f_j(x)$, $1 \le j \le n$, $0 \le x < \infty$ be functions whose Laplace transforms exist, and let c_j, $1 \le j \le n$ be real numbers. Then,

$$
\mathcal{L}[c_1 f_1(x) + \cdots + c_n f_n(x)] = c_1 \mathcal{L}[f_1] + \cdots + c_n \mathcal{L}[f_n]. \tag{45.10}
$$

Proof. Clearly, we have

$$
\begin{aligned}
&\mathcal{L}[c_1 f_1(x) + \cdots + c_n f_n(x)] \\
&= \int_0^\infty e^{-sx} (c_1 f_1(x) + \cdots + c_n f_n(x)) dx \\
&= c_1 \int_0^\infty e^{-sx} f_1(x) dx + \cdots + c_n \int_0^\infty e^{-sx} f_n(x) dx \\
&= c_1 \mathcal{L}[f_1] + \cdots + c_n \mathcal{L}[f_n]. \quad \blacksquare
\end{aligned}
$$

Theorem 45.2 (Uniqueness Property). If $f(x)$ and $g(x)$ are continuous functions for $0 \le x < \infty$ and if $\mathcal{L}[f] = \mathcal{L}[g]$, then $f(x) = g(x)$, and conversely.

In fact, if two functions defined on the positive real axis have the same transform, then these functions cannot differ over an interval of positive length, although they may differ at various isolated points. However, this is not important in applications; we may say that the inverse of a given function is essentially unique. Of course, if two continuous functions have the same transform, they must be identical.

Theorem 45.3 (Inverse Linearity Property). Let $f_j(x)$, $1 \leq j \leq n$, $0 \leq x < \infty$ be continuous functions, and let $F_j(s)$, $1 \leq j \leq n$ be their Laplace transforms. Then

$$\mathcal{L}^{-1}[c_1 F_1(s) + \cdots + c_n F_n(s)] = c_1 \mathcal{L}^{-1}[F_1] + \cdots + c_n \mathcal{L}^{-1}[F_n]$$
$$= c_1 f_1(x) + \cdots + c_n f_n(x),$$

where c_j, $1 \leq j \leq n$ are real numbers.

Theorem 45.3 follows immediately from Theorem 45.2.

Example 45.6. Since $\cos^2 x = (1 + \cos 2x)/2$, we have

$$\mathcal{L}[\cos^2 x] = \mathcal{L}\left[\frac{1}{2} + \frac{1}{2}\cos 2x\right]$$
$$= \frac{1}{2}\mathcal{L}[1] + \frac{1}{2}\mathcal{L}[\cos 2x] = \frac{1}{2} \cdot \frac{1}{s} + \frac{1}{2} \cdot \frac{s}{s^2 + 4}.$$

Example 45.7.

$$\mathcal{L}[4x + 7e^{2x} + 5\cos 3x] = 4\mathcal{L}[x] + 7\mathcal{L}[e^{2x}] + 5\mathcal{L}[\cos 3x]$$
$$= \frac{4}{s^2} + \frac{7}{s - 2} + \frac{5s}{s^2 + 9}.$$

Example 45.8. Let

$$F(s) = \frac{1}{(s - a)(s - b)}, \quad a \neq b.$$

We shall find $\mathcal{L}^{-1}[F]$.

$$\mathcal{L}^{-1}[F] = \mathcal{L}^{-1}\left[\frac{1}{a - b}\left(\frac{1}{s - a} - \frac{1}{s - b}\right)\right]$$
$$= \frac{1}{a - b}\left[\mathcal{L}^{-1}\left[\frac{1}{s - a}\right] - \mathcal{L}^{-1}\left[\frac{1}{s - b}\right]\right] = \frac{1}{a - b}\left(e^{ax} - e^{bx}\right).$$

Example 45.9. Let

$$F(s) = \frac{s}{(s - a)(s - b)}, \quad a \neq b.$$

We shall find $\mathcal{L}^{-1}[F]$.

$$\mathcal{L}^{-1}[F] = \mathcal{L}^{-1}\left[\frac{1}{a - b}\left(\frac{a}{s - a} - \frac{b}{s - b}\right)\right] = \frac{1}{a - b}\left(ae^{ax} - be^{bx}\right).$$

If $f(x)$ is a complex-valued function of the real variable x, i.e., $f(x) = u(x) + iv(x)$, where $u(x)$ and $v(x)$ are continuous real functions, then from (45.10) it follows that

$$\mathcal{L}[f] = \mathcal{L}[u + iv] = \mathcal{L}[u] + i\mathcal{L}[v].$$

Hence,

$$\mathcal{L}[\text{Real part of } f] = \text{Real part of } \mathcal{L}[f]$$

and

$$\mathcal{L}[\text{Imaginary part of } f] = \text{Imaginary part of } \mathcal{L}[f].$$

Example 45.10. Since

$$\mathcal{L}[\cos \omega x + i \sin \omega x] \; = \; \mathcal{L}[e^{i\omega x}] = \frac{1}{s - i\omega} = \frac{s + i\omega}{s^2 + \omega^2}$$

$$= \; \frac{s}{s^2 + \omega^2} + i\frac{\omega}{s^2 + \omega^2}$$

it follows that

$$\mathcal{L}[\cos \omega x] = \frac{s}{s^2 + \omega^2} \quad \text{and} \quad \mathcal{L}[\sin \omega x] = \frac{\omega}{s^2 + \omega^2}.$$

Thus, (45.8) and (45.9) can be obtained from (45.5).

Now we shall provide sufficient conditions which guarantee the existence of the integral (45.1). For this, we need to introduce the following definition.

Definition 45.3. A function $f(x)$ is said to be of *exponential order* α if there exist positive constants X and M such that $|f(x)| \leq Me^{\alpha x}$ for all $x \geq X$.

Theorem 45.4 (Existence Theorem). If $f(x)$ is piecewise continuous on $[0, \infty)$ and of exponential order α, then $\mathcal{L}[f]$ exists for $s > \alpha$.

Proof. Clearly,

$$\int_0^\infty e^{-sx} f(x)dx = \int_0^X e^{-sx} f(x)dx + \int_X^\infty e^{-sx} f(x)dx, \qquad (45.11)$$

where X is the same as in Definition 45.3. The first integral in the right side of (45.11) exists because $f(x)$ and hence $e^{-sx} f(x)$ is piecewise continuous on $[0, X]$ for any fixed s. Now since $f(x)$ is of exponential order α, for $x \geq X$ we have $|f(x)| \leq Me^{\alpha x}$ and hence

$$\left| e^{-sx} f(x) \right| = e^{-sx}|f(x)| \; \leq \; e^{-sx} \cdot Me^{\alpha x} = Me^{-(s-\alpha)x}.$$

Therefore, it follows that for $s > \alpha$,

$$\int_X^\infty \left| e^{-sx} f(x) \right| dx \; \leq \; M \int_X^\infty e^{-(s-\alpha)x} dx = \frac{Me^{-(s-\alpha)X}}{s - \alpha} \; < \; \infty,$$

i.e., the second integral in the right side of (45.11) converges absolutely. But then $\int_X^\infty e^{-sx} f(x)dx$ exists for $s > \alpha$. Finally, since both the integrals in the right side of (45.11) exist for $s > \alpha$, the Laplace transform $\mathcal{L}[f]$ exists for $s > \alpha$. ∎

The conditions in Theorem 45.4 are sufficient for most applications, and it is easy to find whether a given function satisfies an inequality of the form $|f(x)| \leq Me^{\alpha x}$. For example, a bounded function is of exponential order 0. This is clear from the fact that $|f(x)| \leq M = Me^{0x}$. Thus, $\cos bx$ and $\sin bx$ are of exponential order 0. The functions $e^{ax} \cos bx$ and $e^{ax} \sin bx$ are of exponential order a. The function x^n is of exponential order α for any positive α, since by the Maclaurin series

$$e^{\alpha x} = \sum_{n=0}^{\infty} \frac{\alpha^n x^n}{n!} > \frac{\alpha^n x^n}{n!}$$

so that $x^n < (n!/\alpha^n)e^{\alpha x}$. However, the function e^{x^2} is not of exponential order, because, no matter how large we choose M and α, $e^{x^2} > Me^{\alpha x}$ for all sufficiently large x.

It should be noted that the conditions of Theorem 45.4 are only sufficient rather than necessary. For example, the function $1/\sqrt{x}$ is infinite at $x = 0$, but its transform exists. Indeed, we have

$$\int_0^\infty e^{-sx} x^{-1/2} dx = \frac{1}{\sqrt{s}} \int_0^\infty e^{-\tau} \tau^{-1/2} d\tau = \frac{1}{\sqrt{s}} \Gamma\left(\frac{1}{2}\right) = \sqrt{\frac{\pi}{s}}.$$

We also remark that if $f(x)$ is of exponential order, then $f'(x)$ need not be of exponential order, e.g., $f(x) = \sin e^{x^2}$ is of exponential order 0, however, $f'(x) = 2xe^{x^2} \cos e^{x^2}$ is not of exponential order. But, if $f(x)$ is of exponential order α, then $\int_0^x f(\tau)d\tau$ is of exponential order α. For this it suffices to note that

$$\left| \int_0^x f(\tau)d\tau \right| \leq \int_0^x |f(\tau)|d\tau \leq \int_0^x Me^{\alpha\tau}d\tau = \frac{M}{\alpha}(e^{\alpha x} - 1) \leq \frac{M}{\alpha}e^{\alpha x}.$$

Example 45.11. For the piecewise continuous function

$$f(x) = \begin{cases} x, & 0 < x < 2 \\ 1, & x > 2 \end{cases}$$

we have

$$\mathcal{L}[f] = \int_0^2 e^{-sx} x\, dx + \int_2^\infty e^{-sx}\, dx$$

$$= \left[-\frac{x}{s} e^{-sx} - \frac{1}{s^2} e^{-sx} \right]\Big|_0^2 + \left[-\frac{1}{s} e^{-sx} \right]\Big|_2^\infty$$

$$= \frac{1}{s^2} - \left(\frac{1}{s} + \frac{1}{s^2} \right) e^{-2s}, \quad s > 0.$$

Example 45.12. For the piecewise continuous function

$$f(x) = \begin{cases} -2, & 0 \leq x < 1 \\ 1, & 1 \leq x < 3 \\ e^{2x}, & x \geq 3 \end{cases}$$

we have

$$\mathcal{L}[f] = -2 \int_0^1 e^{-sx}\, dx + \int_1^3 e^{-sx}\, dx + \int_3^\infty e^{(2-s)x}\, dx$$

$$= -\frac{2}{s} + \frac{3e^{-s}}{s} - \frac{e^{-3s}}{s} + \frac{e^{3(2-s)}}{s-2}, \quad s > 2.$$

Lecture 46
Laplace Transforms (Cont'd.)

Using the definition of Laplace transforms to get an explicit expression for $\mathcal{L}[f]$ requires the evaluation of the improper integral, which is often difficult. In the previous lecture we have already seen how the linearity property of the transform can be employed to simplify at least some computation. In this lecture we shall develop several other properties that can be used to facilitate the computation of Laplace transforms.

Theorem 46.1 (Transform of the Derivative). Let $f(x)$ be continuous on $[0, \infty)$ and $f'(x)$ be piecewise continuous on $[0, \infty)$, with both of exponential order α. Then,

$$\mathcal{L}[f'] = s\mathcal{L}[f] - f(0), \quad s > \alpha. \tag{46.1}$$

Proof. Clearly, f' satisfies the conditions of Theorem 45.4, and hence $\mathcal{L}[f']$ exists. If $f'(x)$ is continuous for all $x \geq 0$, then we have

$$\mathcal{L}[f'] = \int_0^\infty e^{-sx} f'(x) dx = e^{-sx} f(x) \Big|_0^\infty + s \int_0^\infty e^{-sx} f(x) dx.$$

Since $|f(x)| \leq M e^{\alpha x}$, the integrated portion on the right is zero at the upper limit when $s > \alpha$ and the lower limit gives $-f(0)$. Hence, it follows that

$$\mathcal{L}[f'] = -f(0) + s\mathcal{L}[f].$$

If $f'(x)$ is only piecewise continuous, the proof remains the same, except that now the range of integration in the original integral must be broken up into parts such that $f'(x)$ is continuous in each such case. ∎

Theorem 46.2. Let $f^{(i)}(x)$, $0 \leq i \leq n - 1$ be continuous on $[0, \infty)$ and $f^{(n)}(x)$ be piecewise continuous on $[0, \infty)$, with all $f^{(i)}(x)$, $0 \leq i \leq n$ of exponential order α. Then

$$\mathcal{L}[f^{(n)}] = s^n \mathcal{L}[f] - s^{(n-1)} f(0) - s^{(n-2)} f'(0) - \cdots - f^{(n-1)}(0). \tag{46.2}$$

Proof. From (46.1) it follows that

$$\mathcal{L}[f''] = s\mathcal{L}[f'] - f'(0) = s[s\mathcal{L}[f] - f(0)] - f'(0) = s^2 \mathcal{L}[f] - sf(0) - f'(0).$$

The proof now can be completed by induction. ∎

R.P. Agarwal, D. O'Regan, *Ordinary and Partial Differential Equations*,
Universitext, DOI 10.1007/978-0-387-79146-3_46,
© Springer Science+Business Media, LLC 2009

Example 46.1.   Let $f(x) = \sin^2 x$ so that $f'(x) = 2\sin x \cos x = \sin 2x,\ f(0) = 0$. Thus, from (46.1) and (45.9), we have

$$s\mathcal{L}[\sin^2 x] - 0 = \mathcal{L}[\sin 2x] = \frac{2}{s^2 + 4},$$

and hence

$$\mathcal{L}[\sin^2 x] = \frac{2}{s(s^2 + 4)}. \tag{46.3}$$

Example 46.2.   Let $f(x) = x\sin\omega x$ so that $f'(x) = \sin\omega x + \omega x\cos\omega x$, $f''(x) = 2\omega\cos\omega x - \omega^2 x\sin\omega x = 2\omega\cos\omega x - \omega^2 f(x),\ f(0) = f'(0) = 0$. Thus, from (46.2) for $n = 2$, and (45.8), we have

$$s^2\mathcal{L}[x\sin\omega x] = \mathcal{L}[2\omega\cos\omega x - \omega^2 x\sin\omega x]$$

$$= 2\omega\mathcal{L}[\cos\omega x] - \omega^2\mathcal{L}[x\sin\omega x] = \frac{2\omega s}{s^2 + \omega^2} - \omega^2\mathcal{L}[x\sin\omega x]$$

and hence

$$\mathcal{L}[x\sin\omega x] = \frac{2\omega s}{(s^2 + \omega^2)^2}. \tag{46.4}$$

Theorem 46.3   (Transform of the Integral).   If $f(x)$ is piecewise continuous on $[0, \infty)$ and of exponential order α, then

$$\mathcal{L}\left[\int_0^x f(\tau)d\tau\right] = \frac{1}{s}\mathcal{L}[f], \quad s > \max\{0, \alpha\}. \tag{46.5}$$

Proof.   Clearly, if $f(x)$ is of negative exponential order, it is also of positive exponential order, and hence we can assume that $\alpha > 0$. Now the integral $g(x) = \int_0^x f(\tau)d\tau$ is continuous and as we have seen in the previous lecture it is of exponential order α. Also, $g'(x) = f(x)$ except for points at which $f(x)$ is discontinuous. Hence, $g'(x)$ is piecewise continuous and $g(0) = 0$. Thus, from (46.1) we have

$$\mathcal{L}[f] = \mathcal{L}[g'] = s\mathcal{L}[g] - 0 = s\mathcal{L}\left[\int_0^x f(\tau)d\tau\right], \quad s > \alpha$$

which is the same as (46.5).   ■

From (46.5) and the definition of inverse transform it is clear that

$$\mathcal{L}^{-1}\left[\frac{1}{s}F(s)\right] = \int_0^x f(\tau)d\tau. \tag{46.6}$$

Example 46.3.

$$\mathcal{L}\left[\frac{1}{\omega^2}(1 - \cos\omega x)\right] = \mathcal{L}\left[\frac{1}{\omega}\int_0^x \sin\omega\tau d\tau\right]$$

$$= \frac{1}{\omega}\cdot\frac{1}{s}\cdot\frac{\omega}{s^2 + \omega^2} = \frac{1}{s(s^2 + \omega^2)}.$$

Example 46.4.

$$\mathcal{L}\left[\frac{1}{\omega^2}\left(x - \frac{\sin\omega x}{\omega}\right)\right] = \mathcal{L}\left[\frac{1}{\omega^2}\int_0^x (1 - \cos\omega\tau)d\tau\right] = \frac{1}{s^2(s^2 + \omega^2)}.$$

Theorem 46.4 (s-shifting). If $f(x)$ has the transform $F(s)$ where $s > \alpha$, then $e^{ax}f(x)$ has the transform $F(s - a)$, i.e.,

$$\mathcal{L}\left[e^{ax}f(x)\right] = F(s - a). \tag{46.7}$$

Proof. Since $F(s) = \int_0^\infty e^{-sx}f(x)dx$, we have

$$F(s-a) = \int_0^\infty e^{-(s-a)x}f(x)dx = \int_0^\infty e^{-sx}\left[e^{ax}f(x)\right]dx = \mathcal{L}\left[e^{ax}f(x)\right]. \quad \blacksquare$$

From (46.7) it follows that

$$\mathcal{L}^{-1}[F(s - a)] = e^{ax}f(x) = e^{ax}\mathcal{L}^{-1}[F(s)]. \tag{46.8}$$

Example 46.5.

$$\mathcal{L}\left[e^{ax}x^n\right] = \frac{n!}{(s - a)^{n+1}},$$

$$\mathcal{L}\left[e^{ax}\cos\omega x\right] = \frac{s - a}{(s - a)^2 + \omega^2},$$

$$\mathcal{L}\left[e^{ax}\sin\omega x\right] = \frac{\omega}{(s - a)^2 + \omega^2}.$$

Example 46.6.

$$\mathcal{L}^{-1}\left[\frac{s+2}{s^2 + 6s + 25}\right] = \mathcal{L}^{-1}\left[\frac{(s+3) - 1}{(s+3)^2 + 16}\right] = e^{-3x}\mathcal{L}^{-1}\left[\frac{s - 1}{s^2 + 16}\right]$$

$$= e^{-3x}\left\{\mathcal{L}^{-1}\left[\frac{s}{s^2 + 16}\right] - \frac{1}{4}\mathcal{L}^{-1}\left[\frac{4}{s^2 + 16}\right]\right\}$$

$$= e^{-3x}\left(\cos 4x - \frac{1}{4}\sin 4x\right).$$

Theorem 46.5 (x-shifting). If $f(x)$ has the transform $F(s)$, then the function

$$f_a(x) = \begin{cases} 0 & \text{if } x < a \\ f(x - a) & \text{if } x > a \quad (a \geq 0) \end{cases} \tag{46.9}$$

has the transform $e^{-as}F(s)$.

Proof. Since

$$f(x-a)H(x-a) = \begin{cases} 0 & \text{if } x < a \\ f(x-a) & \text{if } x > a \end{cases}$$

(recall the definition of the Heaviside function $H(x)$) Theorem 46.5 can be reformulated as follows: If $\mathcal{L}[f] = F(s)$, then

$$\mathcal{L}[f(x-a)H(x-a)] = e^{-as}F(s). \qquad (46.10)$$

Now we have

$$e^{-as}F(s) = e^{-as}\int_0^\infty e^{-s\tau}f(\tau)d\tau = \int_0^\infty e^{-s(\tau+a)}f(\tau)d\tau$$

$$= \int_a^\infty e^{-sx}f(x-a)dx \quad (\tau + a = x)$$

$$= \int_0^\infty e^{-sx}f(x-a)H(x-a)dx = \mathcal{L}[f(x-a)H(x-a)]. \qquad \blacksquare$$

From (46.10) it is clear that

$$\mathcal{L}^{-1}\left[e^{-as}F(s)\right] = f(x-a)H(x-a). \qquad (46.11)$$

Example 46.7. The function

$$f(x) = \begin{cases} 1, & 0 < x < \pi \\ 0, & \pi < x < 2\pi \\ \sin x, & x > 2\pi \end{cases}$$

can be written as

$$f(x) = H(x) - H(x-\pi) + H(x-2\pi)\sin x.$$

Thus, it follows that

$$\mathcal{L}[f] = \frac{1}{s} - \frac{e^{-\pi s}}{s} + \frac{e^{-2\pi s}}{s^2+1}.$$

Example 46.8. For the transformed function

$$F(s) = \frac{1}{s^2} - \frac{e^{-2s}}{s^2} - \frac{2e^{-2s}}{s} + \frac{2se^{-\pi s}}{s^2+1}$$

an application of (46.11) gives

$$\begin{aligned} f(x) &= \mathcal{L}^{-1}[F(s)] \\ &= x - (x-2)H(x-2) - 2H(x-2) + 2\cos(x-\pi)H(x-\pi) \\ &= x - xH(x-2) - 2\cos xH(x-\pi) \\ &= \begin{cases} x, & 0 < x < 2 \\ 0, & 2 < x < \pi \\ -2\cos x, & x > \pi. \end{cases} \end{aligned}$$

Remark 46.1. Formula (46.10) applies to translations to the right. For translations to the left, we have

$$
\begin{aligned}
\mathcal{L}[f(x+a)] &= \int_0^\infty e^{-sx} f(x+a) dx = \int_a^\infty e^{-s(\tau-a)} f(\tau) d\tau \\
&= e^{sa} \mathcal{L}[f] - \int_0^a e^{s(a-\tau)} f(\tau) d\tau.
\end{aligned}
\tag{46.12}
$$

The finite integral cannot be neglected unless $f(x) = 0$ for $x < a$, as it accounts for the part of the function which has been "lost" by translation to negative x values where the Laplace transform does not operate.

Remark 46.2. Let $f(x)$ be a function and $F(s)$ its Laplace transform. Then, for $a > 0$, we have

$$
\mathcal{L}[f(ax)] = \int_0^\infty e^{-sx} f(ax) dx = \frac{1}{a} \int_0^\infty e^{-(s/a)\tau} f(\tau) d\tau = \frac{1}{a} F\left(\frac{s}{a}\right).
\tag{46.13}
$$

Theorem 46.6 (Derivatives of Transforms). If $f(x)$ is piecewise continuous on $[0, \infty)$ and of exponential order α, then

$$
\mathcal{L}[x^n f(x)] = (-1)^n F^{(n)}(s), \quad s > \alpha, \quad n = 1, 2, \cdots.
\tag{46.14}
$$

Proof. Since $F(s) = \int_0^\infty e^{-sx} f(x) dx$, a formal differentiation with respect to s (under the integral sign) gives

$$
F'(s) = \int_0^\infty (-x) e^{-sx} f(x) dx = - \int_0^\infty e^{-sx} [x f(x)] dx = -\mathcal{L}[x f(x)],
$$

and hence

$$
\mathcal{L}[x f(x)] = -F'(s),
\tag{46.15}
$$

which is the same as (46.14) for $n = 1$. It is now easy to see that repeated differentiations of (46.15) formally lead to (46.14). ∎

From (46.14) it follows that

$$
\mathcal{L}^{-1}\left[F^{(n)}(s)\right] = (-1)^n x^n f(x).
\tag{46.16}
$$

Example 46.9. From (46.15) it is clear that

$$
\begin{aligned}
\mathcal{L}[x \cos \omega x] &= -\frac{d}{ds}\left(\frac{s}{s^2 + \omega^2}\right) = \frac{s^2 - \omega^2}{(s^2 + \omega^2)^2}, \\
\mathcal{L}[x \sin \omega x] &= -\frac{d}{ds}\left(\frac{\omega}{s^2 + \omega^2}\right) = \frac{2\omega s}{(s^2 + \omega^2)^2}.
\end{aligned}
$$

Theorem 46.7 (Integration of Transforms). If $f(x)$ is piecewise continuous on $[0, \infty)$ and of exponential order α, and if $f(x)/x$ has a finite limit as $x \to 0^+$, then

$$\mathcal{L}\left[\frac{f(x)}{x}\right] = \int_s^\infty F(\tau)d\tau. \qquad (46.17)$$

Proof. Clearly,

$$\mathcal{L}\left[\frac{f(x)}{x}\right] = \int_0^\infty \frac{e^{-sx}}{x} f(x)dx = \int_0^\infty \left(\int_s^\infty e^{-\tau x}d\tau\right) f(x)dx$$

$$= \int_s^\infty \left(\int_0^\infty e^{-\tau x}f(x)dx\right)d\tau = \int_s^\infty F(\tau)d\tau;$$

here changing the order of integration is justified by the absolute convergence of the integral. ∎

From (46.17), we have

$$\mathcal{L}^{-1}\left[\int_s^\infty F(\tau)d\tau\right] = \frac{f(x)}{x}. \qquad (46.18)$$

Example 46.10. From (46.17), we find

$$\mathcal{L}\left[\frac{2(1 - \cos\omega x)}{x}\right] = \int_s^\infty 2\left[\frac{1}{\tau} - \frac{\tau}{\tau^2 + \omega^2}\right]d\tau = [2\ln\tau - \ln(\tau^2 + \omega^2)]\Big|_s^\infty$$

$$= \ln\frac{\tau^2}{\tau^2 + \omega^2}\Big|_s^\infty = -\ln\frac{s^2}{s^2 + \omega^2} = \ln\frac{s^2 + \omega^2}{s^2}.$$

Remark 46.3. Formula (46.17) can be generalized, to obtain

$$\mathcal{L}\left[x^{-n}f(x)\right] = \int_s^\infty \int_{\tau_n}^\infty \cdots \int_{\tau_2}^\infty F(\tau_1)d\tau_1 d\tau_2 \cdots d\tau_n. \qquad (46.19)$$

Remark 46.4. From the relations (46.5) and (46.17), we get

$$\mathcal{L}\left[\int_0^x \frac{f(\tau)}{\tau}d\tau\right] = \frac{1}{s}\mathcal{L}\left[\frac{f(x)}{x}\right] = \frac{1}{s}\int_s^\infty F(\tau)d\tau. \qquad (46.20)$$

Theorem 46.8 (Transform of Periodic Functions). If $f(x)$ is piecewise continuous on $[0, \infty)$ and periodic of period T, then $\mathcal{L}[f]$ exists for $s > 0$ and is given by

$$\mathcal{L}[f] = \frac{1}{1 - e^{-sT}}\int_0^T e^{-sx}f(x)dx. \qquad (46.21)$$

Proof. Since

$$\mathcal{L}[f] = \int_0^\infty e^{-sx} f(x)dx = \int_0^T e^{-sx} f(x)dx + \int_T^\infty e^{-sx} f(x)dx,$$

if we change the variable in the second integral to $\tau = x - T$, we obtain

$$\int_T^\infty e^{-sx} f(x)dx = \int_0^\infty e^{-s(\tau+T)} f(\tau + T)d\tau$$

$$= e^{-sT} \int_0^\infty e^{-s\tau} f(\tau)d\tau = e^{-sT} \mathcal{L}[f]$$

and hence

$$\mathcal{L}[f] = \int_0^T e^{-sx} f(x)dx + e^{-sT} \mathcal{L}[f],$$

which can be solved for $\mathcal{L}[f]$ to yield (46.21). ∎

Example 46.11. We shall find Laplace transform of the half-wave rectifier periodic function of period $2\pi/T$ defined by

$$f(x) = \begin{cases} \sin Tx & \text{if } 0 < x < \pi/T \\ 0 & \text{if } \pi/T < x < 2\pi/T. \end{cases}$$

From (46.21), we have

$$\mathcal{L}[f] = \frac{1}{1 - e^{-2\pi s/T}} \int_0^{\pi/T} e^{-sx} \sin Tx\, dx$$

$$= \frac{1}{1 - e^{-2\pi s/T}} \left\{ \frac{e^{-sx}}{s^2 + T^2} (-s \sin Tx - T \cos Tx) \Big|_0^{\pi/T} \right\}$$

$$= \frac{1}{1 - e^{-2\pi s/T}} \frac{T\left(e^{-\pi s/T} + 1\right)}{s^2 + T^2} = \frac{T}{(1 - e^{-\pi s/T})(s^2 + T^2)}.$$

Theorem 46.9 (Convolution Theorem). Let $f(x)$ and $g(x)$ satisfy the conditions of Theorem 45.4. Then, the product of their transforms $F(s) = \mathcal{L}[f]$ and $G(s) = \mathcal{L}[g]$ is the transform $K(s) = \mathcal{L}[k]$ of the *convolution* $k(x)$ of $f(x)$ and $g(x)$, written as $f \star g$ and defined by

$$k(x) = (f \star g)(x) = \int_0^x f(\tau)g(x - \tau)d\tau. \tag{46.22}$$

Proof. Since for $s > \alpha$,

$$e^{-s\tau} G(s) = \mathcal{L}[g(x - \tau)H(x - \tau)]$$

$$= \int_0^\infty e^{-sx} g(x - \tau)H(x - \tau)dx = \int_\tau^\infty e^{-sx} g(x - \tau)dx,$$

we find

$$F(s)G(s) = \int_0^\infty e^{-s\tau} f(\tau)G(s)d\tau = \int_0^\infty f(\tau)\left[\int_\tau^\infty e^{-sx} g(x-\tau)dx\right] d\tau.$$

Here we integrate with respect to x from τ to ∞ and then over τ from 0 to ∞; this corresponds to the shaded region extending to infinity in the $x\tau$-plane. Our assumptions on f and g allow us to change the order of integration. We then integrate first with respect to τ from 0 to x and then over x from 0 to ∞; thus

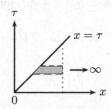

Figure 46.1

$$\begin{aligned}
F(s)G(s) &= \int_0^\infty e^{-sx} \left(\int_0^x f(\tau)g(x-\tau)d\tau\right) dx \\
&= \int_0^\infty e^{-sx} k(x)dx = \mathcal{L}[k] = K(s). \quad \blacksquare
\end{aligned}$$

Example 46.12. Let

$$K(s) = \frac{s}{(s^2+1)^2} = \frac{s}{s^2+1} \cdot \frac{1}{s^2+1}.$$

From (46.22), we find

$$k(x) = \mathcal{L}^{-1}[K(s)] = \cos x \star \sin x = \int_0^x \cos\tau \sin(x-\tau)d\tau = \frac{1}{2}x\sin x.$$

Remark 46.5. Using the definition of convolution $f \star g$ in (46.22) the following properties are immediate

$$\begin{aligned}
f \star g &= g \star f & \text{(commutative law)} \\
(f \star g) \star h &= f \star (g \star h) & \text{(associative law)} \\
f \star (g+h) &= f \star g + f \star h & \text{(distributive law)} \\
f \star 0 &= 0 \star f = 0.
\end{aligned}$$

However, $f \star 1 \neq f$ in general. For example, $1/s^2$ has the inverse x and $1/s$ has the inverse 1, and the convolution theorem confirms that

$$x \star 1 = \int_0^x \tau \cdot 1 d\tau = \frac{x^2}{2}.$$

Laplace Transform Table

	$f(x)$	$Y(s) = \mathcal{L}[f]$
1.	x^n	$\dfrac{\Gamma(n+1)}{s^{n+1}}, \quad n > -1$
2.	$x^n e^{ax}$	$\dfrac{n!}{(s-a)^{n+1}}$
3.	$x^{-1/2}$	$\sqrt{\dfrac{\pi}{s}}$
4.	$e^{ax} \sin bx$	$\dfrac{b}{(s-a)^2 + b^2}$
5.	$e^{ax} \cos bx$	$\dfrac{s-a}{(s-a)^2 + b^2}$
6.	$\sinh ax$	$\dfrac{a}{s^2 - a^2}$
7.	$\cosh ax$	$\dfrac{s}{s^2 - a^2}$
8.	$\dfrac{e^{ax} - e^{bx}}{a - b}$	$\dfrac{1}{(s-a)(s-b)}, \quad a \neq b$
9.	$\dfrac{ae^{ax} - be^{bx}}{a - b}$	$\dfrac{s}{(s-a)(s-b)}, \quad a \neq b$
10.	$x \sin ax$	$\dfrac{2as}{(s^2 + a^2)^2}$
11.	$x \cos ax$	$\dfrac{s^2 - a^2}{(s^2 + a^2)^2}$
12.	$x \sinh ax$	$\dfrac{2as}{(s^2 - a^2)^2}$
13.	$x \cosh ax$	$\dfrac{s^2 + a^2}{(s^2 - a^2)^2}$
14.	$\dfrac{1}{a} e^{-x/a}$	$\dfrac{1}{1 + as}$
15.	$1 - e^{-x/a}$	$\dfrac{1}{s(1 + as)}$
16.	$\dfrac{1}{a^2} x e^{-x/a}$	$\dfrac{1}{(1 + as)^2}$
17.	$e^{-ax}(1 - ax)$	$\dfrac{s}{(s+a)^2}$
18.	$\dfrac{1}{a^2}(1 - \cos ax)$	$\dfrac{1}{s(s^2 + a^2)}$
19.	$\dfrac{1}{a^3}(ax - \sin ax)$	$\dfrac{1}{s^2(s^2 + a^2)}$
20.	$\dfrac{1}{2a^3}(\sin ax - ax \cos ax)$	$\dfrac{1}{(s^2 + a^2)^2}$
21.	$\dfrac{1}{3} e^{-ax} - \dfrac{1}{3} e^{ax/2} \left(\cos \dfrac{\sqrt{3}}{2} ax - \sqrt{3} \sin \dfrac{\sqrt{3}}{2} ax \right)$	$\dfrac{a^2}{s^3 + a^3}$
22.	$\dfrac{1}{2}(\sinh ax - \sin ax)$	$\dfrac{a^3}{s^4 - a^4}$
23.	$\dfrac{e^{-ax}}{\sqrt{\pi x}}$	$\dfrac{1}{\sqrt{s + a}}$

Problems

46.1. Let $f(x)$ and $g(x)$ be piecewise continuous on $[a, b]$. Show that $f(x) + g(x)$ and $f(x)g(x)$ also are piecewise continuous on $[a, b]$.

46.2. Let $f(x)$ and $g(x)$ be of exponential order. Show that $f(x) + g(x)$ and $f(x)g(x)$ also are of exponential order.

46.3. Show that if $\int_0^\infty e^{-sx} f(x)dx$ converges absolutely for $s = s_0$, it converges absolutely for each $s > s_0$.

46.4. Let $\sum_{n=1}^\infty f_n(x)$ be a uniformly convergent series of functions, each of which has a Laplace transform defined for $s \geq \alpha$, i.e., $\mathcal{L}[f_n]$ exists for $s \geq \alpha$. Show that $f(x) = \sum_{n=1}^\infty f_n(x)$ has a Laplace transform for $s \geq \alpha$ defined by $\mathcal{L}[f(x)] = \sum_{n=1}^\infty \mathcal{L}[f_n(x)]$.

46.5. Let $f(x)$ be a piecewise continuous function on $[0, \infty)$ and periodic of period T. Show that $f(x)$ is of exponential order α for any $\alpha > 0$.

46.6. (i) Differentiate (45.7) with respect to a to show that

$$\frac{\Gamma'(a+1) - (\ln s)\Gamma(a+1)}{s^{a+1}} = \int_0^\infty e^{-sx}(\ln x)x^a dx.$$

(ii) Show that $\mathcal{L}[\ln x] = (\Gamma'(1) - \ln s)/s$. The constant $\gamma = \Gamma'(1) = 0.57721566\cdots$ is called Euler's constant.

46.7. Assume that conditions of Theorem 45.4 are satisfied. Show that $\lim_{s \to \infty} F(s) = 0$.

46.8. Use (46.1) to show that

(i) $\lim_{x \to 0^+} f(x) = \lim_{s \to \infty} sF(s)$

(ii) $\lim_{x \to \infty} f(x) = \lim_{s \to 0} sF(s)$.

46.9. Give an example of the function $F(s)$ for which the inverse Laplace transform does not exist.

46.10. Assume that conditions of Theorem 46.1 are satisfied except that $f(x)$ is discontinuous at $x_0 > 0$, but $f(x_0-)$ and $f(x_0+)$ exist. Show that

$$\mathcal{L}[f'] = s\mathcal{L}[f] - f(0) - [f(x_0+) - f(x_0-)]e^{-x_0 s}.$$

46.11. Show that

(i) $\mathcal{L}[\sinh ax] = \dfrac{a}{s^2 - a^2}$

(ii) $\mathcal{L}[\cosh ax] = \dfrac{s}{s^2 - a^2}$

(iii) $\mathcal{L}[\sin 2x \sin 3x] = \dfrac{12s}{(s^2 + 1)(s^2 + 25)}$

(iv) $\mathcal{L}[\sin^3 2x] = \dfrac{48}{(s^2 + 4)(s^2 + 36)}$

(v) $\mathcal{L}\left[e^{-3x}(2\cos 5x - 3\sin 5x)\right] = \dfrac{2s - 9}{s^2 + 6s + 24}$

(vi) $\mathcal{L}\left[\displaystyle\int_0^x \dfrac{\sin \tau}{\tau} d\tau\right] = \dfrac{1}{s}\cot^{-1} s.$

46.12. Let $f(x)$ be the square-wave function defined by

$$f(x) = \begin{cases} 1 & \text{if } 2i < x < 2i + 1 \\ 0 & \text{if } 2i + 1 < x < 2i + 2, \quad i = 0, 1, \cdots. \end{cases}$$

Show that

$$\mathcal{L}[f] = \dfrac{1}{s(1 + e^{-s})}.$$

46.13. Let $f(x)$ be the square-wave function defined by

$$f(x) = \begin{cases} E & \text{if } 2iT < x < (2i + 1)T \\ -E & \text{if } (2i + 1)T < x < (2i + 2)T, \quad i = 0, 1, \cdots. \end{cases}$$

Show that

$$\mathcal{L}[f] = \dfrac{E}{s}\tanh \dfrac{sT}{2}.$$

46.14. Let $f(x)$ be the sawtooth-wave function defined by

$$f(x) = k(x - iT), \quad iT < x < (i + 1)T, \quad i = 0, 1, \cdots.$$

Show that

$$\mathcal{L}[f] = \dfrac{k}{s^2} - \dfrac{kTe^{-sT}}{s(1 - e^{-sT})}.$$

46.15. Let $f(x)$ be the triangular-wave periodic function of period T defined by

$$f(x) = \begin{cases} \dfrac{2x}{T} & \text{if } 0 < x \leq \dfrac{T}{2} \\ 2\left(1 - \dfrac{x}{T}\right) & \text{if } \dfrac{T}{2} \leq x < T. \end{cases}$$

Show that

$$\mathcal{L}[f] = \dfrac{2}{Ts^2}\tanh\left(\dfrac{1}{4}sT\right).$$

46.16. For the Bessel function $J_n(x)$ of order n show that

(i) $\mathcal{L}[J_0(x)] = (s^2 + 1)^{-1/2}$

(ii) $\mathcal{L}[J_0(ax)] = (s^2 + a^2)^{-1/2}, \quad a > 0$

(iii) $\mathcal{L}[J_1(x)] = 1 - s(s^2 + 1)^{-1/2}$ (Hint. use $J_0'(x) = -J_1(x)$)

(iv) $\mathcal{L}[J_0(\sqrt{x})] = e^{-s/4}/s$.

46.17. The *error function* erf is defined by

$$\mathrm{erf}(x) = \frac{2}{\sqrt{\pi}} \int_0^x e^{-u^2}\, du.$$

Show that

$$\mathcal{L}[\mathrm{erf}(\sqrt{x})] = \frac{1}{s\sqrt{s+1}}.$$

46.18. Show that

(i) $\mathcal{L}^{-1}\left[\dfrac{2s}{(s-2)(s-3)(s-6)}\right] = e^{2x} - 2e^{3x} + e^{6x}$

(ii) $\mathcal{L}^{-1}\left[\dfrac{6}{(s^2+1)(s^2+4)}\right] = 2\sin x - \sin 2x$

(iii) $\mathcal{L}^{-1}\left[\dfrac{e^{-s}}{s(s+1)}\right] = H(x-1) - e^{-(x-1)}H(x-1)$

(iv) $\mathcal{L}^{-1}\left[\ln\left(\dfrac{s+1}{s-1}\right)\right] = \dfrac{2\sinh x}{x}$

(v) $\mathcal{L}^{-1}\left[\dfrac{\pi}{2} - \tan^{-1}\dfrac{s}{2}\right] = \dfrac{\sin 2x}{x}.$

46.19. Show that

$$\mathcal{L}\left[\int_0^\infty \frac{\sin xt}{t}\, dt\right] = \int_0^\infty \frac{1}{t^2+s^2}\, dt = \frac{\pi}{2} \cdot \frac{1}{s}$$

and hence deduce that

$$\int_0^\infty \frac{\sin xt}{t}\, dt = \frac{\pi}{2}.$$

46.20. Show that for $x > 0$,

(i) $\displaystyle\int_0^\infty e^{-xt^2}\, dt = \frac{1}{2}\sqrt{\frac{\pi}{x}}$

(ii) $\displaystyle\int_0^\infty \frac{\cos xt}{1+t^2}\, dt = \frac{\pi}{2}e^{-x}$

(iii) $\displaystyle\int_0^\infty \frac{e^{-2x} - e^{-3x}}{x}\, dx = \ln\frac{3}{2}$

(iv) $\displaystyle\int_0^\infty xe^{-2x}\cos x\,dx = \frac{3}{25}.$

46.21. Use convolution theorem to show that

(i) $\displaystyle\mathcal{L}^{-1}\left[\frac{1}{(s^2+1)^2}\right] = -\frac{1}{2}x\cos x + \frac{1}{2}\sin x$

(ii) $\displaystyle\mathcal{L}^{-1}\left[\frac{s}{(s^2-a^2)(s-b)}\right] = \frac{1}{2}\left(\frac{e^{ax}}{a-b} - \frac{e^{-ax}}{a+b} - \frac{2be^{bx}}{a^2-b^2}\right)$

(iii) $\displaystyle\mathcal{L}^{-1}\left[\frac{1}{(s-1)\sqrt{s}}\right] = \frac{2e^x}{\sqrt{\pi}}\int_0^{\sqrt{x}} e^{-\tau^2}d\tau = e^x\,\mathrm{erf}\,(\sqrt{x})$

(iv) $\displaystyle\mathcal{L}^{-1}\left[e^{-\pi s/2}\frac{s}{(s^2+1)(s^2+9)}\right] = \left[\frac{1}{8}\sin 3x + \frac{1}{8}\sin x\right]H\left(x - \frac{\pi}{2}\right).$

Answers or Hints

46.1. Use definition.

46.2. Use definition.

46.3. Compare the integrands.

46.4. Since the series converges uniformly $\int\sum = \sum\int$.

46.5. Consider the function over $[0, T]$.

46.6. Verify directly.

46.7. From Theorem 45.4, we have $F(s) < \int_0^X e^{-sx}f(x)dx + \frac{Me^{-(s-\alpha)X}}{s-\alpha}$.

46.8. (i) In (46.1) let $s \to \infty$ (ii) In (46.1) let $s \to 0$.

46.9. See Problem 46.7. Take $F(s) = 1$.

46.10. Integrate over $[0, x_0-)$ and (x_0+, ∞).

46.12. Use Theorem 46.8.

46.13. Use Theorem 46.8.

46.14. Use Theorem 46.8.

46.15. Use Theorem 46.8.

46.17. Change the order of integration.

46.19. Integrate first with respect to x.

Lecture 47
Laplace Transform Method for Ordinary DEs

Laplace transforms supply an easy, efficient, and quick procedure to find the solutions of differential, difference, and integral equations. Here we summarize this method to solve the second-order initial value problem

$$y'' + ay' + by = r(x), \quad y(0) = y_0, \quad y'(0) = y_1 \qquad (47.1)$$

where a and b are constants. Of course, this method can be extended to higher order initial value problems rather easily. In engineering applications the function $r(x)$ is called the input (driving force) and $y(x)$ is the output (response).

The main steps are as follows:

1.   Take Laplace transform of each side of equation (47.1), i.e.,

$$\mathcal{L}[y'' + ay' + by] = \mathcal{L}[r].$$

2.   Use the linearity property of Laplace transforms, Theorem 46.2 and the initial conditions in (47.1) to obtain a linear algebraic equation; i.e., if we denote $Y = Y(s) = \mathcal{L}[y]$ and $R = R(s) = \mathcal{L}[r]$, then

$$(s^2 Y - sy(0) - y'(0)) + a(sY - y(0)) + bY = R,$$

which is the same as

$$(s^2 + as + b)Y = (s + a)y_0 + y_1 + R.$$

3.   Solve the algebraic equation for Y, i.e.,

$$Y = \frac{(s + a)y_0 + y_1 + R}{(s^2 + as + b)}. \qquad (47.2)$$

4.   Use the table of Laplace transforms to determine the solution $y(x)$ of the initial value problem (47.1). For this, often the partial fraction decomposition of the right–hand side of (47.2) is required.

Example 47.1. For the initial value problem

$$y'' - 2y' + y = e^x + x, \quad y(0) = 1, \quad y'(0) = 0 \qquad (47.3)$$

R.P. Agarwal, D. O'Regan, *Ordinary and Partial Differential Equations*,
Universitext, DOI 10.1007/978-0-387-79146-3_47,
© Springer Science+Business Media, LLC 2009

we have
$$(s^2Y - s) - 2(sY - 1) + Y = \frac{1}{s-1} + \frac{1}{s^2}$$

and hence
$$\begin{aligned}
Y &= \frac{s-2}{(s-1)^2} + \frac{1}{(s-1)^3} + \frac{1}{s^2(s-1)^2} \\
&= \frac{1}{s-1} - \frac{1}{(s-1)^2} + \frac{1}{(s-1)^3} + \frac{1}{(s-1)^2} - \frac{2}{s-1} + \frac{1}{s^2} + \frac{2}{s} \\
&= -\frac{1}{s-1} + \frac{1}{(s-1)^3} + \frac{1}{s^2} + \frac{2}{s}.
\end{aligned}$$

Thus, the solution of the problem (47.3) is
$$y(x) = \mathcal{L}^{-1}[Y] = -e^x + \frac{1}{2}x^2 e^x + x + 2.$$

Example 47.2. For the initial value problem
$$y'' + 4y = \sin 2x, \quad y(0) = 1, \quad y'(0) = 0 \tag{47.4}$$

we have
$$(s^2Y - s) + 4Y = \frac{2}{s^2+4}$$

and hence
$$Y = \frac{s}{(s^2+4)} + \frac{2}{(s^2+4)^2}.$$

Thus, the solution of the problem (47.4) is
$$y(x) = \mathcal{L}^{-1}[Y] = \cos 2x + \frac{1}{8}(\sin 2x - 2x\cos 2x). \tag{47.5}$$

Remark 47.1. A simple observation shows that $y = y(x)$ is a solution of (47.1) if and only if $\phi = \phi(x) = y(x - c)$ is a solution of
$$y'' + ay' + by = r(x - c), \quad y(c) = y_0, \quad y'(c) = y_1. \tag{47.6}$$

Example 47.3. In the initial value problem
$$y'' + y = 2x, \quad y(\pi/4) = \pi/2, \quad y'(\pi/4) = 2 - \sqrt{2}, \tag{47.7}$$

since $r(x) = 2x = 2(x - \pi/4) + \pi/2$, in view of Remark 47.1, first we need to solve the problem
$$y'' + y = 2x + \pi/2, \quad y(0) = \pi/2, \quad y'(0) = 2 - \sqrt{2}. \tag{47.8}$$

For (47.8), we have

$$\left(s^2 Y - \frac{1}{2}\pi s - 2 + \sqrt{2}\right) + Y = \frac{2}{s^2} + \frac{\pi/2}{s}$$

and hence

$$Y = \frac{1}{2}\pi \frac{s}{s^2+1} + (2-\sqrt{2})\frac{1}{s^2+1} + 2\left(\frac{1}{s^2} - \frac{1}{s^2+1}\right) + \frac{\pi}{2}\left(\frac{1}{s} - \frac{s}{s^2+1}\right)$$

$$= \frac{\pi}{2}s + \frac{2}{s^2} - \sqrt{2}\frac{1}{s^2+1},$$

which gives the solution of (47.8) as $y(x) = (\pi/2) + 2x - \sqrt{2}\sin x$. Now once again in view of Remark 47.1, the solution $\phi(x)$ of (47.7) can be obtained from the relation $\phi(x) = y(x - \pi/4)$, i.e.,

$$\phi(x) = \frac{\pi}{2} + 2\left(x - \frac{\pi}{4}\right) - \sqrt{2}\sin\left(x - \frac{\pi}{4}\right) = 2x - \sin x + \cos x.$$

Now we shall consider the initial value problem (47.1) where the function $r(x)$ has discontinuities, is impulsive, or is periodic but not merely a sine or cosine function. The Laplace transform technique shows its real power for these kinds problems.

Example 47.4.   For the initial value problem

$$y'' - 4y' + 3y = r(x), \quad y(0) = 3, \quad y'(0) = 1 \tag{47.9}$$

where

$$r(x) \;=\; \begin{cases} 0, & x < 2 \\ x, & 2 < x < 4 \\ 6, & x > 4 \end{cases}$$

$$= \; xH(x-2) - xH(x-4) + 6H(x-4)$$

$$= \; (x-2)H(x-2) + 2H(x-2) - (x-4)H(x-4) + 2H(x-4)$$

in view of (46.10), we have

$$(s^2 Y - 3s - 1) - 4(sY - 3) + 3Y = \frac{e^{-2s}}{s^2} + \frac{2e^{-2s}}{s} - \frac{e^{-4s}}{s^2} + \frac{2e^{-4s}}{s}$$

and hence

$$Y \;=\; \frac{3s-11}{s^2-4s+3} + \frac{2s+1}{s^2(s^2-4s+3)}e^{-2s} + \frac{2s-1}{s^2(s^2-4s+3)}e^{-4s}$$

$$= \; \frac{4}{s-1} - \frac{1}{s-3} + \left[\frac{10}{9}\frac{1}{s} + \frac{1}{3}\frac{1}{s^2} - \frac{3}{2}\frac{1}{s-1} + \frac{7}{18}\frac{1}{s-3}\right]e^{-2s}$$

$$+ \left[\frac{2}{9}\frac{1}{s} - \frac{1}{3}\frac{1}{s^2} - \frac{1}{2}\frac{1}{s-1} + \frac{5}{18}\frac{1}{s-3}\right]e^{-4s}.$$

Thus, from (46.11) it follows that

$$y(x) = 4e^x - e^{3x} + \left[\frac{10}{9} + \frac{1}{3}(x-2) - \frac{3}{2}e^{(x-2)} + \frac{7}{18}e^{3(x-2)}\right] H(x-2)$$

$$+ \left[\frac{2}{9} - \frac{1}{3}(x-4) - \frac{1}{2}e^{(x-4)} + \frac{5}{18}e^{3(x-4)}\right] H(x-4)$$

$$= \begin{cases} 4e^x - e^{3x}, & 0 \le x \le 2 \\ \left(4 - \dfrac{3}{2e^2}\right)e^x + \left(\dfrac{7}{18e^6} - 1\right)e^{3x} + \dfrac{1}{3}x + \dfrac{4}{9}, & 2 \le x \le 4 \\ \left(4 - \dfrac{3}{2e^2} - \dfrac{1}{2e^4}\right)e^x + \left(\dfrac{7}{18e^6} + \dfrac{5}{18e^{12}} - 1\right)e^{3x} + 2, & x \ge 4. \end{cases}$$

Example 47.5. Consider the initial value problem

$$y'' + \omega^2 y = Ar(x), \quad y(0) = y_0, \quad y'(0) = y_1. \tag{47.10}$$

This problem models undamped oscillations of a spring-mass system, simple pendulum, or LC circuit depending on the interpretation of x, ω, $Ar(x)$, y_0, and y_1. We shall find the solution of (47.10) where $r(x)$ is the square-wave function (see Problem 46.12). We have

$$(s^2 Y - y_0 s - y_1) + \omega^2 Y = \frac{A}{s(1 + e^{-s})}$$

and hence

$$Y = \frac{s}{s^2 + \omega^2} y_0 + \frac{1}{s^2 + \omega^2} y_1 + \frac{A}{s(s^2 + \omega^2)} \sum_{n=0}^{\infty} (-1)^n e^{-ns},$$

which in view of (45.8), (45.9), Example 46.3, and (46.11) gives

$$y(x) = y_0 \cos \omega x + \frac{y_1}{\omega} \sin \omega x + \frac{A}{\omega^2} \sum_{n=0}^{\infty} (-1)^n (1 - \cos \omega(x-n)) H(x-n).$$

In physics and engineering often one encounters forces of very large amplitude that act for a very short period of time, i.e., that are of an impulsive nature. This situation occurs, for example, when a tennis ball is hit, an airplane makes a hard landing, a system is given a blow by a hammer, a ship is hit by a high single wave, and so on. We shall now show that the Laplace transform technique works equally well for problem (47.1) when $r(x)$ is of an impulsive type. For this, we recall that in mechanics, the impulse of a force $f(x)$ over the interval $a \le x \le a + p$ is defined by $\int_a^{a+p} f(x)dx$.

Now consider the function

$$f_p(x) = \begin{cases} 1/p, & a \le x \le a+p \\ 0, & \text{otherwise.} \end{cases} \qquad (47.11)$$

Clearly, the impulse of $f_p(x)$ is 1. The limit of $f_p(x)$ as $p \to 0$ is denoted as $\delta(x-a)$, and is called the *Dirac delta function* after Paul Dirac (1902–1984). Sometimes this function is also termed as *unit impulse function*.

Definition 47.1. The Dirac delta function δ is characterized by the following two properties:

(i). $\delta(x - a) = 0$, $x \ne a$, and

(ii). $\displaystyle\int_{-\infty}^{\infty} f(x)\delta(x - a)dx = f(a)$ for any function $f(x)$ that is continuous on an interval containing $x = a$.

Now we shall show that for $a \ge 0$,

$$\mathcal{L}[\delta(x - a)] = e^{-as}. \qquad (47.12)$$

For this, we note that the function $f_p(x)$ defined in (47.11) can be written as

$$f_p(x) = \frac{1}{p}[H(x - a) - H(x - (a + p))].$$

Thus, from our earlier considerations

$$\mathcal{L}[f_p(x)] = \frac{1}{ps}\left[e^{-as} - e^{-(a+p)s}\right] = e^{-as}\frac{1 - e^{-ps}}{ps};$$

and hence in view of the linearity of Laplace transforms, we have

$$\lim_{p\to 0}\mathcal{L}[f_p(x)] = \mathcal{L}\left[\lim_{p\to 0}f_p(x)\right] = \mathcal{L}[\delta(x - a)] = e^{-as}.$$

Finally, we remark that $\delta(x-a)$ is not a function in the ordinary sense as used in calculus, but a so–called *generalized function*. An ordinary function which is everywhere 0 except at a single point must have the integral 0, but $\int_{-\infty}^{\infty} \delta(x - a)dx = 1$.

Example 47.6. For the initial value problem

$$y'' + y = \delta(x - \pi), \quad y(0) = y'(0) = 0 \qquad (47.13)$$

we have

$$s^2Y + sY = e^{-\pi s}$$

and hence

$$Y = \frac{e^{-\pi s}}{s^2 + 1},$$

which in view of (46.11) gives

$$y(x) = \sin(x - \pi)H(x - \pi) = -\sin x \, H(x - \pi) = \begin{cases} 0, & 0 \le x < \pi \\ -\sin x, & x \ge \pi. \end{cases}$$

Some initial value problems which involve differential equations with variable coefficients can be solved by the method of Laplace transforms. However, for such problems there is no general method. To apply Laplace transforms to specific problems first from (46.2) and (46.14) we note that

$$\mathcal{L}[xf'(x)] = -\frac{d}{ds}[sF(s) - f(0)] = -F(s) - sF'(s) \qquad (47.14)$$

$$\mathcal{L}[xf''(x)] = -\frac{d}{ds}[s^2F(s) - sf(0) - f'(0)] = -2sF(s) - s^2F'(s) + f(0). \tag{47.15}$$

Hence, if a differential equation has coefficients such as $(c_0 x + c_1)$, we get a first-order differential equation for $F(s)$, which can be solved. We illustrate the method in the following example.

Example 47.7.   For Laguerre's DE (8.10) with $a = 0$, (47.14) and (47.15) leads to

$$-2sY - s^2Y' + y(0) + sY - y(0) - (-Y - sY') + nY = 0.$$

Thus, we have

$$(s - s^2)Y' + (n + 1 - s)Y = 0,$$

which on separating the variables gives

$$\frac{dY}{Y} = \left(\frac{n}{s-1} - \frac{n+1}{s} \right)$$

and hence

$$Y(s) = \frac{(s-1)^n}{s^{n+1}};$$

here the integration constant C we have taken as 1 (often C is determined by using the fact that $\lim_{s \to \infty} Y(s) = 0$, or some other properties of the initial value problems).

We shall show that

$$y(x) = \mathcal{L}^{-1}[Y(s)] = L_n(x) = \frac{e^x}{n!} \frac{d^n}{dx^n} \left(x^n e^{-x} \right), \quad n = 0, 1, 2, \cdots.$$

For this, recall that

$$\mathcal{L}\left[x^n e^{-x} \right] = \frac{n!}{(s+1)^{n+1}}$$

and hence

$$\mathcal{L}\left[(x^n e^{-x})^{(n)}\right] = \frac{n!\, s^n}{(s+1)^{n+1}}.$$

Therefore,

$$\mathcal{L}[L_n(x)] = \frac{1}{n!}\, \frac{n!\,(s-1)^n}{s^{n+1}} = \frac{(s-1)^n}{s^{n+1}}.$$

Example 47.8. For Bessel's DE of order zero, i.e., (2.15) with $a = 0$, with the initial condition $y(0) = 1$, we have

$$-2sY - s^2 Y' + 1 + sY - 1 - Y' = 0$$

and hence

$$(s^2 + 1)Y' + sY = 0,$$

which on integration gives

$$Y(s) = \frac{C}{\sqrt{s^2+1}} = \frac{C}{s}\left(1 + \frac{1}{s^2}\right)^{-1/2}.$$

Now expanding the function $Y(s)$ in binomial series for $s > 1$, we obtain

$$\begin{aligned}
Y(s) &= \frac{C}{s}\left[1 - \frac{1}{2}\frac{1}{s^2} - \frac{1}{2}\left(-\frac{3}{2}\right)\frac{1}{2!\,s^4} - \cdots\right] \\
&= \frac{C}{s}\left[1 + \sum_{m=1}^{\infty}(-1)^m \frac{1 \cdot 3 \cdot 5 \cdots (2m-1)}{2^m\, m!\, s^{2m}}\right] \\
&= C\sum_{m=0}^{\infty}\frac{(-1)^m\,(2m)!}{(2^m\, m!)^2 s^{2m+1}}.
\end{aligned}$$

Thus, from the inverse transform it follows that

$$y(x) = C\sum_{m=0}^{\infty}\frac{(-1)^m}{(2^m\, m!)^2}x^{2m}.$$

However, since $y(0) = 1$ it follows that $C = 1$, and hence

$$y(x) = \sum_{m=0}^{\infty}\frac{(-1)^m}{(m!)^2}\left(\frac{x}{2}\right)^{2m} = J_0(x),$$

which is the same as given in (9.8) as it should.

Problems

47.1. Use the Laplace transform technique to solve the following initial value problems:

(i) $y' + 3y = 1$, $y(0) = 2$

(ii) $y'' - 3y' + 2y = e^{-x}$, $y(0) = 3$, $y'(0) = 4$

(iii) $y'' + 2y' + 5y = 5$, $y(0) = 0$, $y'(0) = 0$

(iv) $y'' + y = \cos 3x$, $y(0) = 0$, $y'(0) = 0$

(v) $y'' - 3y' + 2y = H(x - 6)$, $y(0) = y'(0) = 0$

(vi) $y'' - 5y' + 6y = xe^{2x} + e^{3x}$, $y(0) = 0$, $y'(0) = 1$

(vii) $y'' - 3y' - 4y = H(x - 1) + H(x - 2)$, $y(0) = 0$, $y'(0) = 1$

(viii) $y'' + 4y' + 3y = x$, $y(-1) = 0$, $y'(-1) = 2$

(ix) $y'' + 4y' + 5y = \delta(x - \pi) + \delta(x - 2\pi)$, $y(0) = 0$, $y'(0) = 2$

(x) $y'' - 4y' + 3y = 8\delta(x - 1) + 12H(x - 2)$, $y(0) = 1$, $y'(0) = 5$

(xi) $y''' + 4y'' + 5y' + 2y = 10\cos x$, $y(0) = 0$, $y'(0) = 0$, $y''(0) = 3$

(xii) $y''' + 3y'' - y' - 3y = 0$, $y(0) = 1$, $y'(0) = 1$, $y''(0) = -1$

(xiii) $y'''' - 2y''' + 5y'' - 8y' + 4y = 0$, $y(0) = 0$, $y'(0) = 0$,
$$y''(0) = 1, \quad y'''(0) = 3$$

(xiv) $y'''' - k^4 y = 0$, $y(0) = y'(0) = y''(0) = 0$, $y'''(0) = 1$

(xv) $y'''' - k^4 y = 0$, $y(0) = 1$, $y'(0) = y''(0) = y'''(0) = 0$.

47.2. Suppose $y = y(x)$ is the solution of the initial value problem

$$y'' + ay' + by = 0, \quad y(0) = 0, \quad y'(0) = 1.$$

Show that the solution $\phi(x)$ of (47.1) with $y_0 = y_1 = 0$ can be written as

$$\phi(x) = (y \star r)(x) = \int_0^x y(x - \tau)r(\tau)d\tau.$$

47.3. Use integration by parts to show that

$$\int_{-\infty}^{\infty} f(x)\delta'(x)dx = -f'(0).$$

In general prove that

$$\int_{-\infty}^{\infty} f(x)\delta^{(n)}(x)dx = (-1)^n f^{(n)}(0).$$

47.4. Suppose $z = z(x)$ is the solution of the initial value problem

$$y'' + ay' + by = \delta(x), \quad y(0) = y'(0) = 0.$$

Show that the solution $\phi(x)$ of (47.1) with $y_0 = y_1 = 0$ can be written as

$$\phi(x) = (z \star r)(x) = \int_0^x z(x - \tau)r(\tau)d\tau.$$

47.5. Use Laplace transform technique to solve the following initial value problems:

(i) $y'' + 2xy' - 4y = 1, \quad y(0) = y'(0) = 0$

(ii) $y'' + 4xy' - 8y = 4, \quad y(0) = y'(0) = 0$

(iii) $y'' + xy' - 2y = 1, \quad y(0) = y'(0) = 0$

(iv) $y'' - 2xy' + 2y = 0, \quad y(0) = 0, \ y'(0) = 1$

(v) $y'' - xy' + y = 1, \quad y(0) = 1, \ y'(0) = 2$

(vi) $xy'' + y' + 4xy = 0, \quad y(0) = 3, \ y'(0) = 0$

(vii) $xy'' + 2(x-1)y' + 2(x-1)y = 2e^{-x}\cos x, \quad y(0) = 0, \ y'(0) = -1$

(viii) $xy'' - (2+x)y' + 3y = x - 1, \quad y(0) = y'(0) = 0.$

47.6. Consider the mechanical system depicted in Figure 47.1. The system consists of a mass M, a spring with spring constant K, and a viscous damper μ. The mass is subjected to an external force $r(x)$. Let $y(x)$ denote the deviation of the mass from its equilibrium at time x. Then, Newton's and Hooke's laws lead to the differential equation

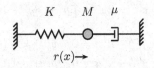

Figure 47.1

$$My'' + \mu y' + Ky = r(x), \quad M > 0, \ K > 0 \ \text{ and } \ \mu \geq 0.$$

Use the Laplace transform technique to obtain $y(x)$ in each of the following cases:

(i) $r(x) \equiv 0, \ \mu^2 - 4MK = 0, \ y(0) = y_0, \ y'(0) = y_1$ (critically damped)

(ii) $r(x) \equiv 0, \ \mu^2 - 4MK > 0, \ y(0) = y_0, \ y'(0) = y_1$ (overdamped)

(iii) $r(x) \equiv 0, \ \mu^2 - 4MK < 0, \ y(0) = y_0, \ y'(0) = y_1$ (underdamped)

(iv) $r(x) = F\sin\omega x, \ \mu = 0, \ K/M \neq \omega^2, \ y(0) = y'(0) = 0$

(simple harmonic motion with sinusoidal force)

(v) $r(x) = F\sin\omega x, \ \mu = 0, \ K/M = \omega^2, \ y(0) = y'(0) = 0$ (resonance).

Answers or Hints

47.1. (i) $\frac{5}{3}e^{-3x} + \frac{1}{3}$ (ii) $\frac{4}{3}e^{2x} + \frac{3}{2}e^x + \frac{1}{6}e^{-x}$ (iii) $1 - e^{-x}\cos 2x - \frac{1}{2}e^{-x}\sin 2x$
(iv) $\frac{1}{8}\cos x - \frac{1}{8}\cos 3x$ (v) $H(x-6)[\frac{1}{2}e^{2(x-6)} - e^{x-6} + \frac{1}{2}]$ (vi) $e^{3x}[x+1] -$
$e^{2x}[1 + x + \frac{x^2}{2}]$ (vii) $H(x-1)[-\frac{1}{4} + \frac{1}{20}e^{4(x-1)} + \frac{1}{5}e^{-(x-1)}] + H(x-2)[-\frac{1}{4} +$
$\frac{1}{20}e^{4(x-2)} + \frac{1}{5}e^{-(x-2)}] + \frac{1}{5}e^{4x} - \frac{1}{5}e^{-x}$ (viii) $2e^{-(x+1)} - \frac{11}{9}e^{-3(x+1)} + \frac{1}{3}x - \frac{4}{9}$
(ix) $[2 - e^{2\pi}H(x-\pi) + e^{4\pi}H(x-2\pi)]e^{-2x}\sin x$ (x) $2e^{3x} - e^x + [4e^{3(x-1)} -$
$4e^{(x-1)}]H(x-1) + [2e^{3(x-2)} - 6e^{(x-2)} + 4]H(x-2).$ (xi) $2\sin x - \cos x -$

$e^{-2x}+2e^{-x}-2xe^{-x}$ (xii) $\frac{3}{4}e^x+\frac{1}{2}e^{-x}-\frac{1}{4}e^{-3x}$ (xiii) $\frac{1}{25}e^x+\frac{2}{5}xe^x-\frac{1}{25}\cos 2x -$ $\frac{11}{50}\sin 2x$ (xiv) $\frac{1}{2k^3}\sin kx$ (xv) $\frac{1}{2}\cosh kx + \frac{1}{2}\cos kx$

47.2. Use Convolution Theorem 46.9.

47.3. Verify directly.

47.3. Use Convolution Theorem 46.9.

47.5. (i) $x^2/2$ (ii) $2x^2$ (iii) $x^2/2$ (iv) x (v) $1+2x$ (vi) $3J_0(2x)$ (vii) $\frac{1}{2}(C-1)e^{-x}\sin x - \frac{1}{2}(C+1)xe^{-x}\cos x$, C is arbitrary (viii) $x/2$.

47.6. (i) $y(x) = e^{-(\mu/2M)x}\left[y_0 + \frac{\mu}{2M}y_0 x + y_1 x\right]$

(ii) $y(x) = e^{-(\mu/2M)x}\left[y_0 \cosh\theta x + \frac{1}{\theta}\left(y_1 + \frac{\mu}{2M}y_0\right)\sinh\theta x\right]$, $\theta = \frac{\sqrt{\mu^2-4MK}}{2M}$

(iii) $y(x) = e^{-(\mu/2M)x}\left[y_0 \cos\phi x + \frac{1}{\phi}\left(y_1 + \frac{\mu}{2M}y_0\right)\sin\phi x\right]$, $\phi = \frac{\sqrt{4MK-\mu^2}}{2M}$

(iv) $y(x) = \frac{F}{(K-M\omega^2)}\left[\sin\omega x - \omega\sqrt{\frac{M}{K}}\sin\sqrt{\frac{K}{M}}x\right]$.

(v) $y(x) = \frac{F}{2M\omega}\left[\frac{1}{\omega}\sin\omega x - x\cos\omega x\right]$.

Lecture 48
Laplace Transform Method for Partial DEs

In this lecture we shall apply the Laplace transform technique to find solutions of partial differential equations. For this, we note that for a given function $u(x,t)$ defined for $a \leq x \leq b$, $t > 0$, we have

$$\mathcal{L}\left[\frac{\partial u}{\partial t}\right] = \int_0^\infty e^{-st}\frac{\partial u}{\partial t}dt$$

$$= e^{-st}u(x,t)\Big|_0^\infty + s\int_0^\infty e^{-st}u(x,t)dt$$

$$= -u(x,0) + s\mathcal{L}[u] = -u(x,0) + sU(x,s),$$

where $\mathcal{L}[u] = U(x,s)$. Similarly, we find

$$\mathcal{L}\left[\frac{\partial u}{\partial x}\right] = \int_0^\infty e^{-st}\frac{\partial u}{\partial x}dt = \frac{d}{dx}\int_0^\infty e^{-st}u(x,t)dt = \frac{dU(x,s)}{dx},$$

$$\mathcal{L}\left[\frac{\partial^2 u}{\partial t^2}\right] = \mathcal{L}\left[\frac{\partial}{\partial t}\left(\frac{\partial u}{\partial t}\right)\right] = s\mathcal{L}\left[\frac{\partial u}{\partial t}\right] - \frac{\partial u(x,0)}{\partial t}$$

$$= s^2 U(x,s) - su(x,0) - \frac{\partial u(x,0)}{\partial t},$$

and

$$\mathcal{L}\left[\frac{\partial^2 u}{\partial x^2}\right] = \frac{d^2 U(x,s)}{dx^2}.$$

Example 48.1. We shall find the solution of

$$u_x = 2u_t + u, \quad u(x,0) = 6e^{-3x} \tag{48.1}$$

which is bounded for all $x > 0$, $t > 0$.

Taking Laplace transforms of the given partial differential equation with respect to t, we obtain

$$\frac{dU}{dx} = 2\{sU - u(x,0)\} + U, \quad U = U(x,s) = \mathcal{L}[u(x,t)],$$

or

$$\frac{dU}{dx} - (2s+1)U = -12e^{-3x}. \tag{48.2}$$

R.P. Agarwal, D. O'Regan, *Ordinary and Partial Differential Equations*,
Universitext, DOI 10.1007/978-0-387-79146-3_48,
© Springer Science+Business Media, LLC 2009

Thus, we find that Laplace transformation has transformed the partial differential equation into an ordinary differential equation. Clearly, the solution of (48.2) can be written as

$$U(x, s) = \frac{6}{s+2}e^{-3x} + C(s)e^{(2s+1)x},$$

where $C(s)$ is an arbitrary constant. Now since $u(x, t)$ must be bounded as $x \to \infty$, we must have $U(x, s)$ also bounded as $x \to \infty$ and so we must choose $C(s) = 0$. Hence,

$$U(x, s) = \frac{6}{s+2}e^{-3x}. \tag{48.3}$$

From (48.3) it immediately follows that $u(x, t) = 6e^{-2t-3x}$.

Example 48.2. We shall solve the problem

$$u_x + xu_t = 0, \quad x > 0, \quad t > 0$$
$$u(x, 0) = 0, \quad u(0, t) = t. \tag{48.4}$$

Taking Laplace transforms of the given partial differential equation with respect to t, we find

$$\frac{dU}{dx} + x[sU(x, s) - u(x, 0)] = 0,$$

or

$$\frac{dU}{dx} + xsU = 0.$$

The general solution of this ordinary differential equation is

$$U(x, s) = C(s)e^{-sx^2/2}.$$

Now since $\mathcal{L}[u(0, t)] = \mathcal{L}[t] = 1/s^2$, we have $U(0, s) = 1/s^2$. Hence,

$$U(x, s) = \frac{1}{s^2}e^{-sx^2/2}.$$

Thus, the solution of (48.4) can be written as

$$u(x, t) = \left(t - \frac{x^2}{2}\right)H\left(t - \frac{x^2}{2}\right) = \begin{cases} 0, & t < x^2/2 \\ t - (x^2/2), & t > x^2/2. \end{cases}$$

Example 48.3. We shall find the solution of the following initial–boundary value problem

$$u_t = u_{xx}, \quad 0 < x < 1, \quad t > 0$$
$$u(x, 0) = 3\sin 2\pi x, \quad 0 < x < 1 \tag{48.5}$$
$$u(0, t) = 0, \quad u(1, t) = 0, \quad t > 0.$$

Taking Laplace transforms of the given partial differential equation with respect to t, we obtain

$$sU - u(x,0) = \frac{d^2U}{dx^2},$$

or

$$\frac{d^2U}{dx^2} - sU = -3\sin 2\pi x.$$

The general solution of this ordinary differential equation is

$$U(x,s) = c_1(s)e^{\sqrt{s}x} + c_2(s)e^{-\sqrt{s}x} + \frac{3}{s + 4\pi^2}\sin 2\pi x.$$

Now taking Laplace transform of the boundary conditions, we have

$$\mathcal{L}[u(0,t)] = U(0,s) = 0 \quad \text{and} \quad \mathcal{L}[u(1,t)] = U(1,s) = 0.$$

Thus, it follows that

$$0 = c_1(s) + c_2(s)$$
$$0 = c_1(s)e^{\sqrt{s}} + c_2(s)e^{-\sqrt{s}}$$

and hence, $c_1(s) = c_2(s) = 0$. Therefore, we have

$$U(x,s) = \frac{3}{s + 4\pi^2}\sin 2\pi x,$$

which gives the solution of (48.5), $u(x,t) = 3e^{-4\pi^2 t}\sin 2\pi x$.

For our next example we recall the definition of the *complementary error function* erfc(t) :

$$\text{erfc}(t) = 1 - \text{erf}(t) = 1 - \frac{2}{\sqrt{\pi}}\int_0^t e^{-u^2}\,du.$$

Example 48.4. We shall find the bounded solution of the problem

$$u_t = u_{xx}, \quad x > 0, \quad t > 0$$
$$u(x,0) = 0, \quad u(0,t) = u_0. \tag{48.6}$$

Taking Laplace transforms of the given partial differential equation and of boundary condition $u(0,t) = u_0$, we find

$$\frac{d^2U}{dx^2} - sU = 0, \quad U(0,s) = \frac{u_0}{s}.$$

The general solution of the above ordinary differential equation is $U(x,s) = c_1(s)e^{\sqrt{s}x} + c_2(s)e^{-\sqrt{s}x}$. Since $u(x,t)$ is bounded as $x \to \infty$, $U(x,s)$ must also be bounded as $x \to \infty$. Thus, we must have $c_1(s) = 0$, assuming $s > 0$, so that $U(x, s) = c_2(s)e^{-\sqrt{s}x}$. This solution satisfies the condition $U(0, s) = u_0/s$ provided $c_2(s) = u_0/s$, and hence

$$U(x, s) = u_0 \frac{e^{-\sqrt{s}x}}{s}. \tag{48.7}$$

Now we shall show that the inverse transform of (48.7), i.e., the solution of (48.6) can be written as

$$u(x, t) = u_0 \frac{2}{\sqrt{\pi}} \int_{x/2\sqrt{t}}^{\infty} e^{-u^2} du = u_0 \operatorname{erfc}\left(\frac{x}{2\sqrt{t}}\right). \tag{48.8}$$

For this first we shall find $\mathcal{L}^{-1}[e^{-\sqrt{s}}]$. Let $Y = e^{-\sqrt{s}}$ so that

$$Y' = -\frac{e^{-\sqrt{s}}}{2\sqrt{s}}, \qquad Y'' = \frac{e^{-\sqrt{s}}}{4s} + \frac{e^{-\sqrt{s}}}{4s^{3/2}}.$$

Thus, it follows that

$$4sY'' + 2Y' - Y = 0, \quad ' = \frac{d}{ds}. \tag{48.9}$$

Now $Y'' = \mathcal{L}[t^2 y]$ so that

$$sY'' = \frac{d}{dt}\mathcal{L}[t^2 y] = \mathcal{L}[t^2 y' + 2ty].$$

Also $Y' = \mathcal{L}[-ty]$ so that (48.9) can be written as

$$4\mathcal{L}[t^2 y' + 2ty] - 2\mathcal{L}[ty] - \mathcal{L}[y] = 0,$$

or

$$4t^2 y' + (6t - 1)y = 0,$$

which can be solved to find the solution

$$y(t) = \frac{C}{t^{3/2}} e^{-1/4t},$$

where C is an arbitrary constant, and hence $ty = (C/\sqrt{t})e^{-1/4t}$. Next we have

$$\mathcal{L}[ty] = -\frac{d}{ds}\mathcal{L}[y] = -\frac{d}{ds}\left(e^{-\sqrt{s}}\right) = \frac{e^{-\sqrt{s}}}{2\sqrt{s}}.$$

Clearly, for large t, $ty \sim c/\sqrt{t}$ and $\mathcal{L}[ty] = C\sqrt{\pi}/\sqrt{s}$. Further, for small s, $(e^{-\sqrt{s}}/2\sqrt{s}) \sim (1/2\sqrt{s})$. Hence, from Problem 46.8(ii) it follows that $C\sqrt{\pi} = 1/2$, or $C = (1/2\sqrt{\pi})$. Thus, we find

$$\mathcal{L}^{-1}\left[e^{-\sqrt{s}}\right] = \frac{1}{2\sqrt{\pi}t^{3/2}} e^{-1/4t}. \tag{48.10}$$

Now using the convolution theorem, and letting $u = 1/(4v^2)$ we obtain

$$\mathcal{L}^{-1}\left[\frac{e^{-\sqrt{s}}}{s}\right] = \int_0^t \frac{1}{2\sqrt{\pi}u^{3/2}} e^{-1/4u} du$$

$$= \frac{2}{\sqrt{\pi}} \int_{1/2\sqrt{t}}^{\infty} e^{-v^2} dv = \operatorname{erfc}\left(\frac{1}{2\sqrt{t}}\right).$$

Finally, to find $\mathcal{L}^{-1}[e^{-x\sqrt{s}}/s]$ we use the change of scale property, i.e.,

$$\mathcal{L}^{-1}\left[\frac{e^{-\sqrt{x^2 s}}}{x^2 s}\right] = \frac{1}{x^2}\operatorname{erfc}\left(\frac{1}{2\sqrt{t/x^2}}\right) = \frac{1}{x^2}\operatorname{erfc}\left(\frac{x}{2\sqrt{t}}\right).$$

Hence, we have

$$\mathcal{L}^{-1}\left[\frac{e^{-x\sqrt{s}}}{s}\right] = \operatorname{erfc}\left(\frac{x}{2\sqrt{t}}\right). \tag{48.11}$$

Example 48.5. We shall find the bounded solution of the problem

$$u_t = c^2 u_{xx}, \quad x > 0, \quad t > 0, \quad c > 0$$
$$u(x,0) = k, \quad x > 0 \tag{48.12}$$
$$u(0,t) = f(t), \quad t > 0,$$

which generalizes (48.6).

Using Laplace transforms, we get the ordinary differential equation

$$sU - u(x,0) = c^2 \frac{d^2 U}{dx^2},$$

which can be solved to obtain

$$U(x,s) = c_1 e^{-px} + c_2 e^{px} + \frac{k}{s}, \quad p = \sqrt{s}/c.$$

Now the fact $|u(x,t)| < \infty$ implies that $c_2 = 0$, and since $U(0,s) = F(s)$, we have

$$F(s) = c_1 + \frac{k}{s}.$$

Thus, it follows that

$$U(x,s) = \frac{k}{s}\left(1 - e^{-px}\right) + F(s)e^{-px}.$$

However, since

$$\mathcal{L}\left[\operatorname{erfc}\left(\frac{x}{2c\sqrt{t}}\right)\right] = \frac{e^{-px}}{s},$$

we obtain

$$\mathcal{L}\left[\frac{d}{dt}\mathrm{erfc}\left(\frac{x}{2c\sqrt{t}}\right)\right] = \mathcal{L}\left[\frac{x}{2c\sqrt{\pi}t^{3/2}}\exp\left(-\frac{x^2}{4c^2t}\right)\right] = e^{-px}.$$

Thus, in view of convolution theorem, the solution of (48.12) can be written as

$$u(x,t) = k\,\mathrm{erf}\left(\frac{x}{2c\sqrt{t}}\right) + \frac{x}{2c\sqrt{\pi}}\int_0^t f(t-\tau)\frac{\exp\left(-x^2/4c^2\tau\right)}{\tau^{3/2}}d\tau. \quad (48.13)$$

The above representation of the solution is due to J.M.C. Duhamel (1797–1872).

Example 48.6. We shall use Laplace transforms to find the bounded solution of the problem

$$\begin{aligned} u_t &= c^2 u_{xx}, \quad x > 0, \quad t > 0, \quad c > 0 \\ u(x,0) &= f(x), \quad x > 0 \\ u(0,t) &= k, \quad t > 0. \end{aligned} \qquad (48.14)$$

As in Example 48.5 it is clear that the corresponding ordinary differential equation is

$$\frac{d^2U}{dx^2} - \frac{s}{c^2}U = -\frac{1}{c^2}f(x),$$

which can be solved to obtain

$$\begin{aligned} U(x,s) &= Ae^{px} + Be^{-px} - \frac{1}{pc^2}\int_0^x \sinh p(x-\tau)f(\tau)d\tau, \quad p = \sqrt{s}/c \\ &= e^{px}\left[A - \frac{1}{2pc^2}\int_0^x e^{-p\tau}f(\tau)d\tau\right] + e^{-px}\left[B + \frac{1}{2pc^2}\int_0^x e^{p\tau}f(\tau)d\tau\right]. \end{aligned}$$

Thus, the condition that $U(x,s)$ is bounded as $x \to \infty$ yields

$$A = \frac{1}{2pc^2}\int_0^\infty e^{-p\tau}f(\tau)d\tau$$

and the condition at $x = 0$ will then produce

$$B = \frac{k}{s} - \frac{1}{2pc^2}\int_0^\infty e^{-p\tau}f(\tau)d\tau.$$

Hence, it follows that

$$
\begin{aligned}
U(x,s) &= \frac{k}{s}e^{-px} - \frac{e^{-px}}{2pc^2}\int_0^\infty e^{-p\tau}f(\tau)d\tau \\
&\quad + \frac{e^{px}}{2pc^2}\int_0^\infty e^{-p\tau}f(\tau)d\tau - \frac{1}{pc^2}\int_0^x \sinh p(x-\tau)f(\tau)d\tau \\
&= \frac{k}{s}e^{-px} + \frac{1}{2c\sqrt{s}}\left[\int_0^x \left[-e^{-px-p\tau} + e^{-px+p\tau}\right]f(\tau)d\tau\right. \\
&\quad \left. + \int_x^\infty \left[-e^{-px-p\tau} + e^{px-p\tau}\right]f(\tau)d\tau\right] \\
&= \frac{k}{s}e^{-px} + \frac{1}{2c\sqrt{s}}\int_0^\infty f(\tau)\left[e^{-|x-\tau|p} - e^{-(x+\tau)p}\right]d\tau
\end{aligned}
$$

and therefore,

$$
u(x,t) = k\,\mathrm{erfc}\left(\frac{x}{2c\sqrt{t}}\right) + \frac{1}{2c}\int_0^\infty f(\tau)[g(|x-\tau|,t) - g(x+\tau,t)]d\tau,
$$

where

$$
\mathcal{L}[g(x,t)] = \frac{1}{\sqrt{s}}e^{-px}.
$$

We shall show that

$$
g(x,t) = \mathcal{L}^{-1}\left[\frac{1}{\sqrt{s}}e^{-px}\right] = \frac{1}{\sqrt{\pi t}}\exp\left(-\frac{x^2}{4c^2 t}\right).
$$

For this it suffices to establish the relation

$$
\mathcal{L}\left[\frac{1}{\sqrt{\pi t}}e^{-a^2/4t}\right] = \frac{e^{-a\sqrt{s}}}{\sqrt{s}}. \tag{48.15}
$$

Since, $\mathcal{L}^{-1}[s^n] = 0$, $n = 0, 1, 2, \cdots$ we have

$$
\begin{aligned}
\mathcal{L}^{-1}\left[\frac{e^{-a\sqrt{s}}}{\sqrt{s}}\right] &= \mathcal{L}^{-1}\left[\frac{1}{\sqrt{s}}\left(1 - \frac{a\sqrt{s}}{1!} + \frac{(a\sqrt{s})^2}{2!} - \frac{(a\sqrt{s})^3}{3!} + \cdots\right)\right] \\
&= \mathcal{L}^{-1}\left[\frac{1}{\sqrt{s}} + \frac{a^2 s^{1/2}}{2!} + \frac{a^4 s^{3/2}}{4!} + \cdots\right] \\
&= \mathcal{L}^{-1}\left[\sum_{m=0}^\infty \frac{a^{2m} s^{(2m-1)/2}}{(2m)!}\right] = \sum_{m=0}^\infty \frac{a^{2m}}{(2m)!}\frac{t^{-m-1/2}}{\Gamma(-m+1/2)} \\
&= \frac{t^{-1/2}}{\Gamma(1/2)} + \frac{a^2 t^{-3/2}}{2\Gamma(-1/2)} + \cdots = \frac{t^{-1/2}}{\sqrt{\pi}} + \frac{a^2 t^{-3/2}}{2\,(-2\sqrt{\pi})} + \cdots \\
&= \frac{1}{\sqrt{\pi t}} - \frac{a^2}{4\sqrt{\pi}\,t^{3/2}} + \cdots = \frac{1}{\sqrt{\pi t}}e^{-a^2/4t}.
\end{aligned}
$$

Example 48.7. We shall solve the following initial-boundary value problem

$$u_{tt} = u_{xx}, \quad 0 < x < a, \quad t > 0$$

$$u(x,0) = b \sin \frac{\pi}{a} x, \quad 0 < x < a$$

$$u_t(x,0) = -b \sin \frac{\pi}{a} x, \quad 0 < x < a \qquad (48.16)$$

$$u(0,t) = 0, \quad u(a,t) = 0, \quad t > 0.$$

Transforming the equation and the boundary conditions yields

$$\frac{d^2U}{dx^2} = s^2U - bs \sin \frac{\pi}{a} x + b \sin \frac{\pi}{a} x$$

$$U(0,s) = U(a,s) = 0,$$

which has the solution

$$U(x,s) = \frac{a^2 b(s-1)}{a^2 s^2 + \pi^2} \sin \frac{\pi}{a} x.$$

Hence, the solution of (48.16) can be written as

$$u(x,t) = b \sin \frac{\pi}{a} x \left[\cos \frac{\pi}{a} t - \frac{a}{\pi} \sin \frac{\pi}{a} t \right].$$

Example 48.8. We shall find the solution of the following initial-boundary value problem

$$u_{tt} - 4u_{xx} + u = 16x + 20 \sin x, \quad 0 < x < \pi, \quad t > 0$$

$$u(x,0) = 16x + 12 \sin 2x - 8 \sin 3x, \quad u_t(x,0) = 0, \quad 0 < x < \pi \qquad (48.17)$$

$$u(0,t) = 0, \quad u(\pi,t) = 16\pi, \quad t > 0.$$

Taking Laplace transforms, we find

$$s^2 U - s u(x,0) - \frac{\partial u}{\partial t}(x,0) - 4 \frac{d^2 U}{dx^2} + U = \frac{16x}{s} + \frac{20 \sin x}{s},$$

which in view of the initial conditions is the same as

$$\frac{d^2 U}{dx^2} - \frac{1}{4}(s^2+1)U = -\frac{4(s^2+1)x}{s} - \frac{5 \sin x}{s} - 3s \sin 2x + 2s \sin 3x. \quad (48.18)$$

We need to solve (48.18) along with the boundary conditions

$$U(0,s) = 0, \quad U(\pi,s) = \frac{16\pi}{s}. \qquad (48.19)$$

The general solution of the differential equation (48.18) can be written as

$$U(x,s) = c_1 e^{-\frac{1}{2}\sqrt{s^2+1}x} + c_2 e^{\frac{1}{2}\sqrt{s^2+1}x}$$

$$+ \frac{16x}{s} + \frac{20 \sin x}{s(s^2+5)} + \frac{12s \sin 2x}{s^2+17} - \frac{8s \sin 3x}{s^2+37}.$$

The boundary conditions (48.19) imply that $c_1 = c_2 = 0$, and hence

$$U(x, s) = \frac{16x}{s} + \frac{20 \sin x}{s(s^2 + 5)} + \frac{12s \sin 2x}{s^2 + 17} - \frac{8s \sin 3x}{s^2 + 37},$$

which gives the required solution

$$u(x,t) = 16x + 4 \sin x (1 - \cos \sqrt{5}t) + 12 \sin 2x \cos \sqrt{17}t - 8 \sin 3x \cos \sqrt{37}t.$$

Problems

48.1. Solve by Laplace transforms:

(i) $u_x + 2xu_t = 2x$, $x > 0$, $t > 0$, $u(x,0) = 1$, $u(0,t) = 1$

(ii) $xu_x + u_t = xt$, $x > 0$, $t > 0$, $u(x,0) = 0$, $u(0,t) = 0$

(iii) $u_x + xu_t = x^3$, $x > 0$, $t > 0$, $u(x,0) = 0$, $u(0,t) = 0$

(iv) $u_x + u_t = x$, $x > 0$, $t > 0$, $u(x,0) = f(x)$, $u(0,t) = 0$
                              (assume that $f'(x)$ exists for all x)

(v) $u_x - u_t = 1 - e^{-t}$, $0 < x < 1$, $t > 0$, $u(x,0) = x$, $|u(x,t)| < \infty$.

48.2. Solve by Laplace transforms:

(i) $u_t = 2u_{xx}$, $0 < x < 5$, $t > 0$, $u(x,0) = 10 \sin 4\pi x - 5 \sin 6\pi x$,
      $u(0,t) = u(5,t) = 0$

(ii) $u_t = 3u_{xx}$, $0 < x < \pi/2$, $t > 0$, $u(x,0) = 20 \cos 3x - 5 \cos 9x$,
      $u_x(0,t) = u(\pi/2,t) = 0$

(iii) $u_t = u_{xx}$, $0 < x < a$, $t > 0$, $u(x,0) = k + b\sin(\pi x/a)$,
      $u(0,t) = u(a,t) = k$

(iv) $u_t = c^2 u_{xx}$, $x > 0$, $t > 0$, $c > 0$, $u(x,0) = 0$,
      $u_x(0,t) = -k$, $\lim_{x \to \infty} u(x,t) = 0$

(v) $u_t = u_{xx} - 4u$, $0 < x < \pi$, $t > 0$, $u(x,0) = 6 \sin x - 4 \sin 2x$,
      $u(0,t) = u(\pi,t) = 0$.

48.3. Solve by Laplace transforms:

$$u_{tt} = c^2 u_{xx}, \quad x > 0, \quad t > 0, \quad c > 0$$
$$u(x,0) = 0, \quad x > 0$$
$$u_t(x,0) = 0, \quad x > 0$$
$$u(0,t) = f(t), \quad t > 0$$
$$\lim_{x \to \infty} u(x,t) = 0, \quad t > 0.$$

In particular, find the solution when

(i) $c = 1$, $f(t) = 10 \sin 2t$

(ii) $f(t) = \begin{cases} \sin t, & 0 \le t \le 2\pi \\ 0, & \text{otherwise.} \end{cases}$

48.4.   Solve by Laplace transforms:

(i) $u_{tt} = 9u_{xx}$, $0 < x < 2$, $t > 0$, $u(x,0) = 20 \sin 2\pi x - 10 \sin 5\pi x$,
$u_t(x,0) = 0$, $u(0,t) = u(2,t) = 0$

(ii) $u_{tt} = c^2 u_{xx}$, $x > 0$, $t > 0$, $c > 0$, $u(x,0) = 0$, $u_t(x,0) = -1$,
$u(0,t) = t^2$, $\lim_{x \to \infty} u(x,t)$ exists

(iii) $u_{tt} = u_{xx}$, $0 < x < 1$, $t > 0$, $u(x,0) = x - x^2$, $u_t(x,0) = 0$,
$u(0,t) = u(1,t) = 0$

(iv) $u_{tt} = u_{xx} + \sin(\pi x/a) \sin \omega t$, $0 < x < a$, $t > 0$
$u(x,0) = u_t(x,0) = u(0,t) = u(a,t) = 0$.

Answers or Hints

48.1. (i) $t + 1 - (t - x^2)H(t - x^2)$ (ii) $x(t - 1 + e^{-t})$ (ii) $x^2 t - t^2 +$
$\left(t - \frac{x^2}{2}\right)^2 H\left(t - \frac{x^2}{2}\right)$ (iv) Take Laplace transform with respect to x,
$f(x - t)H(x - t) + \frac{1}{2}x^2 - \frac{1}{2}(x - t)^2 H(x - t)$ (v) $x + 1 - e^{-t}$.

48.2. (i) $10e^{-32\pi^2 t} \sin 4\pi x - 5e^{-72\pi^2 t} \sin 6\pi x$ (ii) $20e^{-27t} \cos 3x - 5e^{-243t}$
$\times \cos 9x$ (iii) $k + be^{-\pi^2 t/a^2} \sin \frac{\pi x}{a}$ (iv) $k\left[2c\sqrt{\frac{t}{\pi}}e^{-x^2/4t} - x\,\text{erfc}\left(\frac{x}{2c\sqrt{t}}\right)\right]$
(v) $6e^{-5t} \sin x - 4e^{-8t} \sin 2x$.

48.3. $f\left(t - \frac{x}{c}\right)H\left(t - \frac{x}{c}\right)$. (i) $10 \sin 2(t - x)H(t - x)$
(ii) $\begin{cases} \sin\left(t - \frac{x}{c}\right), & \frac{x}{c} < t < \frac{x}{c} + 2\pi \\ 0, & \text{otherwise.} \end{cases}$

48.4. (i) $20 \sin 2\pi x \cos 6\pi t - 10 \sin 5\pi x \cos 15\pi t$ (ii) $-t + \left[\left(t - \frac{x}{c}\right)\right.$
$\left. + \left(t - \frac{x}{c}\right)^2\right] H\left(t - \frac{x}{c}\right)$ (iii) $x - x^2 - t^2 + \sum_{n=0}^{\infty}(-1)^n \left[(t - n - x)^2 H(t - n - x)\right.$
$\left. + (t - n - 1 + x)^2 H(t - n - 1 + x)\right]$ (iv) $\frac{\omega a^2}{\pi^2 - a^2 \omega^2}\left(\frac{1}{\omega} \sin \omega t - \frac{a}{\pi} \sin \frac{\pi t}{a}\right) \sin \frac{\pi x}{a}$.

Lecture 49
Well-Posed Problems

A problem consisting of a partial DE in a domain with a set of initial and/or boundary conditions is said to be *well-posed* if the following three fundamental properties hold:

1. *Existence*: There exists at least one solution of the problem.

2. *Uniqueness*: There is at most one solution of the problem.

3. *Stability*: The unique solution depends continuously on data (initial and boundary conditions): i.e., a slight change in the data leads to only a small change in the solution.

From Problem 34.4 we know that the Neumann problem $u_{xx} + u_{yy} = 0$, $0 < x < a$, $0 < y < b$, $u_y(x,0) = f(x)$, $u_y(x,b) = g(x)$, $u_x(0,y) = 0 = u_x(a,y)$ has an infinite number of solutions, and hence it is not a well posed problem. As another example, consider the problem $u_{xx} + u_{yy} = 0$, $-\infty < x < \infty$, $y > 0$, $u(x,0) = 0$, $u_y(x,0) = (1/n)\sin nx$. It has solution $u \equiv 0$ when $u_y(x,0) = 0$, but for positive values of n the solution is $u(x,y) = (1/n^2)\sin nx \sinh ny$. Clearly, $u_y(x,0) = (1/n)\sin nx \to 0$ as $n \to \infty$; however, $u(x,y) = (1/n^2)\sin nx \sinh ny \not\to 0$ as $n \to \infty$. Thus, the stability property is violated and the problem is not well posed.

In what follows we will only briefly comment on each of the three requirements, because a detailed discussion of the conditions under which a given problem is well-posed requires some deeper concepts. To ensure the existence the series or integral representation of solutions of problems we have obtained earlier will be verified in the next lecture. In this lecture we shall address the uniqueness and stability of the solutions. An important consequence of the uniqueness of solutions is that different methods lead to the same solution; however, there may be distinct representations of the same solution. We begin our discussion with the heat equation.

Since heat flows from a higher temperature to the lower temperature, in the absence of any internal heat source, the hottest and the coldest spots can occur only initially or on one of the two ends of the rod. If the rod is burned at one end and the other end is in a freezer, the heat will flow from the burning end to the end in the freezer. However, the end that is burned will always be hotter than any other point of the rod, and the end in the freezer will always be cooler than any other point on the rod. A mathematical description of this observation is stated as follows:

R.P. Agarwal, D. O'Regan, *Ordinary and Partial Differential Equations*,
Universitext, DOI 10.1007/978-0-387-79146-3_49,
© Springer Science+Business Media, LLC 2009

Theorem 49.1 (Maximum–Minimum Principle for the Heat Equation). Let the function $u(x,t)$ be continuous in a closed rectangle $R = \{(x,t) : 0 \leq x \leq a,\ 0 \leq t \leq T\}$ and satisfy the heat equation (30.1) in the interior of R. Then $u(x,t)$ attains its maximum and minimum on the base $t = 0$ or on the vertical sides $x = 0$ or $x = a$ of the rectangle.

As an application of this principle we shall prove the uniqueness of solutions of the Dirichlet problem for the heat equation

$$
\begin{aligned}
u_t - c^2 u_{xx} &= q(x,t), \quad 0 < x < a, \quad t > 0, \quad c > 0 \\
u(x,0) &= f(x), \quad 0 < x < a \\
u(0,t) &= g(t), \quad t > 0 \\
u(a,t) &= h(t), \quad t > 0.
\end{aligned}
\tag{49.1}
$$

Theorem 49.2. Assume that q, f, g, and h are continuous in their domain of definition, and $f(0) = g(0)$, $f(a) = h(0)$. Then there is at most one solution to the problem (49.1).

Proof. Assume to the contrary that there are two solutions $u_1(x,t)$ and $u_2(x,t)$ of (49.1). Then the function $w = u_1 - u_2$ satisfies $w_t - c^2 w_{xx} = 0$, $w(x,0) = 0$, $w(0,t) = w(a,t) = 0$. Let $T > 0$. In view of the maximum principle, $w(x,t)$ attains its maximum on the base $t = 0$ or on the vertical sides $x = 0$ or $x = a$ of the rectangle $R = \{(x,t) : 0 \leq x \leq a,\ 0 \leq t \leq T\}$. Therefore, it follows that $w(x,t) \leq 0$, $(x,t) \in R$. Similarly, from the minimum principle, we have $w(x,t) \geq 0$, $(x,t) \in R$. Hence, $w(x,t) \equiv 0$, i.e., $u_1(x,t) \equiv u_2(x,t)$, $(x,t) \in R$. Finally, since T is arbitrary, the result follows. ∎

Alternative Proof. We can also prove Theorem 49.2 by a technique known as the *energy method* as follows: Multiplying the equation $w_t - c^2 w_{xx} = 0$ by w, we get

$$
0 = w(w_t - c^2 w_{xx}) = \left(\frac{1}{2}w^2\right)_t - (c^2 w w_x)_x + c^2 w_x^2.
$$

Thus, an integration with respect to x gives

$$
0 = \int_0^a \left(\frac{1}{2}w^2\right)_t dx - c^2 w w_x \Big|_{x=0}^{x=a} + c^2 \int_0^a w_x^2 dx.
$$

However, since $w(0,t) = w(a,t) = 0$ it follows that

$$
\frac{d}{dt}\int_0^a \frac{1}{2}[w(x,t)]^2 dx = -c^2 \int_0^a [w_x(x,t)]^2 dx \leq 0.
$$

Therefore, in view of $w(x,0) = 0$, we have

$$
\int_0^a \frac{1}{2}[w(x,t)]^2 dx \leq \int_0^a \frac{1}{2}[w(x,0)]^2 dx = 0,
$$

and hence $\int_0^a [w(x,t)]^2 dx = 0$ for all $t > 0$. But, this implies that $w(x,t) \equiv 0$, i.e., $u_1(x,t) \equiv u_2(x,t)$. ∎

Clearly, the above proof holds as long as $ww_x|_0^a = 0$. Hence, our proof applies to other boundary conditions also, e.g., in (49.1) we can replace the Dirichlet conditions $u(0,t) = g(t)$, $u(a,t) = h(t)$ by the Neumann conditions $u_x(0,t) = g(t)$, $u_x(a,t) = h(t)$.

The following stability result immediately follows from Theorem 49.1.

Theorem 49.3. Let $u_1(x,t)$ and $u_2(x,t)$ be solutions to the problem (49.1) with data f_1, g_1, h_1 and f_2, g_2, h_2, respectively. Further, let T and ϵ be any positive real numbers. If $\max_{0 \le x \le a} |f_1(x) - f_2(x)| \le \epsilon$, $\max_{0 \le t \le T} |g_1(t) - g_2(t)| \le \epsilon$ and $\max_{0 \le t \le T} |h_1(t) - h_2(t)| \le \epsilon$, then

$$\max_{0 \le x \le a, \ 0 \le t \le T} |u_1(x,t) - u_2(x,t)| \le \epsilon.$$

Next we shall discuss the Laplace equation. If in equation (34.4) we consider $u(x,y)$ as the steady-state temperature distribution in a plate, then the temperature at any interior point cannot be higher than all other surrounding points. In fact, otherwise the heat will flow from the hot point to the cooler points. But, then the temperature will change with time, and would lead to a contradiction to the steady-state condition. Because of the same reasoning at any interior point the temperature cannot be lower than all other surrounding points. Mathematically this result can be stated as follows:

Theorem 49.4 (Maximum–Minimum Principle for the Laplace Equation). Let $D \subset \mathbb{R}^2$ be a bounded and connected open set. Let $u(x,y)$ be a harmonic function in D that is continuous on $\overline{D} = D \cup \partial D$, where ∂D is the boundary of D. Then, the maximum and minimum values of u are attained on ∂D, unless u is identically a constant.

As an application of this principle we shall prove the uniqueness of solutions of the Dirichlet problem for the Laplace equation

$$\begin{aligned} u_{xx} + u_{yy} &= q(x,y) \quad \text{in} \quad D \\ u(x,y) &= f(x,y) \quad \text{on} \quad \partial D. \end{aligned} \tag{49.2}$$

Theorem 49.5. Assume that q and f are continuous in their domain of definition. Then there is at most one solution to the problem (49.2).

Proof. Assume to the contrary that there are two solutions $u_1(x,y)$ and $u_2(x,y)$ of (49.2). Then, the function $w = u_1 - u_2$ satisfies $w_{xx} + w_{yy} = 0$ in D and $w = 0$ on ∂D. Thus, from the maximum–minimum principle w must attain its maximum and minimum values on ∂D. Hence, $w(x,y) \equiv 0$; i.e., $u_1(x,y) \equiv u_2(x,y)$, $(x,y) \in \overline{D}$. ∎

Alternative Proof. We can also prove Theorem 49.5 by using *Green's identity* (energy technique): Let $D \subset \mathbb{R}^2$ be a bounded and open region whose boundary ∂D is a piecewise continuously differentiable curve. Then, for any function $w = w(x, y)$ having continuous second-order partial derivatives in D and continuous first-order partial derivatives on $D \cup \partial D$ the following holds

$$\iint_D w(w_{xx} + w_{yy})dxdy = \oint_{\partial D} w\frac{\partial w}{\partial n}ds - \iint_D (w_x^2 + w_y^2)dxdy,$$

where $\partial w/\partial n$ is the exterior normal derivative.

Green's identity for the difference of two solutions of (49.2), i.e., $w = u_1 - u_2$ reduces to

$$\iint_D (w_x^2 + w_y^2)dxdy = 0,$$

which clearly implies that w is a constant. However, since $w = 0$ on ∂D, this constant must be zero, i.e., $w = u_1 - u_2 \equiv 0$. ∎

The following stability result is a direct consequence of Theorem 49.4.

Theorem 49.6. Let $u_1(x, y)$ and $u_2(x, y)$ be solutions to the problem (49.2) with data f_1 and f_2, respectively. Further, let ϵ be any positive real number. If $\max_{\partial D} |f_1(x, y) - f_2(x, y)| \le \epsilon$, then

$$\max_{\overline{D}} |u_1(x, y) - u_2(x, y)| \le \epsilon.$$

Finally, we shall employ the energy technique to prove the uniqueness of solutions of the initial–boundary value problem for the wave equation

$$
\begin{aligned}
u_{tt} &= c^2 u_{xx} + q(x, t), \quad 0 \le x \le a, \quad t \ge 0 \\
u(x, 0) &= f(x), \quad u_t(x, 0) = g(x), \quad 0 \le x \le a \qquad (49.3) \\
u(0, t) &= \alpha(t), \quad u(a, t) = \beta(t), \quad t \ge 0.
\end{aligned}
$$

We multiply the wave equation by u_t and integrate with respect to x, to get

$$\frac{d}{dt} \int_0^a \frac{1}{2} u_t^2 dx = c^2 \int_0^a u_{xx} u_t dx + \int_0^a q u_t dx.$$

However, since

$$\int_0^a u_{xx} u_t dx = u_x u_t \Big|_0^a - \int_0^a u_x u_{xt} dx$$

and

$$u_x u_{xt} = \frac{\partial}{\partial t}\left(\frac{1}{2} u_x^2\right),$$

it follows that

$$\frac{d}{dt}\int_0^a \frac{1}{2}[u_t^2 + c^2 u_x^2]dx = c^2 u_x u_t\Big|_0^a + \int_0^a qu_t dx. \tag{49.4}$$

The identity (49.4) is called the energy equation for the wave equation.

Theorem 49.7. Assume that q, f, g, α, and β are sufficiently smooth in their domain of definition. Then, there is at most one solution of (49.3) which is continuous together with its first- and second-order partial derivatives for $0 \le x \le a$, $t \ge 0$.

Proof. Assume to the contrary that there are two such solutions $u_1(x,y)$ and $u_2(x,y)$ of (49.3). Then the function $w = u_1 - u_2$ satisfies $w_{tt} = c^2 w_{xx}$, $w(x,0) = 0$, $w_t(x,0) = 0$, $w(0,t) = 0$, $w(a,t) = 0$. But then, from the assumptions on the solutions, we also have $w_x(x,0) = 0$, $w_t(0,t) = 0$ and $w_t(a,t) = 0$. Thus, for the function w, the identity (49.4) reduces to

$$\frac{d}{dt}\int_0^a \frac{1}{2}[w_t^2 + c^2 w_x^2]dx = 0.$$

Therefore, we have

$$\int_0^a \frac{1}{2}[w_t^2(x,t) + c^2 w_x^2(x,t)]dx = \int_0^a \frac{1}{2}[w_t^2(x,0) + c^2 w_x^2(x,0)]dx = 0,$$

which immediately implies that w is a constant. However, since $w(x,0) = 0$ and w is continuous this constant must be zero; i.e., $w = u_1 - u_2 \equiv 0$. ∎

Lecture 50
Verification of Solutions

In our previous lectures the series or integral form of the solutions we have obtained were only formal; there we did not attempt to establish their validity. In this lecture we shall prove a few theorems which verify that these are actually the solutions. For this, we shall need the following results in two independent variables:

(P_1). *Weierstrass's M-test*: If the terms of a series of functions of two variables are majorized on a rectangle by the terms of a convergent numerical series, then the series of functions is absolutely and uniformly convergent on the rectangle.

(P_2). If a series of continuous functions converges uniformly on a rectangle, then its sum is continuous on the rectangle.

(P_3). If the series obtained from a given convergent series by formal term-by-term partial differentiation is a uniformly convergent series of continuous functions on a closed rectangle, then the given series has a continuous derivative which on the rectangle is the sum of the series obtained by term-by-term differentiation.

(P_4). *Abel's test*: The series $\sum_{n=1}^{\infty} X_n(x)Y_n(y)$ converges uniformly with respect to the two variables x and y together, in a closed region R of the xy-plane, provided the series $\sum_{n=1}^{\infty} X_n(x)$ converges uniformly with respect to x in R, and for all y in R the functions $Y_n(y)$, $n = 1, 2, \cdots$ are monotone (nondecreasing or nonincreasing) with respect to n and uniformly bounded; i.e., $|Y_n(y)| < M$ for some M.

First we prove the following theorem for the heat equation.

Theorem 50.1. In the interval $[0, a]$ let $f(x)$ be continuous and $f'(x)$ piecewise continuous, and let $f(0) = f(a) = 0$. Then the series (30.10) represents a unique solution to the problem (30.1)–(30.4) which is continuous for all $\{(x, t) : 0 \leq x \leq a, \ t \geq 0\}$.

Proof. The uniqueness of the solutions has already been proved in Theorem 49.2. Thus, it suffices to show that the function $u(x, t)$ defined by the series (30.10) is a solution. Clearly, this $u(x, t)$ satisfies the boundary conditions (30.3) and (30.4). Next from (30.10), we have $u(x, 0) = \sum_{n=1}^{\infty} c_n \sin n\pi x/a$, and since the function f satisfies the conditions of The-

R.P. Agarwal, D. O'Regan, *Ordinary and Partial Differential Equations*,
Universitext, DOI 10.1007/978-0-387-79146-3_50,
© Springer Science+Business Media, LLC 2009

orem 23.6, it follows that $u(x, 0) = f(x)$. Now from (30.11), we find that

$$|c_n| \leq \frac{2}{a} \int_0^a |f(x)|dx =: C$$

so that, for the n-th term of (30.10), we obtain the majorization

$$\left| c_n e^{-(n^2\pi^2c^2/a^2)t} \sin \frac{n\pi x}{a} \right| \leq Ce^{-(n^2\pi^2c^2/a^2)t_0},$$

for all $0 \leq x \leq a$, $t \geq t_0 > 0$. Next since the numerical series

$$\sum_{n=1}^{\infty} Ce^{-(n^2\pi^2c^2/a^2)t_0}$$

is convergent; from (P_1) it follows that the series (30.10) is absolutely and uniformly convergent for all $0 \leq x \leq a$, $t \geq t_0 > 0$. Thus, from (P_2) and the fact that t_0 is arbitrary, we conclude that the sum $u(x,t)$ of the series is continuous for all $0 \leq x \leq a$, $t > 0$. Now we formally differentiate the series (30.10) term-by-term with respect to t, to obtain

$$u_t(x,t) = -\frac{\pi^2c^2}{a^2} \sum_{n=1}^{\infty} n^2 c_n e^{-(n^2\pi^2c^2/a^2)t} \sin \frac{n\pi x}{a}. \qquad (50.1)$$

This series has the majorizing series

$$\frac{\pi^2c^2}{a^2} \sum_{n=1}^{\infty} Cn^2 e^{-(n^2\pi^2c^2/a^2)t_0}$$

for all $0 \leq x \leq a$, $t \geq t_0 > 0$. By the ratio test this numerical series is convergent. Hence, again by (P_1) it follows that the differentiated series is absolutely and uniformly convergent for all $0 \leq x \leq a$, $t \geq t_0 > 0$. Therefore, (P_3) implies that the sum $u(x,t)$ of the series (30.10) has continuous partial derivative with respect to t for all $0 \leq x \leq a$, $t > 0$ and this derivative can be obtained term-by-term differentiation. In a similar manner we can show that u_x, u_{xx} exist, are continuous, and can be obtained by term-by-term differentiation. In fact, we have

$$u_{xx}(x,t) = -\frac{\pi^2}{a^2} \sum_{n=1}^{\infty} n^2 c_n e^{-(n^2\pi^2c^2/a^2)t} \sin \frac{n\pi x}{a} \qquad (50.2)$$

for all $0 \leq x \leq a$, $t \geq t_0 > 0$.

Finally, from (50.1) and (50.2) it is clear that $u_t(x,t) = c^2 u_{xx}(x,t)$ for all $0 \leq x \leq a$, $t \geq t_0 > 0$. ∎

Now we state the following theorem which confirms that the Gauss–Weierstrass formula (43.3) (see equivalent representation in Problem 44.5) is actually a unique solution of (43.1).

Theorem 50.2. Let $f(x)$, $-\infty < x < \infty$ be a bounded and piecewise continuous function. Then, the Gauss–Weierstrass integral (43.3) defines a unique solution of the problem (43.1) with $\lim_{t \to 0} u(x, t) = [f(x + 0) + f(x - 0)]/2$.

Our next result is for the initial value problem (43.9), where we do not assume the condition that u and u_x are finite as $|x| \to \infty$, $t > 0$.

Theorem 50.3. Let for all $-\infty < x < \infty$ the function $f_1(x)$ be twice continuously differentiable and the function $f_2(x)$ continuously differentiable. Then, the initial value problem (43.9) has a unique twice continuously differentiable solution $u(x, t)$, given by d'Alembert's formula (43.11) (see also Problem 33.10).

Proof. From Lecture 28 (see (28.2)) it is clear that the solution of the wave equation can be written as

$$u(x, t) = F(x + ct) + G(x - ct), \tag{50.3}$$

where F and G are arbitrary functions. This solution $u(x, t)$ is twice continuously differentiable provided that F and G are twice differentiable. Differentiating (50.3) with respect to t, we get

$$u_t(x, t) = cF'(x + ct) - cG'(x - ct). \tag{50.4}$$

Thus, $u(x, t)$ satisfies the initial conditions $u(x, 0) = f_1(x)$, $u_t(x, 0) = f_2(x)$, if and only if,

$$\begin{aligned} F(x) + G(x) &= f_1(x) \\ cF'(x) - cG'(x) &= f_2(x). \end{aligned} \tag{50.5}$$

We integrate the second equation in (50.5), to obtain

$$cF(x) - cG(x) = \int_0^x f_2(\xi)d\xi + K, \tag{50.6}$$

where K is an arbitrary constant. Combining (50.6) with the first equation of (50.5), we can solve for F and G, to find

$$\begin{aligned} F(x) &= \frac{1}{2}f_1(x) + \frac{1}{2c}\int_0^x f_2(\xi)d\xi + \frac{1}{2}\frac{K}{c} \\ G(x) &= \frac{1}{2}f_1(x) - \frac{1}{2c}\int_0^x f_2(\xi)d\xi - \frac{1}{2}\frac{K}{c}. \end{aligned}$$

Using these expressions in (50.3), we obtain d'Alembert's formula (43.11).

The above explicit construction shows that if the problem (43.9) has a solution it must be given by (43.11) and is unique. Conversely, since $f_1(x)$ is twice continuously differentiable and $f_2(x)$ is continuously differentiable, it is trivial to verify that (43.11) is indeed a twice continuously differentiable solution of (43.9). ∎

Now we shall prove a theorem for the boundary value problem (34.4) – (34.8) with $g(x) \equiv 0$. We note that when $g(x) \equiv 0$, the solution (34.15) can be written as

$$u(x,y) = \sum_{n=1}^{\infty} a_n \frac{\sinh \omega_n(b-y)}{\sinh \omega_n b} \sin \omega_n x, \quad \omega_n = \frac{n\pi}{a} \qquad (50.7)$$

where a_n is given in (34.16).

Theorem 50.4. In the interval $[0, a]$ let $f(x)$ be continuously differentiable and $f''(x)$ piecewise continuous, and let $f(0) = f(a) = 0$. Then, the series (50.7) represents a unique solution $u(x,y)$ to the problem (34.4)–(34.8) with $g(x) \equiv 0$. This solution u and u_x, u_y are continuous in the closed rectangle $0 \le x \le a$, $0 \le y \le b$, while u_{xx} and u_{yy} are continuous in the rectangle $0 \le x \le a$, $0 < y \le b$.

Proof. The uniqueness of the solutions has already been proved in Theorem 49.5. Thus, it suffices to show that the function $u(x,y)$ defined by the series (50.7) is a solution. Clearly, this $u(x,t)$ satisfies the boundary conditions (34.7), (34.8) and (34.6), and since the function f satisfies the conditions of Theorem 23.6, it follows that $u(x,0) = f(x) = \sum_{n=1}^{\infty} a_n \sin \omega_n x$ uniformly. Now we consider the sequence of functions

$$Y_n(y) = \frac{\sinh \omega_n(b-y)}{\sinh \omega_n b}.$$

It is clear that $0 \le Y_n(y) \le 1$ for all n and $0 \le y \le b$; i.e., these functions are uniformly bounded. We claim that for all $0 \le y \le b$ the functions $Y_n(y)$ are nonincreasing as n increases. This is immediate for $y = 0$ and $y = b$, and for $0 < y < b$ it suffices to show that the function $S(s) = \sinh qs / \sinh ps$, where $p > q > 0$ is a decreasing function of $s > 0$. Since

$$
\begin{aligned}
2S'(s)\sinh^2 ps &= 2q \sinh ps \cosh qs - 2p \sinh qs \cosh ps \\
&= -(p-q)\sinh(p+q)s + (p+q)\sinh(p-q)s \\
&= -(p^2 - q^2)\left[\frac{\sinh(p+q)s}{p+q} - \frac{\sinh(p-q)s}{p-q} \right] \\
&= -(p^2 - q^2)\sum_{n=0}^{\infty} \left[(p+q)^{2n} - (p-q)^{2n} \right] \frac{s^{2n+1}}{(2n+1)!}
\end{aligned}
$$

it follows that $S'(s) < 0$, and hence $S(s)$ is a decreasing function of $s > 0$. From this and the fact that Fourier sine series of $f(x)$ converges uniformly, (P$_4$) implies that the series (50.7) converges uniformly to $u(x, y)$ with respect to x, y in the closed rectangle $0 \le x \le a$, $0 \le y \le b$. The continuity of u in this rectangle immediately follows from (P$_2$).

Next note that the function f' also satisfies the conditions of Theorem 23.6, and hence the Fourier cosine series $\sum_{n=1}^{\infty} a_n \omega_n \cos \omega_n x$, obtained by differentiating the sine series term-by-term also converges uniformly for $0 \le x \le a$. Thus, as above the series

$$\sum_{n=1}^{\infty} a_n \omega_n \frac{\sinh \omega_n (b - y)}{\sinh \omega_n b} \cos \omega_n x$$

converges uniformly to $u_x(x, y)$ with respect to x, y in the closed rectangle $0 \le x \le a$, $0 \le y \le b$. The continuity of u_x in this rectangle also follows from (P$_2$).

Now in view of Problem 23.14(i) the numerical series $\sum_{n=1}^{\infty} |a_n \omega_n|$ converges, and hence from (P$_1$) the series $\sum_{n=1}^{\infty} a_n \omega_n \sin \omega_n x$ converges uniformly for $0 \le x \le a$. We also note that the sequences of functions

$$\overline{Y}_n(y) = \frac{\cosh \omega_n (b - y)}{\sinh \omega_n b}$$

is nonincreasing as n increases for all $0 \le y \le b$. In fact, this follows from the relation

$$\overline{Y}_n^2(y) = \frac{1}{\sinh^2 \omega_n b} + Y_n^2(y) \tag{50.8}$$

and our earlier considerations. The uniform boundedness of $\overline{Y}_n(y)$ is also immediate from (50.8). Combining these arguments, we find that the series

$$-\sum_{n=1}^{\infty} a_n \omega_n \frac{\cosh \omega_n (b - y)}{\sinh \omega_n b} \sin \omega_n x$$

converges uniformly to $u_y(x, y)$ with respect to x, y in the closed rectangle $0 \le x \le a$, $0 \le y \le b$. The continuity of u_y in this rectangle once again follows from (P$_2$).

Finally, since $|a_n| \le C$, and for $0 \le y \le b$,

$$\sinh \omega_n (b - y) < \frac{1}{2} e^{\omega_n (b - y)}, \quad \sinh \omega_n b \ge \frac{1}{2} e^{\omega_n b} \left(1 - e^{-2\pi b/a}\right)$$

for the series obtained by differentiating twice (50.7) with respect to x or y, the terms have the absolute value less than

$$C \omega_n^2 \frac{\frac{1}{2} e^{\omega_n (b - y)}}{\frac{1}{2} e^{\omega_n b} \left(1 - e^{-2\pi b/a}\right)} = C \omega_n^2 \frac{e^{-\omega_n y}}{\left(1 - e^{-2\pi n/a}\right)}.$$

Now since by the ratio test, the numerical series

$$\sum_{n=1}^{\infty} C\omega_n^2 \frac{e^{-\omega_n y_0}}{\left(1 - e^{-2\pi n/a}\right)},$$

where $0 < y_0 \leq b$ converges, the series of the second derivatives obtained from (50.7) by term-by-term differentiating twice with respect to x or y converges uniformly for $0 \leq x \leq a$, $0 < y \leq b$. The continuity of u_{xx}, u_{yy} in this rectangle again follows from (P$_2$). ■

Finally, we state the following theorem which confirms that the integral (44.2) is actually a solution of (44.1).

Theorem 50.13. Let $f(x)$, $-\infty < x < \infty$ be a bounded and piecewise continuous function. Then, the integral (44.2) defines a unique solution of the problem (44.1) with $\lim_{y \to 0} u(x,y) = [f(x+0) + f(x-0)]/2$.

If we allow unbounded solutions, then the uniqueness is lost. For example, the function $u(x,y) = y$ satisfies Laplace's equation in the half-plane and $\lim_{y \to 0} u(x,y) = 0$. This can be added to any solution.

References for Further Reading

1. R.P. Agarwal and D. O'Regan, *An Introduction to Ordinary Differential Equations*, Springer–Verlag, New York, 2008.

2. L.C. Andrews, *Special Functions for Engineers and Applied Mathematicians*, Macmillan, New York, 1985.

3. N.H. Asmar, *Partial Differential Equations with Fourier Series and Boundary Value Problems*, 2nd edition, Prentice–Hall, Englewood Cliffs, 2005.

4. A. Aziz and T.Y. Na, *Perturbation Methods in Heat Transfer*, Hemisphere Publishing Corporation, New York, 1984.

5. H. Bateman, *Partial Differential Equations of Mathematical Physics*, Cambridge University Press, London, 1959.

6. C.M. Bender and S.A. Orszag, *Advanced Mathematical Methods for Scientists and Engineers*, Springer–Verlag, New York, 1999.

7. P.W. Berg and J.L. McGregor, *Elementary Partial Differential Equations*, Holden–Day, San Francisco, 1966.

8. H.S. Carslaw, *Introduction to the Theory of Fourier's Series and Integrals*, Dover, New York, 1950.

9. R.V. Churchill and J.W. Brown, *Fourier Series and Boundary Value Problems*, 6th edition, McGraw–Hill, New York, 2000.

10. G. Folland, *Introduction to Partial Differential Equations*, Princeton University Press, 1976.

11. A. Friedman, *Partial Differential Equations*, Holt, Rinehart and Wilson, New York, 1969.

12. P. Garabedian, *Partial Differential Equations*, 2nd edition, Chelsea, New York, 1998.

13. D. Gilbarg and N. Trudinger, *Elliptic Partial Differential Equations of Second Order*, 2nd edition, Springer–Verlag, New York, 1983.

14. R.B. Guenther and J.W. Lee, *Partial Differential Equations of Mathematical Physics and Integral Equations*, Prentice–Hall, Englewood Cliffs, 1988.

15. R. Haberman, *Applied Partial Differential Equations with Fourier Series and Boundary Value Problems*, Pearson Prentice–Hall, New Jersey, 2004.

16. H. Hochstadt, *The Functions of Mathematical Physics*, Wiley, New York, 1971.

17. V.G. Jenson and G.V. Jefferys, *Mathematical Methods in Chemical Engineering*, Academic Press, London, 1977.

18. F. John, *Partial Differential Equations*, 4th edition, Springer–Verlag, New York, 1982.

19. J. Kevorkian, *Partial Differential Equations, Analytical Solution Techniques*, Wadsworth & Brooks/Cole, 1990.

20. J.D. Logan, *An Introduction to Nonlinear Partial Differential Equations*, Wiley, New York, 1994.

21. R.E. O'Malley, Jr., *Singular Perturbation Methods for Ordinary Differential Equations*, Springer–Verlag, New York, 1991.

22. D.L. Powers, *Boundary Value Problems*, 4th edition, Academic Press, New York, 1999.

23. M.H. Protter and H.F. Weinberger, *Maximum Principles in Differential Equations*, Prentice-Hall, Englewood Cliffs, 1967.

24. E.D. Rainville, *Special Functions*, Chelsea, New York, 1960.

25. M. Renardy and R.C. Rogers, *An Introduction to Partial Differential Equations*, Springer–Verlag, New York, 1993.

26. H. Sagen, *Boundary and Eigenvalue Problems in Mathematical Physics*, Dover, New York, 1989.

27. S.L. Sobolev, *Partial Differential Equations of Mathematical Physics*, Pergamon Press, Toronto, 1964.

28. I. Stakgold, *Green's Functions and Boundary Value Problems*, 2nd edition, Wiley, New York, 1997.

29. W.A. Strauss, *Partial Differential Equations: An Introduction*, Wiley, New York, 1992.

30. G.P. Tolstov, *Fourier Series*, Prentice–Hall, Englewood Cliffs, 1962.

31. F. Treves, *Basic Linear Partial Differential Equations*, Academic Press, New York, 1975.

32. H.F. Weinberger, *A First Course in Partial Differential Equations*, Blaisdell, Waltham, Mass., 1965.

33. E.C. Young, *Partial Differential Equations*, Allyn and Bacon, Boston, 1972.

34. E.C. Zachmanoglou and D.W. Thoe, *Introduction to Partial Differential Equations with Applications*, Dover, New York, 1989.

35. E. Zauderer, *Partial Differential Equations of Applied Mathematics*, 2nd edition, Wiley, New York, 1989.

Index